Buckling of Thin Metal Shells

Buckling of Thin Metal Shells

J.G. Teng and J.M. Rotter

Spon Press
Taylor & Francis Group

LONDON AND NEW YORK

First published 2004
by Spon Press
11 New Fetter Lane, London EC4P 4EE

Spon Press is an imprint of the Taylor & Francis Group

Typeset in Times New Roman by
Newgen Imaging Systems (P) Ltd, Chennai, India
Printed and bound in Great Britain by
TJ International, Padstow, Cornwall

British Library Cataloguing in Publication Data
A catalogue record for this book is available from the
British Library

Library of Congress Cataloging in Publication Data
Teng, J.G.
    Buckling of thin metal shells / J.G. Teng and J.M. Rotter.
        p. cm.
    1. Metals. 2. Buckling (Mechanics) 3. Thin-walled structures.
    I. Rotter, J. Michael. II. Title.
    TA460.T375 2003
    624.1′82—dc21   2003008011

ISBN 0–419–24190–6

Contents

Figures

Tables

Contributors

P. Ansourian
The University of Sydney, Sydney,
Australia
e-mail: p.ansourian@civil.usyd.edu.au

J. Arbocz
Delft University of Technology, Delft,
The Netherlands
e-mail: j.arbocz@lr.tudelft.nl

A.E. Elwi
University of Alberta, Edmonton,
Canada
e-mail: aeelwi@civil.ualberta.ca

R. Greiner
Technical University of Graz, Graz,
Austria
e-mail: greiner@steel.tu-graz.ac.at

W. Guggenberger
Technical University of Graz, Graz,
Austria
e-mail: werner@steel.tu-graz.ac.at

J.M.F.G. Holst
Ingenieurbüro Peter Holst, Hamburg,
Germany
e-mail: janmark.holst@freenet.de

J.F. Jullien
INSA Lyon, Lyon, France
e-mail: jean-francois.jullien@
insa-lyon.fr

G.L. Kulak
University of Alberta, Edmonton,
Canada
e-mail: glkulak@civil.ualberta.ca

G. Michel
INSA Lyon, Lyon, France
e-mail: gerard.michel@insa-lyon.fr

J.M. Rotter
The University of Edinburgh,
Edinburgh, UK
e-mail: michael@civ.ed.ac.uk

H. Schmidt
University of Essen (GH), Essen,
Germany
e-mail: herbert.schmidt@uni-essen.de

J. Singer
Israel Institute of Technology, Haifa,
Israel
e-mail: j-singer@techunix.technion.ac.il

A.R. Stam
Delft University of Technology, Delft,
The Netherlands

J.G. Teng
The Hong Kong Polytechnic
University, Hong Kong, China
e-mail: cejgteng@polyu.edu.hk

Th.A. Winterstetter
Werner Sobek Ingenieure GmbH,
Stuttgart, Germany
e-mail: twinterstetter@aol.com

Y. Zhao
Zhejiang University, Hangzhou, China
e-mail: 96981930r@polyu.edu.hk

Preface

Thin-shell structures are widely used in many branches of engineering. Examples include aircraft, spacecraft, rockets, submarines, nuclear reactors, cooling towers, roofs, tubular towers, silos, tanks, pressure vessels, pipelines and offshore platforms. The aim of this book is to provide a state-of-the-art summary of existing knowledge of the buckling behaviour of thin metal shells for which failure by buckling is often the controlling design criterion. Due to the complexity and diversity of the buckling phenomena of thin metal shells, no book of a similar nature has been published for almost two decades.

The subject of shell buckling is vast and very complex: applied mathematicians, mechanicians and engineers have all been fascinated by the subject from different perspectives. Indeed, many illustrious names in the history of applied mechanics are associated with this subject, with Love and Rayleigh laying foundations that were built on by Southwell, Timoshenko, von Karman, Novozhilov, Tsien, Koiter, Reissner, Mindlin and Hutchinson. The period from the 1900s to the 1970s may be referred to as the classical period in shell buckling research. There were very strong activities in applied mechanics on shell buckling, which laid a solid foundation for the stability theory of isotropic thin shells covering both elastic and inelastic behaviour. In addition, many classical solutions were derived for the buckling loads of simple shell geometries under simple loads, which illustrated the diverse applications of the theory and clarified key phenomenological aspects of shell buckling, such as the acute sensitivity of shell buckling strength to small geometric imperfections.

Despite many advances in classical applied mechanics research on shell buckling, the impact of this research on practical design has been limited for two reasons. First, the theory of shells leads to high-order partial differential equations that are very difficult to solve, except for a few simple cases. Thus, before the computer era, no theoretical solution could be found for many practical problems. Second, for cases where classical solutions were available, the predictions often greatly overestimated the experimental buckling strength of the shell, due to its acute sensitivity to geometric imperfections in a wide variety of forms that arise naturally in the fabrication process. Classical work on the effect of imperfections is only available for a limited number of problems and focuses on idealised imperfections that may be very different from those in real structures.

From the 1960s onwards, the advent of high-speed digital computers and the associated advances in computational mechanics offered a new approach to shell buckling problems. Since then, computer analysis of shell buckling problems has been a very important area of computational mechanics. Indeed, even today, the development of powerful and trouble-free shell elements remains an active research area in computational solid mechanics.

Engineering differs from science in that solutions must be found for practical design problems before the scientific basis is fully established. This is particularly clear in the engineering design of shell structures. From the earliest days (e.g. the Forth Bridge, 1890), designs for thin metal shells were chiefly based on test results, although theoretical work has always been exploited as much as is practicable. For most shell buckling problems for which design rules exist, these were developed as empirical lower bounds to available test results. This engineering approach brings its own problems: a large amount of reliable test data is required for the development of a design rule, so design rules only exist for a small number of simple design situations.

By the mid-1980s, computer packages for the buckling analysis of shells had been developed to the point where they could deliver realistic buckling loads if realistic imperfections could be modelled. The wide implementation of path-following methods, in particular the arc-length method (Riks, Wempner, Crisfield and Ramm), was an important step forward, as it enabled the tracing of highly non-linear equilibrium paths, so that the prediction of buckling was no longer mingled with divergences in the solution process. By this time, engineering researchers gradually gained confidence in the ability of computer software to deliver reliable buckling loads. The shell buckling research community was one of the first groups of engineering researchers to fully exploit the power of modern computer programs, since buckling experiments are often particularly difficult to conduct and can be very unreliable if not properly undertaken.

By the beginning of the 1990s, a new paradigm emerged for shell buckling research: a given problem which requires a design method is generally subject to extensive parametric studies founded on a good mastery of classical shell buckling theory and numerical techniques, supplemented by a few careful experiments when required. This paradigm seems to have gradually been accepted by other branches of research on metal structures, though hitherto to a lesser degree. Several general-purpose computer packages are now widely used, such as ABAQUS, ADINA, ANSYS and NASTRAN: these are now brand names, as MTS, Schenck and Instron are for structural testing machines. Specialist proprietary computer packages for shells, such as BOSOR5 and NEPAS have also contributed significantly to this numerically driven research. Indeed, computer analysis is now on an equal footing with laboratory experiment, and computer packages play a role comparable to that of testing machines. Most of the recent advances summarised in this book have been obtained mainly within this new research paradigm.

From early in the 1990s researchers began to ponder on the direct use of numerical buckling analyses in the design process. In this area, they lagged behind design philosophies in other parts of structural, mechanical and aeronautical engineering design, but for a very good reason. It is extremely difficult to specify imperfections in the shell structure that can be guaranteed to be adequately deleterious to strength under all conditions and for all geometries, and which are coupled to tolerance restrictions that can ensure that the assumptions of the design are achieved in practice. Moreover, the range of different analyses that could be used for shell buckling is considerable, so design processes are required to identify how each might be used safely. The new European standard on the Strength and Stability of Shell Structures has presented the first attempt to codify these matters, with a comprehensive conceptual framework that should permit the next generation of shell buckling research studies to be exploited more effectively in design.

This book consists of 15 chapters written by leading international experts on the buckling of thin metal shells. The first chapter provides an extended introduction to the field, and indicates the place of each of the later chapters within the whole framework. The chapters are grouped together, relating to the different fundamental loading types for a cylinder: axial compression, circumferential compression and membrane shear.

The first group of chapters is concerned with cylinders under axial compression. Chapter 2 provides a comprehensive treatment of isotropic cylindrical shell buckling under uniform axial compression. Chapter 3 examines the local elastic or plastic failure above a local support at its lower end, introducing local axial compression into the cylinder. Chapter 4 reviews the whole subject of non-uniform settlement beneath cylindrical tanks and reaches a new point in developing our understanding of that subject.

Chapters 5–7 deal with cylinders under external pressure, beginning with uniform pressure (or vacuum) in Chapter 5, and continuing with general non-uniform pressures in Chapter 6. Chapter 7 deals with the practical and challenging case of buckling of empty tanks and silos under wind pressures.

Chapters 8–10 deal with problems in which membrane shear stresses play a strong role, beginning in Chapter 8 with torsion and transverse shear. One of the key problems involving shear is the cantilever cylinder subject to cyclic shear caused by earthquake, and this is dealt with in Chapter 9. Finally, Chapter 10 considers combined load cases of different types.

The above chapters are all concerned with isotropic shells. Chapter 11 extends the discussion to stiffened shells used in aerospace, with many different types of stiffening considered. Chapter 12 describes the behaviour of thicker cylinders under shear and bending.

Chapters 13 and 14 are concerned with conditions where the cylindrical shell meets another shell segment. Chapter 13 considers the junction itself, which often needs strengthening with a ring. Chapter 14 considers the ring at a junction, a simple structure that can display quite complicated behaviour.

The foregoing chapters have all presented a deterministic view of shell buckling: there is a defined strength for a given set of conditions. The final Chapter 15 gives us a probabilistic perspective, so necessary in structural design, and outlines how the huge scatter associated with shell buckling strengths may be safely, economically and reliably accommodated into the design conceptual framework.

The two editors are very pleased that they have now come to the end of a long road which started soon after the publication of a review article by the first editor in *Applied Mechanics Reviews*. It was felt then that a book of this kind would provide a useful consolidated source of information for researchers and engineers alike. The preparation of this book has, however, taken much longer than initially expected, for which the editors would like to record their apologies to the contributors. Nevertheless, in this complex field of research, the information contained in this book, being of such high quality, depth and thoroughness, will be an indispensable source of reference to interested readers for many years to come. It is hoped that when looking back many years later, all contributors will feel proud of their involvement in this project.

The preparation of a book of this size would not have been possible without the support of many individuals and organisations. The editors would like to thank all contributors for their meticulous preparation of manuscripts, their patience and their cooperation. They are also indebted to the publishers and their staff: their gentle reminders and persuasions were important to the successful completion of this book.

The first editor would like to thank the Research Grants Council of Hong Kong Special Administrative Region and The Hong Kong Polytechnic University for financial support over many years to his research on steel shell structures and for the excellent contributions to this research made by his research students and collaborators, particularly Drs T. Hong, Y.F. Luo, H.W. Ma, C.Y. Song and Y. Zhao, Messrs B. Gabriel and X. Lin, and Miss F. Chan. The first editor is also indebted to Prof. J.M. Rotter, the second editor of this book, who supervised his PhD research on steel shell buckling in the second half of the 1980s, and provided strong and continuous support and encouragement since then. Finally, he would like to thank his family for their support and understanding.

The second editor would like to thank his many colleagues on the Project Team that developed the European standard for Shell Strength and Stability, and the Committee for the ECCS Recommendations on Shell Buckling. In particular, Prof. Richard Greiner of Graz and Prof. Herbert Schmidt of Essen contributed enormously to his understanding of shell behaviour and its formulation into design methodology. Special thanks are also due to Prof. Jacques Heyman of Cambridge who taught him a love of structures and Prof. Nick Trahair of Sydney, Prof. Jean-Francois Jullien of Lyon and Prof. Chris Calladine of Cambridge, each of whom gave most valuable support and encouragement which greatly promoted this work. He would also like to thank his former research students and collaborators Profs J.G. Teng and W. Guggenberger and Drs J.Y. Ooi, P.T. Jumikis,

M.J. Blackler, J. Zhang, H.Y. Li, J.M.F.G. Holst, M. Gillie and M. Cai. Finally, the support and understanding of his long-suffering family is most gratefully acknowledged.

Jin-Guang Teng
The Hong Kong Polytechnic University

J. Michael Rotter
The University of Edinburgh

Chapter 1

Buckling of thin shells

An overview

J.G. Teng and J.M. Rotter

Introduction

Thin-shell structures are widely used in many branches of engineering. Examples include aircraft, spacecraft, cooling towers, nuclear reactors, steel silos and tanks for storage of bulk solids and liquids, pressure vessels, pipelines and offshore platforms, though the latter are largely deemed to be outside the scope of this book. The class of shells covered here is very thin, so failure by buckling is often the controlling design criterion. It is therefore essential that the buckling behaviour of these shells is properly understood so that suitable design methods can be established.

This chapter, which is an expansion and updating of the survey paper of Teng (1996a), provides a review of recent research advances and trends in the area of thin-shell buckling. The chapter is intended as an overview in preparation for the special chapters devoted to particular topics which follow. Furthermore, this chapter covers matters that are outside the scope of the later chapters, where that is appropriate. In particular, the following topics are given particular emphasis: (a) imperfections in real structures, and their influence; (b) the use of computer buckling analysis in the stability design of complex thin shell structures; and (c) the buckling of shells under local or non-uniform loads and localised compressive stresses. Because buckling in these thin structures often occurs at very low stress levels, the phenomena are often elastic. Wherever the buckling is not identified as being in the plastic range, elastic buckling is being discussed.

Brief historical outline

General

Shell structures are widely used in many fields and have been studied scientifically for more than a hundred years. The first theoretical shell buckling problem to be solved was the cylindrical shell under axial compression (Lorenz 1908; Timoshenko 1910; Southwell 1914). This analysis determined the linear bifurcation stress for a cylinder with simply supported ends and a uniform membrane (in-plane) prebuckling stress distribution. It is almost universally referred to as the 'classical elastic buckling stress'. Early tests (Robertson 1929; Flugge 1932; Lundquist 1933; Wilson and Newmark 1933) indicated that real cylinders buckle at much lower loads, with experimental values even below 30% of the classical load being common. Moreover, the test strengths were found to be very scattered, even when great care was taken with the testing. The search for the reasons for this major discrepancy led to an enormous amount of research in the following decades. Similar discrepancies are found for externally pressurised spherical shells, but cylinders under other loading conditions and shells of other geometries display the same behaviour in less marked form. However, because of its economic importance, high sensitivity to geometric imperfections and simplicity of testing, the axially compressed cylinder has been studied more extensively than any other shell buckling problem.

The big discrepancy between the experimental and simple theoretical strengths has commonly been attributed to one of four factors:

1   prebuckling deformations and their associated changes in stress,
2   boundary conditions,
3   eccentricities and non-uniformities in the applied load or support, and
4   geometric imperfections and associated residual stresses.

Whilst the effects of each of these factors has been investigated for many shell buckling problems, their effects are briefly outlined here for axially compressed isotropic and stringer-stiffened cylinders. The outline provides a brief glimpse of the development of ideas on thin-shell buckling and indicates the boundary of current knowledge.

Axially compressed isotropic cylinders

Introduction

Most shell buckling research has, in the past, concentrated on simple 'uniform' loads. These are load cases that, in principle, produce a constant membrane prebuckling stress state throughout the shell. They include axial compression, external pressure, torsion and their combinations. The following historical description therefore relates principally to these conditions: more realistic loading conditions are discussed later in the chapter.

It is probably fair to say that the foundations of shell stability theory were almost all laid in the study of cylinders under uniform axial compression, so this topic provides a useful starting point. A much fuller description of the different phenomena is given in Chapter 2.

Prebuckling stresses and nonlinear changes before buckling

The classical linear buckling theory assumes that the state of stress before buckling is perfectly uniform and consists of membrane stresses alone. In the case of a cylindrical shell under axial compression, this implies that the compressed cylinder is free to expand laterally under Poisson effects. However, the ends of most shells are not free to expand in either experiments or real structures, so local bending stresses and deformations arise near the ends, and these can reduce the buckling load of the axially compressed cylinder. These stresses increase in a non-linear manner as the axial load increases and they depend on both the shell length and the boundary conditions, so their complete effect took some time to be fully established. They were first investigated by Fischer (1962, 1963) and Stein (1962, 1964), who reported that the buckling strength could be half the classical value. However, it was not recognised until the later work of Hoff and his co-workers (Hoff 1966) that this strength loss was caused by the relaxed boundary condition of freedom to displace circumferentially, and similar reductions were obtained using the assumption of a simple membrane prebuckling stress state. More thorough studies (Fischer 1965; Almroth 1966; Gorman and Evan-Iwanowski 1970; Yamaki and Kodama 1972) showed that the effect of prebuckling deformations is generally small ($\sim$15%) and cannot be the primary reason for the dramatic difference between theory and experiment, or for the great scatter in test results.

Boundary conditions

The effect of different boundary conditions on the buckling strength of cylindrical shells was explored extensively, with particular attention to the boundary conditions that affect the displacements and the membrane stresses during the buckling process. It should be noted that the ends of the shell have three translational and one rotational degrees of freedom, and that restraint of any of these during buckling induces corresponding stresses which affect the buckling strength. As noted above, under membrane prebuckling stresses (Ohira 1961, 1963; Stein 1962; Hoff 1965, 1966; Hoff and Rehfield 1965; Hoff and Soong 1965; Almroth 1966), it was discovered that the greatest sensitivity to boundary conditions occurs if the shell ends are free to displace in the circumferential direction during buckling, halving of the buckling stress. It was initially thought that this might be a cause of the low buckling strengths achieved in tests, but it was soon evident that this effect is both insufficient and an unrealistic explanation for the difference between the classical prediction and experimental strengths, since such boundary conditions rarely exist in either laboratory or constructed shells. Other boundary condition changes have a much smaller effect, and it is particularly notable that a change from simply supported to built-in ends has little effect for most practical lengths of cylinder. Extensive information on the effects of different boundary conditions may be found in Yamaki (1984).

Loading eccentricities and non-uniformities

For axially compressed isotropic cylinders, small eccentricities in the line of action of the total load do not have a major influence on the buckling strength

(Simitses *et al.* 1985), and eccentricities of the applied load relative to the middle surface of the shell at the boundary are usually eliminated both in tests and in practice by appropriate detailing. However, where the local value of the applied load is subject to variation around the circumference, a very marked effect can ensue (Calladine 1983a). Studies of this problem are not common (Knödel and Schulz 1988, 1992; Holst and Rotter 2001), but considerable effort is commonly expended to ensure that both test cylinders (e.g. Berry *et al.* 2000) and full-scale shells are uniformly supported. Further relevant information on this question can be found in Chapters 3 and 4.

Geometric imperfections and co-existent residual stresses

The single most important factor contributing to the discrepancy between theory and experiment for axially compressed isotropic cylinders is now widely accepted to be initial imperfections in the shell geometry. An enormous amount of research has therefore been carried out on the imperfection sensitivity of shell buckling in different configurations. The most notable contributors to this research include von Karman and Tsien (1941), Koiter (1945), Donnell and Wan (1950), Budiansky and Hutchinson (1966) and Yamaki (1984). The latter gives a very thorough description of the buckling behaviour of cylinders with many different imperfection forms and amplitudes under different types of uniform load.

Very few studies have explored the effects of residual stresses in shells on the buckling strength, even though these are usually present where geometric imperfections exist, and can be quite large. Recent studies have begun to explore the mechanics of the development of residual stresses (Guggenberger 1996a; Holst *et al.* 1999), and a few studies have explored their effect on buckling strength (Bornscheuer *et al.* 1983; Rotter 1988, 1996b; Ravn-Jensen and Tvergaard 1990; Guggenberger 1996a; Holst *et al.* 1996a, 2000; Hübner *et al.* 2003). In general, it seems that residual stresses can account for some of the discrepancies between experimental and theoretical calculations, but they do not have a systematically deleterious effect, as they do in columns and beams.

Finally, thickness variations may cause significant imperfections in some thin shells. These were first studied by Koiter *et al.* (1994), and their effect on external pressure buckling has recently been comprehensively solved (Jullien *et al.* 1999).

Axially compressed stringer-stiffened cylinders

Although geometric imperfections are responsible for discrepancies between simple buckling theory and experiment in most shells, when stringer stiffeners are present this effect is less pronounced (Brush 1968). The imperfection sensitivity of stringer-stiffened cylinders depends on the geometry of the stiffeners, particularly the area ratio A_s/bt, where A_s is the cross-sectional area of a stringer, b is the circumferential separation of stringers and t is the thickness of the shell skin (Arbocz and Sechler 1976; Weller and Singer 1977). The boundary conditions

play a stronger role in stringer-stiffened cylinders (Weller 1978). Axial restraint at
the boundaries may increase the buckling load by more than 50%. Singer and his
colleagues (Weller *et al.* 1974; Singer and Rosen 1976; Singer and Abramovich
1979; Singer 1982a,b, 1983) contributed greatly to understanding the buckling of
stiffened cylinders under axial compression, with special emphasis on boundary
conditions and load eccentricities. Load eccentricities in stringer-stiffened cylinders were first studied by Stuhlman *et al.* (1966), followed by extensive work by
Singer's group (Weller *et al.* 1974; Singer 1983). These studies showed that the
buckling load can be reduced by up to 50% in some practical configurations, but
that prebuckling deformations are only significant if the shell is short enough for
the boundary condition to affect stresses throughout the shell length.

The many studies from Singer's group demonstrated conclusively that the
consequences of geometric imperfections should be considered only after the
effects of boundary conditions and load eccentricities have been used to find
the perfect shell strength accurately. They developed an experimental technique
to determine the real achieved boundary conditions and load eccentricities using
non-destructive vibration methods. This technique also gives good assessments
of experimental buckling loads (Singer and Abramovich 1979; Singer 1982a,b,
1983). More recent studies have been undertaken by Byskov and Hansen (1980),
Dowling *et al.* (1982), Dowling (1991) and Düsing (1994). Further discussion of
stiffened cylinders is given in Chapter 11.

Summaries of research: books and review articles

Many other shell buckling problems have been studied in recent decades, and
several books and review articles have been written on the subject. Noor (1990)
provides an extensive list of books, conference proceedings and survey articles
on shell structures. Several books were published around 15 years ago (Calladine
1983a; Kollar and Dulacska 1984; Yamaki 1984; Bushnell 1985) which provide
a wealth of information on the buckling behaviour and strength of thin shells.
Conference proceedings (Koiter 1960; Budinansky 1974; Koiter and Mikhailov
1980; Ramm 1982; Thompson and Hunt 1983; Jullien 1991; Rotter 1996a; Krupka
and Schneider 1997; Drew and Pellegrino 2002), journal special issues (Galletly
1995; Teng 1998) and paper collections (Fung and Sechler 1974; Zamrik and
Dietrich 1982; Harding *et al.* 1982; Narayanan 1985) are also sources of very
useful information.

Many review articles on thin shell buckling have also been written. Nash (1960)
summarised early achievements in shell buckling research. Hoff (1966) discussed
the buckling behaviour of cylinders with various boundary conditions. Work on
elastic postbuckling and imperfection sensitivity was reviewed in Budiansky and
Hutchinson (1966, 1979), Hutchinson and Koiter (1970), Tvergaard (1976) and
Citerley (1982). Babcock (1974) and Singer (1980, 1982b) surveyed experimental
research on shell buckling. Babcock (1983) addressed imperfection sensitivity,
dynamic buckling, plastic buckling, experiments and computer buckling analysis

in his general review of shell buckling research. Bushnell (1982) described the computer analysis of shell buckling and examined many plastic buckling problems. Further important contributions to plastic buckling were made by Ore and Durban (1989, 1992), Galletly *et al.* (1990), Giezen *et al.* (1991) and Combescure (1991). More recent review articles include those of Rotter (1985a), Simitses (1986), Teng (1996a), Schmidt (2000) and Rotter (2002a,b), with the practical applications being discussed in the review articles by Blachut (1998), Öry *et al.* (1998), Rotter (1998b), Smith (1998) and Teng (2000). A handbook for shell stability design was produced by Samuelson and Eggwertz (1992) which also considers the practical aspects of many problems which have not yet been researched properly.

Design standards for the buckling of thin shells

The above summaries of research cover knowledge of the mechanics, behaviour and modelling of shell buckling. However, there are many aspects that are needed for safe and economic design that are not covered in research summaries. For safe design, it is necessary to find a lower bound that covers the uncertainties of fabrication and consequent imperfections, non-uniformities of loading, simplified loading and analysis assumptions and stochastic variability (Greiner 1997; Rotter 1997a). This has led to a wide gap between the extensive research output and its adoption into practical design (Rotter 2002a).

Most standards for shell structures deal with very specific applications (NASA SP8007 1968; AWWA D100–79 1979; API 620 1982; Trahair *et al.* 1983a; Rotter 1985c; API RP2A 1989) and use very simplified assessments for the buckling strength, taking no formal account of the imperfection amplitude. They generally cover only very basic and application-limited geometries of cylinders with conical and spherical closures, subject to simple loading conditions such as uniform axial compression, external pressure and torsion. These rules have almost all been devised as empirical lower bound approximations to collected test data, and have only a passing relationship to the theoretical research described above.

The first generic standard on the buckling of shells (not devised for a particular application) appears to have been the ECCS Recommendations (1988), the first edition of which appeared in 1980. The three succeeding editions expanded the scope of the original document considerably, and at present this document covers the largest range of specific problems in shell buckling. The later DIN standard (DIN 18800 1990) was a more consistent document, but covered fewer specific problems. The rules of these standards were also all devised as empirical lower bounds on collected test data (with different choices on which data to accept), and consequently relate poorly to the theoretical research on shell buckling.

The ECCS and DIN standards were superseded by the later Eurocode (ENV 1993-1-6 1999), which is the first generic standard to deal with both strength and stability of shells. It represents a major advance in defining the use and interpretation of computer calculations of all kinds in the design of shells, and permits each kind of calculation to be entered into some part of the hand calculation process (Rotter 2002a,b). In particular, in the context of the discussion above, it defines different qualities of fabrication, with corresponding tolerances, and deduces the imperfection amplitude and consequent buckling strength in a manner that is consistent with the extensive research on imperfection sensitivity. However, it should be noted that it was calibrated by Rotter (1998a) against the earlier ECCS standard, and that attempts to correlate the empirical lower bound data with elastic–plastic imperfect shell buckling calculations do not necessarily produce the precise match that might be hoped for (Rotter 1997b). More research is needed to bring safe design procedures and quality research findings into a common outcome.

The current situation

Difficulties in transforming research into practice

The choice of an empirical basis for design standards has chiefly been made because of two major difficulties encountered previously in shell buckling research. First, the buckling phenomenon in shells is highly complex, described by nonlinear partial differential equations that, before the computer era, were too difficult to solve except for a few simple cases. Second, unlike beams and plates, the buckling of a shell is generally sensitive to small geometric imperfections induced in the fabrication process. Theoretical buckling loads obtained assuming a perfect geometry often greatly overestimate the actual strength of a shell, so design methods rely heavily on empirical interpretations of experimental data, but this is only available for a limited number of cases. Indeed, in the second edition of 'Stability of Metal Structures: World View' (Beedle 1991), it was noted that the greatest need for more experimental data was in the area of shell buckling.

Computational analysis of shell buckling

The availability of powerful computers and the development of sophisticated finite element and other numerical techniques in recent years have changed the situation drastically (Bushnell 1985; Yang et al. 1990). It is no longer impossible to solve a specific complicated nonlinear buckling problem with accuracy and confidence using numerical methods: indeed there are many proprietary and commercial codes (e.g. Bushnell 1976; Almroth and Brogan 1978; Esslinger et al. 1984; Wunderlich et al. 1985; Combescure et al. 1987; Teng and Rotter 1989a,b, 1991c; Teng and Luo 1997; Hibbit et al. 2001; Hong and Teng 2002) that can perform the task: most are based on the finite element method, but some use other numerical methods. Thus, the first difficulty, that of solving nonlinear equations, seems to have substantially disappeared, although the application of a reliable computer program to a nonlinear buckling analysis of a shell is a considerable undertaking itself. The recent study of Teng and Song (2001) illustrates well the kinds of pitfalls and challenges in such analyses.

The difficulty of solving a shell buckling problem is, however, replaced by another. Numerical predictions have much less value in design than comprehensive algebraic relationships, but algebraic formulations do not arise naturally from the numerical calculations. Considerable effort is required to identify the controlling parameters of any new problem, so that although the 'correct' answer to a problem may be formally obtained using numerical methods, the underlying controlling physics is not apparent, and the manner in which the results should be expressed still requires a major research effort. The complexity of practical problems means that there continues to be great need for research on shell buckling. Nevertheless, numerical modelling opens the door to great strides to improve our understanding of shell buckling under a much wider range of conditions.

The second difficulty, the sensitivity to geometric imperfections in unknown forms, remains. It requires urgent attention if numerical buckling analyses are either to be applied directly in design or to be used to provide predictions that can replace the experimental studies that were always previously required to develop design methods. The establishment of a procedure to convert numerically obtained buckling loads into a reliable design strength predictions is one of the most important challenges facing the shell buckling research community (Samuelson 1991a). The new European Standard on Strength and Stability of Shells (ENV 1993-1-6 1999) has made a substantial first step in this direction (Rotter 2002a,b), but much more research is needed to verify its procedures and to discover shortcomings that may be present in the process.

Imperfections and their characterisation

Imperfections in shell buckling design

It is evident that the buckling strength of imperfect cylindrical shells remains an area of active research. Lower bound strength estimates give no reward for good quality fabrication, and often seriously underestimate the reliable strength that can be achieved. To exploit the knowledge of imperfection sensitivity, the design must explicitly define the imperfection that is adopted. The key problem for the design process is to identify and characterise appropriate practically relevant geometric imperfections, including both their form and amplitude.

The imperfection forms that are most deleterious to shell strength are often unlikely to arise in practice, and the imperfection amplitudes are often difficult to predict in advance. The imperfection form that is most deleterious to strength also changes substantially with the load case, so where several different load cases must be addressed in design, it may be that different imperfection forms and amplitudes must be assumed for each load case. Moreover, the assumptions made about imperfections in design must be transformed into tolerances in execution, and the tolerance measurement systems must allow for the substantial differences that there may be between the assumed imperfection form and the real imperfections.

All these considerations make the task of reliable and economic adoption of imperfection sensitivity into the design process a very considerable challenge.

Research on imperfection measurement

Most early research on imperfection sensitivity was concerned with idealised imperfection forms and imperfections in small-scale laboratory models, but it has now been well recognised that these are generally not representative of real imperfections in full-scale structures. In aerospace structures, Babcock, Arbocz, Singer and their colleagues (Arbocz and Babcock 1981; Arbocz 1982, 1991; Singer 1982b; Weller *et al.* 1986; Elishakoff *et al.* 1987; Arbocz and Hol 1991; Singer and Abramovich 1995) pioneered the precise measurement of imperfections in both laboratory and full-scale shells. Different techniques were required for full-scale civil engineering shells, and these were developed later by Clarke and Rotter (1988), Coleman *et al.* (1992) and Ding *et al.* (1996a,b).

Others later applied similarly thorough measurement techniques to laboratory models (Blachut *et al.* 1991; Chryssanthopoulos *et al.* 1991a,b; Chryssanthopoulos and Poggi 1995; Berry *et al.* 1996, 2000; Teng *et al.* 2001; Zhao and Teng 2001; Lin and Teng 2002). Their work naturally demonstrated that both the form and the amplitude of imperfections are dependent on the fabrication process and quality.

For aeronautical shells, an International Imperfection Data Bank was established with branches in Delft and Haifa for the evaluation of imperfection measurements and correlation studies (Arbocz 1982; Singer 1982b). They also developed statistically based design methods using measured imperfections, but these generally only apply to a defined manufacturing process, where the data relates to structures that are in production and can be measured.

The task set out at the beginning of this section remains substantially un-researched: although the substantial effect of imperfections on shell strength is well known, and measurement techniques have been devised and used, much further development is needed to devise reliable methods that can adopt this knowledge into the design process. In particular, the power of computational methods will remain untapped until these questions can be properly resolved.

Studies of specific problems involving uniform loads

Studies of cylindrical shells

Many other interesting studies have been conducted on shells under uniform loads. Examples include the interaction between different basic load cases on both unstiffened cylinders (e.g. Galletly *et al.* 1987; Shen and Chen 1991) and stiffened cylinders (e.g. Agelidis *et al.* 1982; Miller and Vojta 1984; Croll 1985; Abramovich *et al.* 1991; Dowling 1991). Most of these interaction studies involve compressive stresses in more than one direction. However, the effect of internal pressure leads to an enhancement of the buckling strength under axial compression and has been widely studied (Lo *et al.* 1947; Harris *et al.* 1957; Weingarten *et al.* 1965; Rotter and Teng 1989b; Limam *et al.* 1991; Teng and Rotter 1992a; Knödel *et al.* 1995; Knöbel and Schweizerhof 1995; Rotter 1996b). At high internal pressures, elastic–plastic elephant's foot collapse buckling occurs. This was studied by Rotter (1990), whose design equation has been adopted into the Eurocodes on shells (ENV 1993-1-6 1999), silos (ENV 1993-4-1 1999) and tanks (ENV 1993-4-2 1999).

Systematic imperfections caused by the fabrication process have also been explored: the effect of lap joints (Esslinger and Geier 1977; Rotter and Teng 1989a; Teng 1994a; Greiner and Yang 1996) and weld depressions (Rotter and Teng 1989b; Berry and Rotter 1996) both produce rather well-defined strength reductions because they cause local circumferential stresses in the shell. Corrugated shells (e.g. Rotter 1986a; Rotter *et al.* 1987; Yeh *et al.* 1992; Ross and Humphries 1993) offer more complexities but can be similarly described. Systematic imperfections in the form of holes or cutouts are described below because of their local character.

Studies of conical and spherical shells and shell closures

Further studies have explored the buckling of liquid-filled conical shells (e.g. Vandepitte *et al.* 1982), paraboloidal shells (e.g. Chen and Xu 1992; Ishakov 1993), externally pressurised steel domes (e.g. Blachut *et al.* 1991; Galletly and Blachut 1991; Blachut and Galletly 1995) and externally pressurised toriconical heads (e.g. Wunderlich *et al.* 1987) and torispherical heads (e.g. Galletly 1982, 1985; Galletly and Blachut 1985b; Lu *et al.* 1995).

The range of additional problems

It is beyond the scope of this chapter to identify even the range of thin-shell buckling problems that have been explored in recent years. However, a few topics should not be omitted: shells containing solids and on elastic foundations (e.g. Seide 1962; Trahair *et al.* 1983b; Rotter *et al.* 1989; Rotter and Zhang 1990; Knödel and Schulz 1992; Luo and Teng 1998); the optimisation of shell forms to resist various loads (e.g. Blachut 1987; Jullien and Araar 1991; Reitinger and Ramm 1995); buckle propagation in submarine pipelines (e.g. Kyriakides and Babcock 1981; de Winter *et al.* 1985; Kamalarasa and Calladine 1988; Hahn *et al.* 1992; Lin *et al.* 1993); buckling of buried pipelines (e.g. Moore and Selig 1990; Yun and Kyriakides 1990); thermal buckling (e.g. Combescure and Brochard 1991); creep buckling (e.g. Arnold *et al.* 1989; Miyazaki 1988, 1992; Sammari and Jullien 1995); dynamic buckling (e.g. Saigal *et al.* 1987; Birch and Jones 1990; Florence *et al.* 1991; Lindberg 1991; Wang *et al.* 1993a,b; Pédron and Combescure 1995); non-destructive testing (e.g. Singer 1982b; Nicholls and Karbhari 1989; Souza and Assaid 1991), lower bound buckling loads (e.g. Croll 1985, 1995;

Yamada and Croll 1993) and the development of design codes and guidelines (e.g. Akiyama *et al*. 1991; Odland 1991; Samuelson 1991a; Schmidt 1991; Greiner 1997; Rotter 1997a, 2002a,b).

Cylinders under more complex load cases

Introduction

The above section dealt with a huge range of shell buckling problems involving relatively simple load cases. However, most practical structures are subject to complex load cases that lead to local high stresses that may cause buckling. Sometimes the maximum stress is adjacent to a boundary or local support or nozzle or other penetration, but in other cases, the high stresses are internal to the shell. These two conditions tend to produce rather different consequences for the buckling of the shell. The following is a brief outline of a few problems that have been explored in recent years.

Cylindrical shells under transverse shear or non-uniform torsion

Shear stresses may cause buckling in cylinders, and the reference 'uniform' load case is uniform torsion. However, many cylinders are subject to transverse loading (the complete cylinder being subject to non-uniform bending). Buckling under shear stresses was first studied by Lundquist (1935) who produced the classical solution for buckling under torsion. The corresponding elastic buckling strength under transverse shear was found by Schroeder (1972) for perfect cylinders. Yamaki *et al*. (1979) conducted experiments on polyester cylinders subject to transverse shear (including external pressure). Yamaki (1984) describes both tests and new analyses of these cylinders and shows a close match with the theory of Schroeder (1972). Perfect cylinders under non-uniform torsion, which leads to a shear stress gradient in only one direction, were studied by Jumikis and Rotter (1986).

Further extensive studies of imperfect cylinder buckling under bending and shear have been conducted over many years by Jullien's group (Antoine 2000; Jullien and Limam 2002), clearly defining the interactions and zones of relevance of buckling in bending and in shear, including the effect of internal pressure. Further information may be found in Chapters 8 and 9.

Thicker cylindrical shells under transverse shear and global bending

Thicker cylinders may buckle in the plastic range. Galletly and Blachut (1985a) performed plastic shear buckling experiments on steel cylinders and proposed a simple design equation. Further plastic shear buckling tests were conducted by Dostal *et al*. (1987). The plastic shear buckling problem for thicker cylinders has been investigated further for large diameter fabricated tubes in a number of experimental (Bailey and Kulak 1984; Obaia *et al.* 1992a) and finite element (Mok and Elwi 1987; Roman and Elwi 1989) studies, and an improved design equation was devised (Obaia *et al.* 1992b). Both numerical and experimental studies of the effects of imperfections showed that these were moderate. Roman and Elwi (1989) showed that residual stresses due to a cold-bending fabrication process can lead to a large reduction in the ultimate load, but the residual stresses due to a longitudinal seam-weld seem unimportant. The postbuckling load carrying mechanism for a cylindrical shell with end ring stiffeners may be regarded as a tension field anchored in the ring stiffeners (Bailey and Kulak 1984) and modelled as an equivalent truss (Roman and Elwi 1988). Plastic buckling of cylindrical shells under transverse shear in combination with other loads has also been explored (e.g. Akiyama *et al.* 1987; Kokubo *et al.* 1993). A fuller description of this work is given in Chapter 12.

The plastic buckling of thicker cylinders in bending has also received much attention in recent years (Bushnell 1981; Calladine 1983b; Kyriakides and Shaw 1987; Ju and Kyriakides 1991a,b, 1992; Kyriakides and Ju 1992; Murray and Bilston 1992; Li and Molyneaux 1994). Very high strength thicker tubes are also now presenting important shell buckling phenomena (Elchalakani *et al.* 2002).

Cylindrical shells subject to wind and earthquake loads

Thin-walled silos and tanks, when empty or partially filled, are susceptible to buckling failure under wind pressure. Several such buckling failures have occurred in the past, leading to a substantial research effort in this area (e.g. Brendel *et al.* 1981; Resinger and Greiner 1982; Johns 1983; Jerath and Sadid 1985; Uchiyama *et al.* 1987; Blackler and Ansourian 1988; Greiner and Derler 1995; Greiner 1998). Simple design methods have also been developed (e.g. BS 2654 1989; ENV 1993-1-6 1999). Some interesting aspects considered include the effect of a rectangular cutout on the buckling strength of wind-loaded cylinders (Jerath 1987), the buckling of tanks with an unrestrained upper edge which occurs during the construction stage (Saal and Schrufer 1991), buckling of cylinders under combined wind and snow loads (Kapania 1990), exploitation of the postbuckling responses (Schmidt *et al.* 1998) and the buckling of taller cylinders for which axial and shear stresses also become important (Greiner and Derler 1995). Interesting new findings for chimney structures were obtained by Schneider *et al.* (1997). Further discussion may be found in Chapters 5–7.

Seismic action and its potential to cause buckling of unanchored liquid-filled storage tanks has received a great deal of research in recent years (e.g. Niwa and Clough 1982; Fischer *et al.* 1985; Chiba *et al.* 1987; Peek 1989; Fujita *et al.* 1990; Haroun and Mourad 1990; Uras and Liu 1990; Peek and El-Bkaily 1991; Zhou *et al.* 1992; Manos 1994). Rammerstorfer and Scharf (1990) presented a comprehensive survey of research on storage tanks subject to earthquake loading, whilst Liu *et al.* (1991) reviewed research advances in dynamic buckling analysis of liquid-filled shells. Elastic–plastic elephant's foot buckling was studied as a quasi-static phenomenon by Rotter (1990), whose design equation has been adopted into the New Zealand (NZS 2654 1989) and Eurocode (ENV 1998–4 1999) seismic standards.

Ground-supported tanks and silos under earthquake loads are subject to unbalanced horizontal loads that lead to membrane shear stresses which vary approximately linearly in the axial direction (e.g. Rotter and Hull 1989). The elastic buckling of cylinders subject to shear stresses of linear longitudinal variation (non-uniform torsion) was studied by Jumikis and Rotter (1986) whose design equation has been adopted into the Eurocodes on silos (ENV 1993-4-1 1999) and tanks (ENV 1993-4-2 1999).

Nuclear containment vessels represent another problem in which seismic action can induce cyclic dynamic shear buckling (Murakami and Yoguchi 1991). This problem has been extensively studied by Michel et al. (2000a,b) and is discussed in detail in Chapter 9.

Local axial compression

General stress conditions

Many loading conditions can lead to locally elevated axial compressive stresses in cylinders. The buckling strength under stresses that vary around the circumference is therefore of interest. Only a few algebraic studies examined the linear bifurcation buckling of perfect cylinders under circumferentially varying axial loads (Abir and Nardo 1958; Bijlaard and Gallagher 1959; Hoff et al. 1964; Johns 1966; Libai and Durban 1973, 1977; Peter 1974). A simple conclusion from this work might be that buckling of a perfect shell occurs under a circumferentially non-uniform distribution of axial stress when the maximum stress is slightly higher than the classical elastic critical value for uniform axial compression. Libai and Durban (1977) gave simple expressions which describe the increase in buckling stress above this simple rule, but the strength gains are generally small for thin shells. Moreover, the above studies only considered linear bifurcation in perfect elastic cylinders, so the questions of imperfection sensitivity and local plasticity were not addressed.

Some very recent studies have begun to address problems of buckling under very local high stresses (Holst et al. 1996b; Cai et al. 2002; Song et al., submitted). These have the potential to greatly increase our understanding of buckling under more general stress states.

Buckling above local supports

In recent years, several studies have addressed the special problem of buckling above a support at the lower edge of a cylinder in a column-supported silo or tank. As noted above, many of the situations that give rise to high local stresses occur at a boundary or fixture which can itself substantially affect the strength, especially if it restrains the shell against certain buckling displacements. The column support beneath a cylinder is such a problem. Teng and Rotter (1991a) appear to have been the first to investigate the nonlinear buckling behaviour of column-supported cylinders numerically, quickly followed by several other studies (Ramm and Butcher 1991; Rotter *et al.* 1991; Guggenberger 1991; Teng and Rotter 1992b; Dhanens *et al.* 1993; She and Rotter 1993; Rotter 1993). Later extensions of these studies by Guggenberger (1996b), Li (1994), Li and Rotter (1996) and Guggenberger *et al.* (2000) have expanded our understanding considerably. A more complete discussion is given in Chapter 3.

Foundation settlement

Uneven foundation settlement beneath the wall of a ground-supported tank or silo gives rise to a different local stress condition, with displacement-controlled high local stresses if the tank is anchored, but relatively small stresses if it is unanchored (Palmer 1992; Hornung and Saal 1997). Several research groups have contributed to the literature on this problem (Greiner 1980; Holst *et al.* 1996b; Jonaidi and Ansourian 1997, 1998; Lancaster *et al.* 2000). A more complete discussion with a new interpretation is given in Chapter 4.

Silos under unsymmetrical patch loads or eccentric discharge

The loads exerted by stored solids on silo walls are particularly complicated and difficult to predict (Nielsen 1998). These pressures can be very unsymmetrical and are often represented by local patch loads. The most severe form of pressure distribution is that due to eccentric discharge. Many buckling failures have been caused in metal silos in service due to this loading condition, which may occur unexpectedly. Relatively few research studies have been undertaken of buckling under eccentric discharge, the first attempt that involved only a linear shell analysis being that of Rotter (1986b). Experiments were reported by Rotter *et al.* (1989) and further development of the theoretical basis was produced by Rotter (2001). A buckling criterion proposed by Rotter (1986b) was adopted into the Eurocode for silos (ENV 1993-4-1 1999) as a method of assessing the buckling strength in zones of local high axial compression.

A remarkable study of silo shells subject to the patch load pressures defined in different standards was performed by Song and Teng (2002), who showed that nonlinear analysis led to a substantial increase in the assessed buckling strength. They also found large discrepancies between the effects of the patch loads defined by different standards to represent the same phenomena, and large discrepancies between the results of different methods of assessing the buckling strength. They concluded that patch loads give a poor representation of the effects of eccentric discharge. In this context, it is fortunate that the forthcoming revised Eurocode on silo pressures (EN 1991–4 2003) will, at last, provide a standard that provides a relatively realistic representation of eccentric discharge pressures.

Much further work is urgently needed on the stress distributions and buckling conditions associated with unsymmetrical silo pressures and eccentric discharge.

Cylinders with local geometric disturbances or subject to local loads

Systematic disturbances to the geometry

Local disturbances to the simple geometry of an axially compressed cylinder have been explored in many recent studies. Cylindrical shells with cutouts or holes under axial compression have been extensively explored (Tennyson 1968; Almroth and Holmes 1972; Almroth *et al.* 1973; Starnes 1974; Montague and Horne 1981; Miller 1982; Toda 1983; Knödel and Schulz 1985; Allen *et al.* 1990; Velickov and Schmidt 1998). Some design procedures have been formulated (Eggwertz and Samuelson 1991a; Samuelson and Eggwertz 1992). The effect of a local vertical stiffener (Eggwertz and Samuelson 1991b), and of single or multiple localised deep dents caused by collision damage (Krishnakumar and Foster 1991a,b) have also been addressed.

Local and general unsymmetrical loads

Early studies of the buckling of cylinders subject to non-uniform external pressure were carried out by Almroth (1962), Weingarten (1962) and Uemura and Morita (1971). More recent investigations of cylinders under non-uniform or partial external pressure were undertaken by Wei and Shun (1988), Chiba *et al.* (1989), Ramm and Butcher (1991) and Ansourian and Sengupta (1994). The latter study considered external pressures that are circumferentially non-uniform but longitudinally uniform. For this case, they derived a simple design formula from the finite element predictions. The effect of small local radial loads on the buckling strength under axial compression has also been explored by Samuelson (1985, 1991b).

A significant number of studies have examined stability problems in horizontal storage vessels and some simple design methods have been developed (Saal 1982; Tooth and Susatijo 1983; Krupka 1987, 1991a,b; Ansourian and Sengupta 1993).

Rings and shell junctions

Ring stiffeners

Ring stiffeners are often an integral part of a shell structure. These are used both as stiffening elements to give greater buckling resistance (especially under external pressure) and as strengthening elements at shell junctions. The ring itself is also susceptible to buckling failure.

One common mode of buckling failure for a ring is out-of-plane buckling (Bushnell 1977), especially when it is attached to a shell junction or intersection. Several recent studies have been conducted on rings at shell junctions (Jumikis and Rotter 1983; Rotter and Jumikis 1985; Rotter 1987; Sharma *et al.* 1987; Teng and Rotter 1988, 1989d, 1991b, 1992c; Greiner 1991; Teng and Lucas 1994). Esslinger and Geier (1993) and Louca and Harding (1994) investigated the torsional buckling strength of ring stiffeners on externally pressurised cylinders.

In the finite element modelling of ring-stiffened shells, each ring can be modelled either as an additional shell segment using shell theory or as a discrete ring (Bushnell 1985), but discrepancies often arise between the two treatments. This can arise either because the discrete ring model is poor, or because the shell theory fails to capture all behaviours. The most widely used nonlinear shell theory is that of Sanders (1963), which has been generally accepted as being accurate for thin shells for all practical purposes. However, Rotter and Jumikis (1988) cast doubt on the ability of Sanders' theory in modelling buckling phenomena dominated by nonlinear strains arising from in-plane displacements and derived a new nonlinear theory for thin shells of revolution which includes nonlinear strain terms arising from in-plane displacements. Their theory, originally derived in 1982 (Jumikis and Rotter 1983), was later found to be identical to those independently derived by Combescure (1986), Su *et al.* (1987) and Yin *et al.* (1987). Teng and Hong (1998) later presented a set of strain–displacement relations of the same kind for shells of general form, which reduces to the above theory for shells of revolution. Furthermore, Teng and Hong showed that Sanders' theory with the rotation about the normal retained can provide the correct buckling load for in-plane buckling of an annular plate that is modelled using shell elements, but that the simplified Sanders' theory in which this rotation is omitted cannot.

Local circumferential compression and shell junctions

There is a class of shell structures in which localised circumferential compressive stresses arise under loads that give rise to tensile membrane stresses throughout most of the structure. The local compressive stresses can clearly cause buckling. Examples include torispherical and toriconical pressure vessel heads, circular plates under a transverse load, and hemispheres under axial tension. These problems are normally not very sensitive to initial imperfections because the circumferential compression is localised, and the two principal membrane stresses are both tensile at other points in the shell.

Torispherical and toriconical shells have been studied extensively (e.g. Adachi and Benicek 1964; Galletly 1982, 1985; Galletly and Blachut 1985b; Roche and Autrusson 1986; Wunderlich *et al.* 1987; Galletly *et al.* 1990; Soric 1990, 1995; Lu *et al.* 1995) and design methods have been developed and included into several standards including the ECCS (1988) code. Hagihara *et al.* (1991) analysed the bifurcation buckling of torispherical heads dynamically loaded by internal pressure.

Studies of the buckling and collapse of internally pressurized cone–cylinder junctions under internal pressure have been undertaken (Rotter 1985b, 1987; Greiner and Ofner 1991; Teng 1994b, 1995, 1996b) and simple strength equations have been established. Fluid-filled cones supported at the small end on a cylindrical shell were extensively tested by Vandepitte's group (Vandepitte *et al.* 1982, 1988), and have been more recently studied theoretically by Wunderlich and Albertin (1997). In large chimneys, the diameter of a cylindrical shell is often increased by the use of a conical transition between the two sizes. This leads to two cone–cylinder junctions, subject to a high axial compressive load which induces circumferential tension and compression at the two junctions, which can cause both buckling and plastic collapse. This problem has recently been studied by Schmidt and Swadlo (1997).

Other problems in this general class that have been studied include truncated hemispheres under axial tension (Yao 1963), spherical cargo tanks for liquid natural gas (Pedersen and Jensen 1976), spherical caps with a movable edge under internal pressure (Shilkrut 1983), circular plates under a central point load (Adams 1993) and plate-end pressure vessels (Teng and Rotter 1989c). Further discussion of problems at shell junctions is given in Chapters 13 and 14.

Shell stability design by numerical buckling analysis

Modern computer software is extremely powerful, and can provide numerical predictions for most shell buckling problems. These analyses can be divided into different classes, as used in the Eurocode for shell strength and stability (ENV 1993-1-6 1999). The classes are linear elastic (LA), which includes linear bifurcation analysis (LBA), geometrically nonlinear elastic (GNA), materially nonlinear (MNA), geometrically and materially nonlinear (GMNA) and geometrically and materially nonlinear analysis with explicit modelling of geometric imperfections (GMNIA). Formal definitions of these different types of analysis and the manner in which they can be adopted into the design process according to this standard were set out by Rotter (2002a,b). This process comes after considerable discussion and debate on how numerical analyses should be used, as noted below.

The key difficulty in using sophisticated software is that current design methods are based on hand calculations coupled with safety margins that are established using test results. If a computer prediction is to be adopted into this design methodology, then it is necessary to identify those parts of the process that can be replaced by the computer analysis, and to separate them from those parts which cannot. Which parts can be so separated depends strongly on the type of analysis that has been used. For example, a linear bifurcation analysis takes no account of shell nonlinear geometric effects, or plasticity, or imperfection sensitivity, or the statistical scatter present in all construction and design representations. The result of such an analysis therefore needs careful interpretation before it can be adopted into the design process.

Several approaches have been considered by various researchers and code writing committees (Schmidt and Krysik 1991; Speicher and Saal 1991; Samuelson and Eggwertz 1992; Teng and Rotter 1995). According to the simplest proposal, a linear elastic bifurcation buckling analysis may be conducted to obtain the bifurcation load of the perfect structure. Reduction factors are then applied to account for the effects of geometric imperfections and plasticity. This approach can be closely linked to hand calculations, since these also begin with linear bifurcation (classical) buckling stresses as a starting point. The chief difficulty lies in deriving appropriate reduction factors for different geometries, loading cases and support conditions (Schmidt and Krysik 1991; Samuelson and Eggwertz 1992), since it is known that all these factors influence the imperfection sensitivity markedly. The above descriptions also fail to address the problem of plasticity in a rigorous manner, and this also leads to uncertainty about the equivalent slenderness of the structure.

At the other extreme, a fully nonlinear analysis with explicit modelling of geometric imperfections (GMNIA) can be performed, using large deflection theory and properly modelled plasticity. Although this should yield a 'correct' answer to the problem, there is considerable difficulty in identifying the form and amplitude of imperfections to be used. Speicher and Saal (1991) proposed four alternative concepts for determining the design strength, using LBA, GMNA and GMNIA analyses. For the latter, they proposed that an equivalent imperfection in the same form as the first bifurcation mode should be used. They deduced the magnitude required of this imperfection for safe design, based on existing tests of axially compressed cylinders. A similar calibration exercise was undertaken by Rotter (1997b) using weld depression imperfections. However, both of these proposals have only been calibrated for axially compressed cylinders, and other load cases and shell geometries may require quite different approaches.

The simple axisymmetric weld depression proposed by Rotter and Teng (1989b) has been shown to represent the dominant components of imperfections in full-scale civil engineering shells and to produce buckling strengths comparable to those from current design criteria when its amplitude is chosen appropriately for a given slenderness (Rotter 1997b). It is expected that this weld depression, or its extension as developed by Winterstetter (2000) may be used to obtain good approximations in other shells of revolution under predominantly meridional compression of uniform or non-uniform distribution. This approach, including both plasticity and imperfections, can lead to a close estimate of the ultimate strength directly. The assumed amplitude of each weld depression still needs to be carefully chosen, and related to tolerance requirements. The value can be based on past measurements or deduced from existing experimental results or design criteria.

As described in Chapter 2, there is a continuing struggle between the need for generality (expressed in terms of a linear eigenmode imperfection) and realistic imperfections (as in the weld depression and similar realistic modes). If the imperfection is defined as an eigenmode, then this may not appear strongly in the constructed shell (e.g. for shells buckling under shear or torsion) and the tolerance control measurements may fail to detect imperfection modes that are more serious. Alternatively, if the characterisation of real imperfections is too simplistic (as a single weld depression), more deleterious modes may go undetected by the tolerance measurement system.

If no information is available on the amplitude or form of realistic imperfections in the structure, and the geometry and load case are complex, the eigenmode method of choosing the imperfections for numerical analysis may be the only satisfactory approach. A better concept, however, would be to develop a statistically based imperfection model for a particular class of shell structures, fabricated using the same process, based on extensive measurements of real geometric imperfections on full-scale structures (Arbocz 1991; Rotter *et al.* 1992; Ding *et al.* 1996b). As in the Eurocode (ENV 1993-1-6 1999), this imperfection model should distinguish between high quality and low quality shells, and define the strength in terms of reliability measures. Although this approach is desirable in the long term, it can only be developed after much more measurement of real structures, since data currently only exists for very few constructed shells (Rotter 1996b; Ding *et al.* 1996b).

The new Eurocode (ENV 1993-1-6 1999) separates the different components of the overall shell buckling calculation into its distinct parts, and permits all the above categories of analysis to be used where appropriate. It carefully defines the manner in which the results of these analyses should be interpreted. The underlying philosophy may be read in Rotter (2002a,b).

Conclusions

This chapter has given an overview of the field of the buckling of thin shell structures, focussing mostly on civil engineering shells, but with some features of shells used in other fields also noted. It has given an outline of the historical development of ideas about shell buckling, using the axially compressed cylinder as a paradigm. It has also indicated the huge range of problems in shell buckling that have been and are being investigated, pointing to the following chapters in which far more complete and detailed information will be set out.

Finally, some discussion of the reliable use of numerical modelling of shell buckling problems has been given, indicating the challenges that still face the research community and designers alike, and setting out the need for additional data collection, imperfection characterisation and tolerance measurements. It is evident that this research field is active and challenging, and will continue to remain so for a long time to come.

Acknowledgement

The authors wish to thank the ASME International for permitting them to make use of material contained in Teng (1996a) in the writing of this chapter.

References

Abir, D. and Nardo, S.V. (1958). Thermal buckling of circular cylindrical shells under circumferential temperature gradients. *Journal of the Aerospace Sciences* **26**(12), 803–808.

Abramovich, H., Singer, J. and Weller, T. (1991). The influence of initial imperfections on the buckling of stiffened cylindrical shells under combined loading. In *Buckling of Shell Structures on Land, in the Sea and in the Air* (ed. J.F. Jullien). Elsevier Applied Science, London, pp. 1–10.

Adachi, J. and Benicek, M. (1964). Buckling of torispherical shells under internal pressure. *Experimental Mechanics* **August**, 217–222.

Adams, G.G. (1993). Elastic wrinkling of a tensioned circular plate using von Karman plate theory. *Journal of Applied Mechanics, ASME* **60**, 520–525.

Agelidis, N., Harding, J.E. and Dowling, P.J. (1982). Buckling tests on stringer-stiffened cylinder models subject to load combination. Det Norske Veritas, Report 82–098, Imperial College.

Akiyama, H., Takahashi, M. and Hashimoto, S. (1987). Buckling tests of steel cylindrical shells subjected to combined bending and shear. *Transactions of Architectural Institute of Japan* No. 371, p. 44 (in Japanese).

Akiyama, H., Ohtsubo, H., Nakamura, H., Matsuura, S., Hagiwara, Y., Yuhara, T., Hirayama, H., Kokubo, K. and Ooka, Y. (1991). Outline of the seismic buckling guideline of FBR. *Proceedings of the 11th International Conference on Structural Mechanics in Reactor Technology*, Vol. E, p. 239.

Allen, L.R., Hutchinson, G.L. and Stevens, L.K. (1990). Buckling considerations in the design of elevated steel water tanks. *Thin-Walled Structures* **9**(1–4), 389–406.

Almroth, B.O. (1962). Buckling of a cylindrical shell subjected to non-uniform external pressure. *Journal of Applied Mechanics, ASME* **29**, 675–682.

Almroth, B.O. (1966). Influence of edge conditions on the stability of axially compressed cylindrical shells. *AIAA Journal* **4**(1), 134–140.

Almroth, B.O and Brogan, F.A. (1978). The STAGS computer codes. NASA CR 2950, February.

Almroth, B.O. and Holmes, A.M.C. (1972). Buckling of shells with cutouts, experiment and analysis. *International Journal of Solids and Structures* **8**(8), 1057–1071.

Almroth, B.O., Brogan, F.A. and Marlowe, M.B. (1973). Stability analysis of cylinders with circular cutouts. *AIAA Journal* **11**(11), 1582–1584.

Ansourian, P. and Sengupta, M. (1993). Analysis and design of thin-walled horizontal storage tanks. *Proceedings of the 13th Australasian Conference on Mechanics of Structures and Materials*, University of Wollongong, Australia, pp. 25–32.

Ansourian, P. and Sengupta, M. (1994). Behaviour of cylindrical tanks subjected to non-uniform pressure. *Proceedings of the Australasian Structural Engineering Conference-1994*, Sydney, Australia, October, pp. 957–979.

Antoine, P.A. (2000). Comportement des coques cylindriques minces sous chargements combinés: vers une amélioration du dimensionnement sous flexion et pression interne'. PhD Thesis, INSA de Lyon, Lyon, France.

API 620 (1982). *Recommended Rules for Design and Construction of Large Welded, Low Pressure Storage Tanks*, 7th edn. American Petroleum Institute, Texas.

API RP2A (1989). *Recommended Practice for Planning, Designing and Constructing Fixed Offshore Platforms*, 18th edn. American Petroleum Institute, Texas.

Arbocz, J. (1982). The imperfection data bank, a mean to obtain realistic buckling loads. In *Buckling of Shells* (ed. E. Ramm). Springer-Verlag, Berlin, pp. 535–567.

Arbocz, J. (1991). Towards an improved design procedure for buckling critical structures. In *Buckling of Shell Structures on Land, in the Sea and in the Air* (ed. J.F. Jullien). Elsevier Applied Science, London, pp. 270–276.

Arbocz, J. and Babcock, C.D., Jr. (1981). Computerised stability analysis using measured initial imperfections. *Proceedings of the 12th Congress of the International Council of the Aeronautical Sciences*, Munchen, 12–17 October, pp. 688–701.

Arbocz, J. and Hol, J.M.A.M. (1991). Collapse of axially compressed cylindrical shells with random imperfections. *AIAA Journal* **29**(12), 2247–2256.

Arbocz, J. and Sechler, E.E. (1976). On the buckling of stiffened imperfect cylindrical shells. *AIAA Journal* **14**(11), 1611–1617.

Arnold, S.M., Robinson, D.N. and Saleeb, A.F. (1989). Creep buckling of cylindrical shell under variable loading. *Journal of Engineering Mechanics, ASCE* **115**(5), 1054–1074.

AWWA D100–79 (1979). *Standard for Welded Steel Tanks for Water Storage*. American Water Works Association Standard, Denver, CO.

Babcock, C.D. (1974). Experiments in shell buckling. In *Thin-Shell Structures, Theory, Experiment and Design* (eds Y.C. Fung and E.E. Sechler). Prentice-Hall, Englewood Cliffs, NJ, pp. 345–369.

Babcock, C.D. (1983). Shell stability. *Journal of Applied Mechanics, ASME* **50**, 935–940.

Bailey, R. and Kulak, G.L. (1984). Flexural and shear behaviour of large diameter steel tubes. *Report No. 119*, Department of Civil Engineering, University of Alberta, Edmonton.

Beedle, L.S. (1991). *Stability of Metal Structures: A World View*, 2nd edn. Stability Research Council, USA.

Berry, P.A. and Rotter, J.M. (1996). Partial axisymmetric imperfections and their effect on the buckling strength of axially compressed cylinders. *Proceedings of the International Workshop on Imperfections in Metal Silos: Measurement, Characterisation and Strength Analysis*, CA-Silo, Lyon, France, 19 April, pp. 35–48.

Berry, P.A., Bridge, R.Q. and Rotter, J.M. (1996). Imperfection measurement of cylinders using automated scanning with a laser displacement meter. *Strain* **32**(1), 3–7.

Berry, P.A., Rotter, J.M. and Bridge, R.Q. (2000). Compression tests on cylinders with circumferential weld depressions. *Journal of Engineering Mechanics, ASCE* **126**(4), 405–413.

Bijlaard, D.L. and Gallagher, R.H. (1959). Elastic instability of a cylindrical shell under arbitrary circumferential variation of axial stresses. *Journal of the Aerospace Sciences* **27**(11), 854–858, 866.

Birch, R.S. and Jones, N. (1990). Dynamic and static axial crushing of axially stiffened cylindrical shells. *Thin-Walled Structures* **9**, 29–60.

Blachut, J. (1987). Combined axial and pressure buckling of shells having optimal positive Gaussian curvature. *Computers and Structures* **26**(3), 513–519.

Blachut, J. (1998). Some recent developments in strength and buckling of pressure vessel components. *Progress in Structural Engineering and Materials* **1**(4), 418–427.

Blachut, J. and Galletly, G.D. (1995). Buckling strength of imperfect steel hemispheres. *Thin-Walled Structures* **23**(1–4), 1–20.

Blachut, J. Galletly, G.D. and Moffat, D.G. (1991). An experimental and numerical study into the collapse strength of steel domes. In *Buckling of Shell Structures, on Land, in the Sea and in the Air* (ed. J.F. Jullien). Elsevier Applied Science, London, pp. 344–358.

Blackler, M.J. and Ansourian, P. (1988). Design against wind Induced instability of silos. *Proceedings of the 11th Australasian Conference on Mechanics of Structures and Materials*, Auckland, pp. 427–430.

BS 2654: (1989). Manufacture of vertical steel welded non-refrigerated storage tanks with butt-welded shells for the petroleum industry. British Standards Institution.

Bornscheuer, F.W., Hafner, L. and Ramm, E. (1983). Zur Stabilitat eines Kreiszylinders mit einer Rundschweissnaht unter Axialbelastung. *Der Stahlbau* **52**(10), 313–318.

Brendel, B., Ramm, E., Fischer, F.D. and Rammerstorfer, F.G. (1981). Linear and nonlinear stability analysis of thin cylindrical shells under wind loads. *Journal of Structural Mechanics* **9**(1), 91–113.

Brush, D.O. (1968). Imperfection sensitivity of of stringer siffened cylinders. *AIAA Journal* **6**(12), 2445–2447.

Budiansky, B. (ed.) (1974). *Buckling of Structures, Proceedings of an IUTAM Symposium*, Harvard University. Springer, New York, 1976.

Budiansky, B. and Hutchinson, J.W. (1966). A survey of some buckling problems. *AIAA Journal* **4**(9), 1505–1510.

Budiansky, B. and Hutchinson, J.W. (1979). Buckling: progress and challenge. In *Trends in Solid Mechanics 1979*, Proceedings of the Symposium Dedicated to the 65th Birthday of W.T. Koiter (eds J.F. Besseling and A.M.A. van der Heijden). Delft University Press, Delft, pp. 93–116.

Bushnell, D. (1976). BOSOR5 – program for buckling of elastic–plastic shells of revolution including large deflections and creep. *Computers and Structures* **6**, 221–239.

Bushnell, D. (1977). BOSOR4: program for stress, buckling and vibration of complex shells of revolution. In *Structural Mechanics Software Series* (eds N. Perrone and W. Pilkey), Vol. 1. University Press of Virginia, Charlottesville, VA, pp. 11–143.

Bushnell, D. (1981). Elastic–plastic bending and buckling of pipes and elbows. *Computers and Structures* **13**, 241–248.

Bushnell, D. (1982). Plastic buckling of various shells. *Journal of Pressure Vessel Technology, ASME* **104**, 51–72.

Bushnell, D. (1985). *Computerised Buckling Analysis of Shells*. Martinus Nijhoff Publishers, Dordrecht.

Byskov, E. and Hansen, J.C. (1980). Postbuckling and imperfection sensitivity analysis of axially stiffened cylindrical panels with mode interaction. *Journal of Structural Mechanics* **8**(2), 205–224.

Cai, M., Holst, J.M.F.G. and Rotter, J.M. (2002). Buckling strength of thin cylindrical shells under localised axial compression. *Proceedings of the 15th ASCE Engineering Mechanics Conference*, 2–5 June, Columbia University, New York, NY.

Calladine, C.R. (1983a). *Theory of Shell Structures*. Cambridge University Press, Cambridge, UK.

Calladine, C.R. (1983b). Plastic buckling of tubes in pure bending. In *Collapse: The Buckling of Structures in Theory and Practice* (eds J.M.T. Thompson and G.W. Hunt). Cambridge University Press, Cambridge, UK, 111–124.

Chen, J.F. and Xu, Z.A. (1992). Buckling of compound hyperbolic paraboloidal shells. *Thin-Walled Structures* **13**(3), 245–257.

Chiba, M., Tani, J. and Yamaki, N. (1987). Dynamic stability of liquid-filled cylindrical shells under vertical excitation. Part I: Experimental results. Part II: Theoretical results. *Earthquake Engineering and Structural Dynamics* **15**(1), 23–36 (Part I): 37–51 (Part II).

Chiba, M., Yamashida, T. and Yamauchi, M. (1989). Buckling of circular cylindrical shells partially subjected to external liquid pressure. *Thin-Walled Structures* **8**(3), 217–233.

Chryssanthopoulos, M.K. and Poggi, C. (1995). Stochastic imperfection modelling in shell buckling studies. *Thin-Walled Structures* **23**(1–4), 179–200.

Chryssanthopoulos, M.K., Baker, M.J. and Dowling, P.J. (1991a). Statistical analysis of imperfections in stiffened cylinders. *Journal of Structural Engineering, ASCE* **117**(7), 1979–1997.

Chryssanthopoulos, M.K., Baker, M.J. and Dowling, P.J. (1991b). Imperfection modelling for buckling analysis of stiffened cylinders. *Journal of Structural Engineering, ASCE* **117**(7), 1998–2017.

Citerley, R.L. (1982). Imperfection sensitivity and postbuckling behaviour of shells. In *Pressure Vessels and Piping: Design Technology-1981 – A Decade of Progress* (eds S.Y. Zamrik and D. Dietrich). ASME, New York, pp. 27–46.

Clarke, M.J. and Rotter, J.M. (1988). A technique for the measurement of imperfections in prototype silos and tanks. *Research Report R565*, School of Civil and Mining Engineering, University of Sydney, Australia.

Coleman, R., Ding, X.L. and Rotter, J.M. (1992). The measurement of imperfections in full-scale steel silos. *Proceedings of the 4th International Conference on Bulk Materials Storage, Handling and Transportation* IEAust, pp. 467–472.

Combescure, A. (1986). Static and dynamic buckling of large thin shells. *Nuclear Engineering and Design* **92**, 339–354.

Combescure, A. (1991). Upon the different theories of plastic buckling: elements for a choice. In *Buckling of Shell Structures, on Land, in the Sea and in the Air* (ed. J.F. Jullien). Elsevier Applied Science, London, pp. 448–457.

Combescure, A. (1994). Etude de la stabilité non linéaire géométrique et nonlinéaire matériau des coques minces, Habilitation, INSA de Lyon, Lyon, 122 pp. Habilitation HDR 940–12.

Combescure, A. and Brochard, J. (1991). Recent advances on thermal buckling, new results obtained at C.E.A. In *Buckling of Shell Structures, on Land, in the Sea and in the Air* (ed. J.F. Jullien). Elsevier Applied Science, London, pp. 381–390.

Combescure, A., Pernette, E. and Reynouard, J.M. (1987). Linear and nonlinear buckling of a discretely supported cooling tower using special axisymmetric shell element. *Proceeding of the International Colloquium on Stability of Plate and Shell Structures* Ghent, Belgium, 6–8 April, pp. 283–291.

Croll, J.A. (1985). Stiffened cylindrical shells under axial and pressure loading. In *Shell Structures: Stability and Strength* (ed. R. Narayanan). Elsevier Applied Science, London, pp. 19–56.

Croll, J.G.A. (1995). Towards a rationally based elastic–plastic shell buckling design methodololory. *Thin-Walled Structures* **23**(1–4), 67–84.

de Winter, P.E., Stark, J.W.B. and Witteveen, J. (1985). Collapse behaviour of submarine pipelines. In *Shell Structures: Stability and Strength* (ed. R. Narayanan). Elsevier Applied Science, London, pp. 221–246.

Dhanens, F., Lagae, G., Rathe, J. and Impe, R.V. (1993). Stresses in and buckling of unstiffened cylinders subjected to local axial loads. *Journal of Constructional Steel Research* **27**, 89–106.

DIN 18800 (1990). Stahlbauten: Stabilitätsfälle, Schalenbeulen, DIN 18800 Part 4, Deutsches Institut für Normung, Berlin, November.

Ding, X.L., Coleman, R.D. and Rotter, J.M. (1996a). Surface profiling system for measurement of engineering structures. *Journal of Surveying Engineering, ASCE* **122**(1), 3–13.

Ding, X.L., Coleman, R.D. and Rotter, J.M. (1996b). Technique for precise measurement of large-scale silos and tanks. *Journal of Surveying Engineering, ASCE* **122**(1), 14–25.

Donnell, L.H. and Wan, C.C. (1950). Effect of imperfections on buckling of thin cylinders and columns under axial compression. *Journal of Applied Mechanics, ASME* **17**(1), 73–83.

Dostal, M., Austin, N., Combescure, A., Peano, A. and Angeloni, P. (1987). Shear buckling of cylindrical vessels, benchmark exercise. *Proceedings of the 9th International Conference on Structural Mechanics in Reactor Technology* Vol. E, pp. 199–208.

Dowling, P.J. (1991). Strength and reliability of stringer-stiffened cylinders in offshore structures. In *Buckling of Shell Structures, on Land, in the Sea and in the Air* (ed. J.F. Jullien). Elsevier Applied Science, London, pp. 242–250.

Dowling, P.J., Harding, J.E., Agelidis, N. and Fahy, W. (1982). Buckling of orthogonally stiffnened cylindircal shells used in offshore engineering. In *Buckling of Shells* (ed. E. Ramm). Springer Verlag, Berlin, pp. 239–273.

Drew, H.R. and Pellegrino, S. (eds) (2002). *New Approaches to Structural Mechanics, Shells and Biological Structures* Celebration volume for the 60th birthday of Prof. C.R. Calladine, University of Cambridge. Kluwer Academic Publishers, London.

Düsing, H.E. (1994). Stabilität längsversteifter stählerner Kreiszylinderschalen unter zentrischem Axialdruck – theoretische Grundlagen und baupraktischer Beulsicherheitsnachweis. Dr-Ing. Dissertation, Universität GH Essen, Fachbereich Bauwesen.

ECCS (1988). *Buckling of Steel Shells: European Recommendations*, 4th edn. European Convention for Constructional Steelwork, Brussels.

Eggwertz, S. and Samuelson, L.A. (1991a). Design of shell structures with openings subjected to buckling. *Journal of Constructional Steel Research* **18**(2), 155–163.

Eggwertz, S. and Samuelson, L.A. (1991b). Buckling of shells with local reinforcements. In *Buckling of Shell Structures, on Land, in the Sea and in the Air* (ed. J.F. Jullien). Elsevier Applied Science, London, pp. 401–408.

Elchalakani, M., Zhao, X.L. and Grzebieta, R. (2002). Bending tests to determine slenderness limits for cold-formed circular hollow sections. *Journal of Constructional Steel Research* **58**(11), 1407–1430.

Elishakoff, I., Manen, S. van, Vermeulen, P.G. and Arbocz, J. (1987). First order secondmoment analysis of the buckling of shells with random imperfections. *AIAA Journal* **25**(8), 1113–1117.

EN 1991-4 (2003). Eurocode 1: Basis of Design and Actions on Structures, Part 4 – Silos and Tanks. CEN, Brussels.

ENV 1993-1-6 (1999). Eurocode 3: Design of Steel Structures, Part 1.6: General Rules – Supplementary Rules for the Strength and Stability of Shell Structures. CEN, Brussels.

ENV 1993-4-1 (1999). Eurocode 3: Design of Steel Structures, Part 4.1: Silos. CEN, Brussels.

ENV 1993-4-2 (1999). Eurocode 3: Design of Steel structures, Part 4.2: Tanks. CEN, Brussels.

ENV 1998-4 (1999). Eurocode 8: Design Provisions for Earthquake Resistance of Structures – Part 4: Silos, Tanks and Pipelines. CEN, Brussels.

Esslinger, M. and Geier, B. (1977). Buckling loads of thin-walled circular cylinders with axisymmetric irregularities. *Steel Plated Structures, Proceedings of an International Symposium* (eds P.J. Dowling, J.E. Harding and P.A. Frieze). Crosby Lockwood Staples, London, pp. 865–888.

Esslinger, M. and Geier, B. (1993). Flat bar steel ring stiffeners on cylinders subjected to external pressure. *Thin-Walled Structures* **15**, 249–269.

Esslinger, M., Geier, B. and Wendt, U. (1984). Berechnung der spannungen und deformationen von rotationsschalen im elasto-plastischen bereich. *Stahlbau* **1**, 17–25.

Fischer, F.D., Rammerstorfer, F.G. and Auli, W. (1985). Strength and stability of uplifting earthquake-loaded liquid filled tanks. *Transactions of the 8th International Conference on Structural Mechanics in Reactor Technology* Vol. B, pp. 475–480.

Fischer, G. (1962). Uber die berechnung der kritischen axiallasten gelenkig gelagerter kreiszylinderschalen mit hilfe des mehrstellenverfahrens. *Zeitschrift für Angewandte Mathematik und Mechanik* **42**.

Fischer, G. (1963). Ueber den Einfluss der gelenkigen Lagerung auf die Stabilitat dunnwandiger Kreiszylinderschalen unter Axiallast und Innendruck. *Zeitschrift für Flugwissenschaften* **11**, 111–119.

Fischer, G. (1965). Influence of boundary conditions on stability of thin-walled cylindrical shells under axial load and internal pressure. *AIAA Journal* **3**, 736–738.

Florence, A.L., Gefken, P.R. and Kirkpatrick, S.W. (1991). Dynamic plastic buckling of copper cylindrical shells. *International Journal of Solids and Structures* **27**(1), 89–103.

Flugge, W. (1932). Die Stabilität der Kreiszylinderschalen. *Ingenieur-Archiv* **3**, 463–506.

Fujita, K., Ito, T. and Wada, H. (1990). Experimental study on the dynamic buckling of cylindrical shell due to seismic excitation. In *Flow-Structure Vibration and Sloshing – 1990*, ASME Pressure Vessels and Piping Division Publication, Vol. 191. New York, pp. 31–36.

Fung, Y.C. and Sechler, E.E. (eds) (1974). *Thin-Shell Structures, Theory, Experiment and Design*, Prentice-Hall, Englewood Cliffs, NJ.

Galletly, G.D. (1982). The buckling of fabricated torispherical shells under internal pressure. In *Buckling of Shells* (ed. E. Ramm). Springer, Berlin, pp. 429–466.

Galletly, G.D. (1985). Torispherical shells. In *Shell Structures: Stability and Strength* (ed. R. Narayanan). Elsevier Applied Science, London, pp. 281–310.

Galletly, G.D. (ed.) (1995). Special Issue on Buckling Strength of Imperfection-Sensitive Shells. *Thin-Walled Structures* **23**(1–4).

Galletly, G.D. and Blachut, J. (1985a). Plastic buckling of short cylindrical shells subject to horizontal edge shear loads. *Journal of Pressure Vessel Technology, ASME* **107**, 101–107.

Galletly, G.D. and Blachut, J. (1985b). Torispherical shells under internal pressure-failure due to asymmetric buckling or axisymmetric yielding. *Proceedings of the Institution of Mechanical Engineers, Part C* **199**(C3), 225–238.

Galletly, G.D. and Blachut, J. (1991). Buckling design of imperfect welded hemispherical shells subjected to external pressure. *Proceedings of the Institution of Mechanical Engineers, Part C* **205**(3), 175–188.

Galletly, G.D., James, S., Kruzelecki, J. and Pemsing, K. (1987). Interactive buckling tests on cylinders subjected to external pressure and axial compression. *Journal of Pressure Vessel Technology, ASME* **109**, 10–18.

Galletly, G.D., Blachut, J. and Moreton, D.N. (1990). Internally pressurised machined domed ends – a comparison of the plastic buckling predictions of the deformation and flow theories. *Proceedings of the Institution of Mechanical Engineers, Part C* **204**, 169–186.

Giezen, J.J., Babcock, C.D. and Singer, J. (1991). Plastic buckling of cylindrical shells under biaxial loading. *Experimental Mechanics* **31**(4), 337–343.

Gorman, D.J. and Evan-Iwanowski, R.M. (1970). An analytical and experimental investigation of the effects of large prebuckling deformations on the buckling of clamped thin-walled circular cylindrical shells subjected to axial loading and internal pressure. *Developments in Theoretical and Applied Mechanics* **4**, 415–426.

Greiner, R. (1980). Ingenieurmäßige Berechnung dünnwandiger Kreiszylinderschalen. Veröffentlichung des Instituts für Stahlbau, Holzbau und Flächentragwerke der Technischen Universität Graz.

Greiner, R. (1991). Elastic plastic buckling at cone cylinder junctions of silos. In *Buckling of Shell Structures, on Land, in the Sea and in the Air* (ed. J.F. Jullien). Elsevier Applied Science, London, pp. 304–312.

Greiner, R. (1997). A concept for the classification of steel containments due to safety considerations. *Containment Structures: Risk, Safety and Reliability* (ed. B. Simpson). E & FN Spon, London, pp. 65–76.

Greiner, R. (1998). Cylindrical shells: wind loading. In *Silos: Fundamentals of Theory, Behaviour and Design* (eds C.J. Brown and J. Nielsen), chap. 17. E & FN Spon, London, pp. 378–399.

Greiner, R. and Derler, P. (1995). Effect of imperfections on wind-loaded cylindrical shells. *Thin-Walled Structures* **23**(1–4), 271–281.

Greiner, R. and Ofner, R. (1991). Elastic plastic buckling at cone cylinder junctions of silos. In *Buckling of Shell Structures on Land, in the Sea and in the Air* (ed. J.F. Jullien). Elsevier, London, pp. 304–312.

Greiner, R. and Yang, Y. (1996). Effect of imperfections on the buckling strength of cylinders with stepped wall thickness under axial loads. *Proceedings of the International Workshop on Imperfections in Metal Silos: Measurement, Characterisation and Strength Analysis*, CA-Silo, Lyon, France, 19 April, pp. 77–86.

Guggenberger, W. (1991). Buckling of cylindrical shells under concentrated loads. In *Buckling of Shell Structures, on Land, in the Sea and in the Air* (ed. J.F. Jullien). Elsevier Applied Science, London, pp. 323–333.

Guggenberger, W. (1996a). Effect of geometric imperfections taking into account the fabrication process and consistent residual stress fields of cylinders under local axial loads. *Proceedings of the International Workshop on Imperfections in Metal Silos: Measurement, Characterisation and Strength Analysis*, CA-Silo, Lyon, France, 19 April, pp. 217–228.

Guggenberger, W. (1996b) Proposal for design rules of axially loaded steel cylinders on local supports. *Proceedings of the International Conference on Advances in Steel Structures*, ICASS '96, Hong Kong, pp. 1225–1230.

Guggenberger, W., Greiner, R. and Rotter, J.M. (2000). The behaviour of locally-supported cylindrical shells: unstiffened shells. *Journal of Constructional Steel Research* **56**, 175–197.

Hagihara, S., Miyazaki, N. and Munakata, T. (1991). Bifurcation buckling analysis of a torispherical shell subjected to dynamically loaded internal pressure. *Transactions of the Japan Society of Mechanical Engineers, Part A* **57**, 2116–2121.

Hahn, G.D., She, M. and Carney, J.F., III. (1992). Buckle propagation in submarine pipelines. *Journal of Engineering Mechanics, ASCE* **118**(11), 2191–2206.

Harding, J.E., Dowling, P.J. and Agelidis, N. (eds) (1982). *Buckling of Shells in Offshore Structures*. Granada, London.

Haroun, M.A. and Mourad, S.A. (1990). Buckling behavior of liquid filled shells under lateral seismic shear. *Flow-Structure Vibration and Sloshing – 1990*, Vol. 191. ASME Pressure Vessels and Piping Division Publication, New York, pp. 11–17.

Harris, L.A., Suer, H.S., Skene, W.T. and Benjamin, R.J. (1957). The stability of thin-walled unstiffened circular cylinders under axial compression including the effects of internal pressure. *Journal Aeronautical Science* **24**(8), pp. 587–596.

Hibbit, Karlsson and Sorensen (2001). *ABAQUS User's Manual Ver 6.2*. Hibbit, Karlsson and Sorensen Inc., Pawtucket, Rhode Island, USA.

Hoff, N.J. (1965). Low buckling stresses of axially compressed circular cylindrical shells of finite length. *Journal of Applied Mechanics, ASME* **32**, 533–541.

Hoff, N.J. (1966). The perplexing behaviour of thin cylindrical shells in axial compression. *Israel Journal of Technology* **4**(1), 1–28.

Hoff, N.J. and Rehfield, W. (1965). Buckling of axially compressed cylindrical shells at stresses smaller than the classical critical value. *Journal of Applied Mechanics, ASME* **32**, 542–546.

Hoff, N.J. and Soong, T.C. (1965). Buckling of circular cylindrical shells in axial compression. *International Journal of Mechanical Sciences* **7**, 489–520.

Hoff, N.J., Chao, C.C. and Madsen, W.A. (1964). Buckling of a thin-walled circular cylindrical shell heated along an axial strip. *Journal of Applied Mechanics, ASME* **31**, 253–258.

Holst, J.M.F.G. and Rotter, J.M. (2001). Geometric imperfections, residual stresses and buckling in cylindrical shells with local support settlement. *Proceedings of the European Mechanics Conference: Euromech 424*, Kerkrade, Netherlands, pp. 36–37.

Holst, J.M.F.G., Rotter, J.M. and Calladine, C.R. (1996a). Geometric imperfections and consistent residual stress fields in elastic cylinder buckling under axial compression. *Proceedings of the International Workshop on Imperfections in Metal Silos: Measurement, Characterisation and Strength Analysis*, CA-Silo, Lyon, France, 19 April, pp. 199–216.

Holst, J.M.F.G, Rotter, J.M. and Calladine, C.R. (1996b). Modelling of experiments on buckling under localised uplift in compressed cylindrical shells. Report No. R96–012, University of Edinburgh.

Holst, J.M.F.G., Rotter, J.M. and Calladine, C.R. (1999). Imperfections in cylindrical shells resulting from fabrication misfits. *Journal of Engineering Mechanics, ASCE* **125**(4), 410–418.

Holst, J.M.F.G., Rotter, J.M. and Calladine, C.R. (2000). Imperfections and buckling in cylindrical shells with consistent residual stresses. *Journal of Constructional Steel Research* **54**, 265–282.

Hong, T. and Teng, J.G. (2002). Non-linear analysis of shells of revolution under arbitrary loads. *Computers and Structures* **80**(18–19), 1547–1568.

Hornung, U. and Saal, H. (1997). Stresses in unanchored tank shells due to settlement of the tank foundation. *Proceedings of the International Conference on Carrying Capacity of Steel Shell Structures*, Brno, 1–3 October pp. 157–163.

Hübner, A., Teng, J.G. and Saal, H. (2003). Buckling behaviour of extensively welded cylinders. *Proceedings of the International Conference on Design, Inspection and Maintenance of Cylindrical Steel Tanks and Pipelines*, Kralupy, Prague, Czech Republic, 9–11 October.

Hutchinson, J.W. and Koiter, W.T. (1970). Postbuckling theory. *Applied Mechanics Reviews, ASME* **23**(12), 1353–1366.

Ishakov, V.I. (1993). Effect of large deflections and initial imperfections on buckling of flexible shallow hyperbolic paraboloid shells. *International Journal of Mechanical Sciences* **35**(2), 103–115.

Jerath, S. (1987). Stability of cylindrical shells with and without cutouts and subjected to wind load. *Dynamics of Structures, Proceedings of the Sessions at Structures Congress '87 Related to Dynamics of Structures, ASCE* (ed. J.M. Roesset). pp. 325–335.

Jerath, S. and Sadid, H. (1985). Buckling of orthotropic cylinders due to wind load. *Journal of Structural Engineering, ASCE* **111**(5), 610–622.

Johns, D.J. (1966). On the linear buckling of circular cylindrical shells under asymmetric axial compressive stress distribution. *Journal of the Royal Aeronautical Society* **December**, 1095–1097.

Johns, D.J. (1983). Wind induced static instability of cylindrical shells. *Journal of Wind Engineering and Industrial Aerodynamics* **13**, 261–270.

Jonaidi, M. and Ansourian, P. (1997). Non-linear behaviour of storage tank shells under harmonic edge settlement. *Proceedings of the International Conference on Carrying Capacity of Steel Shell Structures*, Brno, 1–3 October, pp. 164–170.

Jonaidi, M. and Ansourian, P. (1998). Harmonic settlement effects on uniform and tapered tank shells. *Thin-Walled Structures* **31**, 237–255.

Ju, G.T. and Kyriakides, S. (1991a). On the effect of local imperfections on the stability of elastic-plastic shells in bending. In *Buckling of Shell Structures, on Land, in the Sea and in the Air* (ed. J.F. Jullien). Elsevier Applied Science, London, pp. 370–380.

Ju, G.T. and Kyriakides, S. (1991b). Bifurcation buckling versus limit load instabilities of elastic–plastic tubes under bending and external pressure. *Journal of Offshore Mechanics and Arctic Engineering* **113**(1), 43–52.

Ju, G.T. and Kyriakides, S. (1992). Bifurcation and localization instabilities in cylindrical shells under bending. II. Predictions. *International Journal of Solids and Structures* **29**(9), 1143–1171.

Jullien, J.F. (ed.) (1991). *Buckling of Shell Structures on Land, in the Sea and in the Air*. Elsevier Applied Science, London.

Jullien, J.F. and Araar, M. (1991). Towards an optimal shape of cylindrical shell structures under external pressure. In *Buckling of Shell Structures, on Land, in the Sea and in the Air* (ed. J.F. Jullien). Elsevier Applied Science, London, pp. 21–32.

Jullien, J-F. and Limam, A. (2002). Buckling of thin pressurised cylindrical shells under bending load. *Proceedings of the Third International Conference on Advances in Steel Structures, ICASS '02*, Vol. 2, Hong Kong, pp. 675–682.

Jullien, J-F., Limam, A. and Gusic, G. (1999). Cylindrical shells buckling under external pressure-influence of localised thickness variation. *Proceedings of the Second International Conference on Advances in Steel Structures, ICASS '99*, Vol. 2, Hong Kong, pp. 613–620.

Jumikis, P.T. and Rotter, J.M. (1983). Buckling of simple ringbeams for bins and tanks. *Proceedings of the International Conference on Bulk Materials Storage, Handling and Transportation*, IEAust, Newcastle, August, pp. 323–328.

Jumikis, P.T. and Rotter, J.M. (1986). Buckling of cylindrical shells in non-uniform torsion. *Proceedings of the 10th Australasian Conference on the Mechanics of Structures and Materials*, Adelaide, 20–22 August, pp. 211–216.

Kamalarasa, S. and Calladine, C.R. (1988). Buckle propagation in submarine pipelines. *International Journal of Mechanical Sciences* **30**(3/4), 217–228.

Kapania, R.K. (1990). Stability of cylindrical shells under combined wind and snow loads. *Journal of Wind Engineering and Industrial Aerodynamics* **36**, 937–948.

Knödel, P. and Schulz, U. (1985). Das Beulverhalten von biegebeanspruchten Zylinderschalen mit grossen Mantelöffnungen. *Berichte der Versuchsanstalt für Stahl, Holz und Steine* Karlsruhe, 4, Folge Heft 12.

Knödel, P. and Schulz, U. (1988). Buckling of silo bins loaded by granular solids. *Proceedings of the International Conference: 'Silos-Forschung und Praxis'*, University of Karlsruhe, Germany, October, pp. 287–302.

Knödel, P. and Schulz, U. (1992). Buckling of cylindrical bins – recent results. *Proceedings of the International Conference: 'Silos-Forschung und Praxis'*, University of Karlsruhe, Germany, October, pp. 75–82.

Knödel, P., Ummenhofer, T. and Schulz, U. (1995). On the modelling of different types of imperfections in silo shells. *Thin-Walled Structures* **23**(1–4), 283–293.

Knebel, K. and Schweizerhof, K. (1995). Buckling of cylindrical shells containing granular solids. *Thin-Walled Structures* **23**(1–4), 295–312.

Koiter, W.T. (1945). On the stability of elastic equilibrium. PhD Thesis, University of Delft (in Dutch).

Koiter, W.T. (ed.) (1960). *Proceedings of the Symposium on the Theory of Thin Elastic Shells*. North Holland, Amsterdam.

Koiter, W.T. and Mikhailov, G.K. (eds) (1980). *Theory of Shells, Proceedings of the Third IUTAM Symposium on Shell Theory, 1978*. North-Holland, Amsterdam.

Koiter, W.T., Elishakoff, I., Li, Y.W. and Starnes, J.H., Jr. (1994). Buckling of an axially compressed cylindrical shell of variable thickness. *International Journal of Solids and Structures* **31**(6), 797–805.

Kokubo, K., Nagashima, H., Takayanagi, M. and Mochizuki, A. (1993). Analysis of shear buckling of cylindrical shells. *JSME International Journal, Series A: Mechanics and Material Engineering* **36**(3), 259–266.

Kollar, L. and Dulacska, E. (1984). *Buckling of Shells for Engineers*. Wiley & Sons, New York.

Krishnakumar, S. and Foster, C.G. (1991a). Axial load capacity of cylindrical shells with local geometric defects. *Experimental Mechanics* **31**, 104–110.

Krishnakumar, S. and Foster, C.G. (1991b). Multiple geometric defects, their effect on stability of cylindrical shells. *Experimental Mechanics* **31**(3), 213–219.

Krupka, V. (1987). Buckling and limit carrying capacity of saddle loaded shells. *Proceedings of the Inernational Colloquium on Stability of Plate and Shell Structures*, Gent, Belgium, April, pp. 617–622.

Krupka, V. (1991a). Plastic squeeze of circular shell due to saddle or lug. In *Contact Loading and Local Effects in Thin-Walled Plated and Shell Structures* (eds V. Krupka and M. Drdacky). Springer-Verlag, Berlin, pp. 28–33.

Krupka, V. (1991b). Buckling and plastic punching of circular cylindrical shell due to saddle or lug loads. In *Buckling of Shell Structures, on Land, in the Sea and in the Air* (ed. J.F. Jullien). Elsevier Applied Science, London, pp. 11–20.

Krupka, V. and Schneider, P. (eds) (1997). *Carrying Capacity of Steel Shell Structures, Proceedings of the International Conference*, Brno, 1–3 October.

Kyriakides, S. and Babcock, C.D. (1981). Experimental determination of the propagation pressure of circular pipes. *Journal of Pressure Vessel Technology, ASME* **103**, 328–336.

Kyriakides, S. and Ju, G.T. (1992). Bifurcation and localization instabilities in cylindrical shells under bending. I. Experiments. *International Journal of Solids and Structures*, **29**(9), 1117–1142.

Kyriakides, S. and Shaw, P.K. (1987). Inelastic buckling of tubes under cyclic bending. *Journal of Pressure Vessel Technology, ASME* **109**, 169–178.

Lancaster, E.R., Calladine, C.R. and Palmer, S.C. (2000). Paradoxical buckling behaviour of a thin cylindrical shell under axial compression. *International Journal of Mechanical Sciences* **42**, 843–865.

Li, H.Y. (1994). Analysis of steel silo structures on discrete supports. PhD Thesis, Department of Civil and Environmental Engineering, University of Edinburgh.

Li, H.Y. and Rotter, J.M. (1996). Algebraic analysis of elastic circular cylindrical shells under local loadings (Parts 1 and 2). *Proceedings of the International Conference on Advances in Steel Structures, ICASS '96*, Hong Kong, pp. 801–807, 808–814.

Li, L.Y. and Molyneaux, T.C.K. (1994). Elastic–plastic dynamic instability of long circular cylindrical shells under pure bending. *International Journal of Mechanical Sciences* **36**(5), 431–437.

Libai, A. and Durban, D. (1973). A method for approximate stability analysis and its application to circular cylindrical shells under circumferentially varying loads. *Journal of Applied Mechanics, ASME* **40**, 971–976.

Libai, A. and Durban, D. (1977). Buckling of cylindrical shells subjected to nonuniform axial loads. *Journal of Applied Mechanics, ASME* **44**, 714–720.

Limam, A., Jullien, J.F., Greco, E. and Lestrat, D. (1991). Buckling of thin-walled cylinders under axial compression and internal pressure. In *Buckling of Shell Structures, on Land, in the Sea and in the Air* (ed. J.F. Jullien). Elsevier Applied Science, London, pp. 359–369.

Lin, X. and Teng, J.G. (2002). Buckling behaviour of extensively-welded steel cylinders under axial compression. *Proceedings of the Third International Conference on Advances in Steel Structures, ICASS '02*, Hong Kong, pp. 737–744.

Lin, Z.Q., Shen, H.S. and Chen, T.Y. (1993). Buckling propagation in marine pipelines. *China Ocean Engineering* **7**(1), 31–44.

Lindberg, H. E. (1991). Dynamic pulse buckling of imperfection-sensitive shells. *Journal of Applied Mechanics, ASME* **58**(3), 743–748.

Liu, W.K., Chen, Y.J., Tsukimori, K. and Uras R.A. (1991). Recent advances in dynamic buckling analysis of liquid-filled shells. *Journal of Pressure Vessel Technology, ASME* **113**(2), 314–320.

Lo, H., Crate, H. and Schwartz, E.B. (1947). Buckling of thin-walled cylinders under axial compression and internal pressure. NACA Report 874.

Lorenz, Z. (1908). Achsensymmetrische Verzerrungen in dunwandigen Hohlzylinder. *Zeitschrift des Vereines Deutscher Ingenieure* **52**, 1766–1793.

Louca, L.A. and Harding, J.E. (1994). Torsional buckling of ring-stiffeners in cylindrical shells subjected to external pressure. *Proceedings of the Institution of Civil Engineers, Structures and Buildings* **104**(2), 219–230.

Lu, Z., Obrecht, H. and Wunderlich, W. (1995). Imperfection sensitivity of elastic and elastic–plastic torispherical pressure vessel heads. *Thin-Walled Structures* **23**(1–4), 21–39.

Lundquist, E.E. (1933). Strength test of thin-walled duralumin cylinders in compression. NACA Technical Note, No. 473.

Lundquist, E.E. (1935). Strength test of thin-walled duralumin cylinders in combined transverse shear and bending. NACA Technical Note, No. 523.

Luo, Y.F. and Teng, J.G. (1998). Stability analysis of shells of revolution on nonlinear elastic foundations. *Computers and Structures* **69**(4), 499–511.

Manos, G.C. (1994). Shell buckling of metal model cylindrical storage tanks subjected to static lateral loads. In *Sloshing, Fluid-Structure Interaction and Structural Response due to Shock and Impact Loads*, Vol. 272. ASME Pressure Vessels and Piping Division Publication, New York, pp. 59–66.

Michel, G., Combescure, A. and Jullien, J.F. (2000a). Finite element simulation of dynamic buckling of cylinders subjected to periodic shear. *Thin-Walled Structures* **36**(2), 111–135.

Michel, G., Limam, A. and Jullien, J.F. (2000b). Buckling of cylindrical shells under static and dynamic shear loading. *Engineering Structures* **22**, 535–543.

Miller, C.D. (1982). Experimental study of the buckling of cylindrical shells with reinforced openings. Chicago Bridge and Iron Company, CBI-5388. Presented to *ASME/ANS Nuclear Engineering Conference*, Portland, OR, July.

Miller, C.D. and Vojta, J.F. (1984). Strength of stiffened cylinders subjected to combinations of axial compression and external pressure. *Proceedings of the Structural Stability Research Council Annual Technical Sessions*, San Francisco, CA, April.

Miyazaki, N. (1988). Creep buckling of circular cylindrical shell under both axial compression and internal pressure. *Computers and Structures* **28**(4), 437–441.

Miyazaki, N., Hagihara, S. and Munakata, T. (1992). Bifurcation creep buckling analysis of circular cylindrical shell under axial compression. *International Journal of Pressure Vessels and Piping* **52**(1), 1–10.

Mok, J. and Elwi, A.E. (1987). Shear behaviour of large diameter fabricated steel tubes. *Canadian Journal of Civil Engineering* **14**(3), 97–102.

Montague, P. and Horne, M.R. (1981). The behaviour of circular tubes with large openings subject to axial compression. *Journal of Mechanical Engineering Science* **23**(5), 225–242.

Moore, I.D. and Selig, E.T. (1990). Use of continuum buckling theory for evaluation of buried plastic pipe stability. *ASTM Special Technical Publication*, No. 1093. Philadelphia, PA, pp. 344–359.

Murakami, T. and Yoguchi, H. (1991). Static and dynamic buckling characteristics of imperfect cylindrical shells under transverse shearing loads. In *Buckling of Shell Structures, on Land, in the Sea and in the Air* (ed. J.F. Jullien). Elsevier Applied Science, London, 391–400.

Murray, N.W. and Bilston, P. (1992). Local buckling of thin-walled pipes being bent in the plastic range. *Thin-Walled Structures* **14**(5), 411–434.

NASA SP8007 (1968). Buckling of thin-walled circular cylinders. National Aeronautics and Space Administration, Special Publication 8007, Langley, p. 47.

Narayanan, R. (ed.) (1985). *Shell Structures: Stability and Strength*. Elsevier Applied Science, London.

Nash, W.A. (1960). Recent advances in the buckling of thin shells. *Applied Mechanics Reviews* **13**(3), 161–164.

Nicholls, R. and Karbhari, V. (1989). Nondestructive load predictions of concrete shell buckling. *Journal of Structural Engineering, ASCE* **115**(5), 1191–1211.

Nielsen, J. (1998). Pressures from flowing granular solids in silos. *Philosophical Transactions of the Royal Society of London: Mathematical, Physical and Engineering Sciences, Series A* **356**(1747), 2667–2684.

Niwa, A. and Clough, R.W. (1982). Buckling of cylindrical liquid-storage tanks under earthquake loading. *Earthquake Engineering and Structural Dynamics* **10**, 107–122.

Noor, A.K. (1990). Bibliography of monographs and surveys on shells. *Applied Mechanics Review* **43**(9), 223–234.

NZS 2654:1989 (1989). Specification for manufacture of vertical steel welded non-refrigerated storage tanks with butt-welded shells for the petroleum industry. Standards New Zealand, Wellington, New Zealand.

Obaia, K.H., Elwi, A.E. and Kulak, G.L. (1992a). Ultimate shear strength of large diameter fabricated steel tubes. *Journal of Constructional Steel Research* **22**, 115–132.

Obaia, K.H., Elwi, A.E. and Kulak, G.L. (1992b). Tests of fabricated steel cylinders subjected to transverse loads. *Journal of Constructional Steel Research* **22**, 21–37.

Odland, J. (1991). Design codes for offshore structures, buckling of cylindrical shells. In *Buckling of Shell Structures, on Land, in the Sea and in the Air* (ed. J.F. Jullien). Elsevier Applied Science, London, pp. 277–285.

Ohira, H. (1961). Local buckling theory of axially compressed cylinders. *Proceedings of the 11th Japanese National Congress on Applied Mechanics*, pp. 37–41.

Ohira, H. (1963). Linear local buckling theory of axially compressed cylinders and various eigenvalues. *Proceedings of the 5th International Symposium on Space Technology and Science*, Tokyo, pp. 511–526.

Ore, E. and Durban, D. (1989). Elastoplastic buckling of annular plate in pure shear. *Journal of Applied Mechanics, ASME* **56**, 644–651.

Ore, E. and Durban, D. (1992). Elastoplastic buckling of axially compressed circular cylindrical shells. *International Journal of Mechanical Sciences* **34**(9), 727–742.

Öry, H. Reimerdes, H.-G. and Gómez García, J. (1998). Some practical aspects of the design of shells and tanks in the aerospace industry. *Progress in Structural Engineering and Materials* **1**(4), 405–417.

Palmer, S.C. (1992). Structural effects of foundation tilt on storage tanks. *Proceedings of the Institution of Mechanical Engineers: Part E* **206**, 83–92.

Pedersen, P.T. and Jensen, J.J. (1976). Buckling of spherical cargo tanks for liquid natural gas. *Transactions of the Royal Institution of Naval Architects* **118**, 193–205.

Pédron, C. and Combescure, A. (1995). Dynamic buckling of stiffened cylindrical shells of revolution under a transient lateral pressure shock wave. *Thin-Walled Structures* **23**(1–4), 85–105.

Peek, R. (1989). Buckling criteria for earthquake-induced overturning stresses in cylindrical liquid storage tanks. In *Seismic Engineering: Research and Practice*. ASCE, New York, pp. 428–437.

Peek, R. and El-Bkaily, M. (1991). Postbuckling behaviour of unanchored steel tanks under lateral loads. *Journal of Pressure Vessel Technology, ASME* **113**, 423–428.

Peter, J. (1974). Zur Stabilität von Kreiszylinderschalen unter ungleichmäßig Verteilten axialen Randbelastungen. Dissertation, Technical University of Hanover, Germany.

Ramm, E. (ed.) (1982). *Buckling of Shells*. Springer-Verlag, Berlin.

Ramm, E. and Butcher, N. (1991). Buckling of cylindrical and conical shells under concentrated loading. In *Buckling of Shell Structures, on Land, in the Sea and in the Air* (ed. J.F. Jullien). Elsevier Applied Science, London, pp. 313–322.

Rammerstorfer, F.G. and Scharf, K. (1990). Storage tanks under earthquake loading. *Applied Mechanics Reviews* **43**(11), 261–282.

Ravn-Jensen, K. and Tvergaard, V. (1990). Effect of residual stresses on plastic buckling of cylindrical shell structures. *International Journal of Solids and Structures* **26**(9–10), 993–1004.

Reitinger, R. and Ramm, E. (1995). Buckling and imperfection sensitivity in the optimization of shell Structures. *Thin-Walled Structures* **23**(1–4), 159–177.

Resinger, F. and Greiner, R. (1982). Buckling of wind loaded cylindrical shells – applications to unstiffened and ring stiffened tanks. In *Buckling of Shells* (ed. E. Ramm). Springer-Verlag, Berlin, pp. 305–331.

Robertson, A. (1929). The strength of tubular struts. ARC Report and Memorandum No. 1185.

Roche, R.L. and Autrusson, B. (1986). Experimental tests on buckling of torispherical heads and methods of plastic bifurcation analysis. *Journal of Pressure Vessel Technology ASME* **108**, 138–145.

Roman, V.G. and Elwi, A.E. (1988). Postbuckling shear capacity of thin steel tubes. *Journal of Structural Engineering, ASCE* **114**, 2512–2524.

Roman, V.G. and Elwi, A.E. (1989). Analysis of the postbuckling behaviour of large diameter fabricated steel tubes. *Journal of Engineering Mechanics, ASCE* **115**, 2587–2600.

Ross, C.T.F. and Humphries, M. (1993). Buckling of corrugated circular cylinders under uniform external pressure. *Thin-Walled Structures* **17**(4), 259–271.

Rotter, J.M. (1985a). Buckling of ground-supported cylindrical steel bins under vertical compressive wall loads. *Proceedings of the Metal Structures Conference*, Institution of Engineers Australia, Melbourne, May, pp. 112–127.

Rotter, J.M. (1985b). Analysis and design of ringbeams. In *Design of Steel Bins for the Storage of Bulk Solids* (ed. J.M. Rotter). University of Sydney, Sydney, pp. 164–183.

Rotter, J.M. (ed.) (1985c). *Design of Steel Bins for the Storage of Bulk Solids*. University of Sydney, Sydney, Australia.

Rotter, J.M. (1986a). Recent studies of the stability of light gauge steel silo structures. *Proceedings of the Eighth International Specialty Conference on Cold-Formed Steel Structures*, St Louis, MO, pp. 543–562.

Rotter, J.M. (1986b). The analysis of steel bins subject to eccentric discharge. *Proceedings of the 2nd International Conference on Bulk Materials Storage, Handling and Transportation*, IEAust, pp. 264–271.

Rotter, J.M. (1987). The buckling and plastic collapse of ring stiffeners at cone/cylinder junctions. *Proceedings of the International Colloquium on Stability of Plate and Shell Structures*, Gent, Belgium, April, pp. 449–456.

Rotter, J.M. (1988). Calculated buckling strengths for the cylindrical wall of 10 000 tonne silos at Port Kembla. Investigation Report S663, School of Civil and Mining Engineering, University of Sydney, Australia.

Rotter, J.M. (1990). Local inelastic collapse of pressurised thin cylindrical steel shells under axial compression. *Journal of Structural Engineering, ASCE* **116**(7), 1955–1970.

Rotter, J.M. (1993). The design of circular metal silos for strength. *Proceedings of an International Symposium – Reliable Flow of Particulate Solids II*, Oslo, Norway, August, pp. 219–234.

Rotter, J.M. (ed.) (1996a). *Proceedings of the International Workshop on Imperfections in Metal Silos: Measurement, Characterisation and Strength Analysis*, CA-Silo, Lyon, France, 19 April.

Rotter, J.M. (1996b). Elastic plastic buckling and collapse in internally pressurised axially compressed silo cylinders with measured axisymmetric imperfections: interactions between imperfections, residual stresses and collapse. *Proceedings of the International Workshop on Imperfections in Metal Silos: Measurement, Characterisation and Strength Analysis*, CA-Silo, Lyon, France, 19 April, pp. 119–140.

Rotter, J.M. (1997a). Challenges for the future in the design of bulk solid storages. *Containment Structures: Risk, Safety and Reliability* (ed. B. Simpson). E & Fn Spon, London, pp. 11–34.

Rotter, J.M. (1997b). Pressurised axially compressed cylinders. *Proceedings of the International Conference on Carrying Capacity of Steel Shell Structures*, Brno, pp. 354–360.

Rotter, J.M. (1998a). Development of proposed European design rules for buckling of axially compressed cylinders. *Advances in Structural Engineering* **1**(4), 273–286.

Rotter, J.M. (1998b). Metal silos. *Progress in Structural Engineering and Materials* **1**(4), 428–435.

Rotter, J.M. (2001). Pressures, stresses and buckling in metal silos containing eccentrically discharging solids. Festschrift Richard Greiner, Celebration volume for the 60th birthday of Prof. Richard Greiner, TU Graz, Austria, October, pp. 85–104.

Rotter, J.M. (2002a). Shell buckling and collapse analysis for structural design: the new framework of the European standard. In *New Approaches to Structural Mechanics, Shells and Biological Structures* (eds H.R. Drew and S. Pellegrino). Celebration volume for the 60th birthday of Prof. C.R. Calladine, Kluwer Academic Publishers, London, pp. 355–378.

Rotter, J.M. (2002b). Advanced computer calculations in the design of shell structures. *Proceedings of the Third International Conference on Advances in Steel Structures*, ICASS '02, Vol. 1, Hong Kong, pp. 27–42.

Rotter, J.M. and Hull, T.S. (1989). Wall loads in squat steel silos during earthquakes. *Engineering Structures* **11**(3), 139–147.

Rotter, J.M. and Jumikis, P.T. (1985). Elastic buckling of stiffened ringbeams for large elevated bins. *Proceedings of the Metal Structures Conference*, IEAust, Melbourne, May, pp. 104–111.

Rotter, J.M. and Jumikis, P.T. (1988). Non-linear strain–displacement relations for thin shells of revolution. Research Report R563, School of Civil and Mining Engineering, University of Sydney, Australia, November.

Rotter, J.M. and Teng, J.G. (1989a). Elastic stability of lap-jointed cylinders. *Journal of Structural Engineering, ASCE* **115**(3), 683–697.

Rotter, J.M. and Teng, J.G. (1989b). Elastic stability of cylindrical shells with weld depressions. *Journal of Structural Engineering, ASCE* **115**(5), 1244–1263.

Rotter, J.M. and Zhang, Q. (1990). Elastic buckling of imperfect cylinders containing granular solids. *Journal of Structural Engineering, ASCE* **116**(8), 2253–2271.

Rotter, J.M., Zhang, Q. and Teng, J.G. (1987). Corrugation collapse in circumferentially corrugated steel cylinders'. *Proceedings of the First National Structural Engineering Conference*, Melbourne, Institution of Engineers, Australia, pp. 377–383.

Rotter, J.M., Jumikis, P.T., Fleming, S.P. and Porter, S.J. (1989). Experiments on the buckling of thin-walled model silo structures. *Journal of Constructional Steel Research* **13**(4), 271–299.

Rotter, J.M., Teng, J.G. and Li, H.Y. (1991). Buckling in thin elastic cylinders on column supports. In *Buckling of Shell Structures, on Land, in the Sea and in the Air* (ed. J.F. Jullien). Elsevier Applied Science, London, pp. 334–343.

Rotter, J.M., Coleman, R., Ding, X.L. and Teng, J.G. (1992). The measurement of imperfections in cylindrical silos for buckling strength assessment. *Proceedings of the 4th International Conference on Bulk Materials Storage, Handling and Transportation*, July, pp. 473–479.

Saal, H. (1982). Buckling of long liquid-filled cylindrical shells. In *Buckling of Shells* (ed. E. Ramm), Springer-Verlag, Berlin.

Saal, H. and Schrufer, W. (1991). Stability of wind-loaded shells of cylindrical tanks with unrestrained upper edge. In *Buckling of Shell Structures, on Land, in the Sea and in the Air* (ed. J.F. Jullien). Elsevier Applied Science, London, pp. 223–232.

Saigal, S., Yang, T.Y. and Kapania, R.K. (1987). Dynamic buckling of imperfection-sensitive shell structures. *Journal of Aircraft* **24**(10), 719–724.

Sammari, A. and Jullien, J.F. (1995). Creep buckling of cylindrical shells under external lateral pressure. *Thin-Walled Structures* **23**(1–4), 255–269.

Samuelson, L. (1985). Buckling of cylindrical shells under axial compression and subjected to localised load. In *EuroMech Colloquium 200: Postbuckling of Elastic Structures*, Matrafured.

Samuelson, L. (1991a). The ECCS recommendations on shell stability, design philosophy and practical applications. In *Buckling of Shell Structures, on Land, in the Sea and in the Air* (ed. J.F. Jullien). Elsevier Applied Science, London, pp. 261–264.

Samuelson, L. (1991b). Effect of local loads on the stability of shells subjected to uniform pressure distribution. In *Contact Loading and Local Effects in Thin-Walled Plated and Shell Structures* (eds V. Krupka and M. Drdacky). Springer-Verlag, Berlin, pp. 42–51.

Samuelson, L. and Eggwertz, S. (1992). *Shell Stability Handbook*. Elsevier Applied Science, London.

Sanders, J.L. (1963). Nonlinear theories for thin shells. *Quarterly Journal of Applied Mathematics* **20**(1), 21–36.

Schmidt, H. (1991). The German code DIN 18800 Part 4: Stability of shell-type steel structures, design philosophy and practical applications. In *Buckling of Shell Structures, on Land, in the Sea and in the Air* (ed. J.F. Jullien). Elsevier Applied Science, London, pp. 265–269.

Schmidt, H. (2000). Stability of steel shell structures – general report. *Journal of Constructional Steel Research* **55**(1–3), 159–181.

Schmidt, H. and Krysik, R. (1991). Towards recommendations for shell stability design by means of numerically determined buckling loads. In *Buckling of Shell Structures, on Land, in the Sea and in the Air* (ed. J.F. Jullien). Elsevier Applied Science, London, pp. 508–519.

Schmidt, H. and Swadlo, P. (1997). Strength and stability design of unstiffened cylinder/cone/cylinder and cone/cone shell assemblies under axial compression. *Proceedings of the International Conference on Carrying Capacity of Steel Shell Structures*, Brno, pp. 361–367.

Schmidt, H. Binder, B. and Lange, H. (1998). Postbuckling strength design of open thin-walled cylindrical tanks under wind load. *Thin-Walled Structures* **31**, 203–220.

Schneider, W., Thiele, R. and Bohm, S. (1997). Carrying capacity of slender wind-loaded cylindrical shells. *Proceedings of the International Conference on Carrying Capacity of Steel Shell Structures*, Brno, pp. 171–177.

Schroeder, V.P. (1972). Uber die stabilitat der querkraftbelasteten dunnwandigen kreiszylinderschale. *Zeitschrift für Angewandte Mathematik and Mechanik* T145–T148.

Seide, P. (1962). The stability under axial compression and lateral pressure of circular–cylindrical shells with a soft elastic core. *Journal of the Aerospace Sciences* **29**(7), 851–862.

Sharma, U.C., Rotter, J.M. and Jumikis, P.T. (1987). Shell restraint to ringbeam buckling in elevated steel silos. *Proceedings of the 1st National Structural Engineering Conference*, Melbourne, IEAust, Australia, August, pp. 604–609.

She, K.M. and Rotter, J.M. (1993). Nonlinear and stability behaviour of discretely supported cylinders. Research Report 93-01, Department of Civil Engineering and Building Science, University of Edinburgh, UK.

Shen, H.S. and Chen, T.Y. (1991). Buckling and postbuckling behaviour of cylindrical shells under combined external pressure and axial compression. *Thin-Walled Structures* **12**(4), 321–334.

Shilkrut, D.I. (1983). Bifurcation in tension of nonlinear spherical caps. *Journal of Structural Engineering, ASCE* **109**(1), 289–295.

Simitses, G.J. (1986). Buckling and postbuckling of imperfect cylindrical shells: a review. *Applied Mechanics Reviews* **39**(10), 1517–1524.

Simitses, G.J., Shaw, D, Sheinman, I. and Giri, J. (1985). Imperfection sensitivity of fibre-reinforced, composite, thin cylinders. *Composites Science and Technology* **22**, 259–276.

Singer, J. (1980). Buckling experiments on shells a review of recent developments. TAE Report 403, Department of Aeronautical Engineering, Technion, Israel Institute of Technology.

Singer, J. (1982a). Vibration correlation techniques for improved buckling predictions of imperfect stiffened shells. In *Buckling of Shells in Offshore Structures* (eds J.E. Harding, P.J. Dowling and N. Agelidis). Granada, London, pp. 285–329.

Singer, J. (1982b). The status of experimental buckling investigations of shells. In *Buckling of Shells* (ed. E. Ramm). Springer-Verlag, New York, 501–534.

Singer, J. (1983). Vibration and buckling of imperfect stiffened shells – recent developments. In *Collapse: The Buckling of Structures in Theory and Practice* (eds J.M.T. Thompson and G.W. Hunt), Cambridge University Press, Cambridge, pp. 443–479.

Singer, J. and Abramovich, H. (1979). Vibration techniques for definition of practical boundary conditions in stiffened shells. *AIAA Journal* **17**(7), 762–769.

Singer, J. and Abramovich, H. (1995). The development of shell imperfection measurement techniques. *Thin-Walled Structures* **23**(1–4), 379–398.

Singer, J. and Rosen, A. (1976). Influence of boundary conditions on the buckling of stiffened cylindrical shells. In *Buckling of Structures, Proceedings of an IUTAM Symposium*, Harvard University, Cambridge, MA, June 1974, Springer-Verlag, Berlin, pp. 227–250.

Smith, B.W. (1998). Chimneys, towers and masts. *Progress in Structural Engineering and Materials* **1**(4), 400–404.

Song, C.Y. and Teng, J.G. (2002). Buckling of circular steel silos subject to eccentric discharge pressures. *Proceedings of the Second International Conference on Advances on Steel Structures*, Hong Kong, Vol. I, Part I, pp. 593–702; Part II, pp. 703–712.

Song, C.Y., Teng, J.G. and Rotter, J.M. Imperfection sensitivity of thin elastic cylindrical shells subject to partial axial compression. Submitted for publication.

Soric, J. (1990). Stability analysis of a torispherical shell subjected to internal pressure. *Computers and Structures* **36**(1), 147–156.

Soric, J. (1995). Imperfection sensitivity of internally-pressurized torispherical shells. *Thin-Walled Structures* **23**(1–4), 57–66.

Southwell, R.V. (1914). On the general theory of elastic stability. *Philosophical Transactions of the Royal Society, London, Series A* **213** 187–202.

Souza, M.A. and Assaid, L.M.B. (1991). A new technique for the prediction of buckling loads from nondestructive vibration tests. *Experimental Mechanics* June, 93–97.

Speicher, G. and Saal, H. (1991). Numerical calculation of limit loads for shells of revolution with particular regard to the applying equivalent initial imperfections. In *Buckling of Shell Structures, on Land, in the Sea and in the Air* (ed. J.F. Jullien). Elsevier Applied Science, London, pp. 466–475.

Starnes, J.H. (1974). The effect of cutouts on the buckling of thin shells. In *Thin-Shell Structures, Theory, Experiment and Design* (eds Y.C. Fung and E.E. Sechler). Prentice-Hall, Englewood Cliffs, NJ, pp. 289–304.

Stein, M. (1962). The effect on the buckling of perfect cylinders of prebuckling deformations and stresses induced by edge support. In *Collected Papers on Instability of Shell Structures*, NASA Technical Note, D-1510, pp. 217–226.

Stein, M. (1964). The influence of prebuckling deformations and stresses on the buckling of perfect cylinders. NASA Technical Report TR R-190.

Stuhlman, C.E., DeLuzio, A. and Almroth, B. (1966). Influence of stiffener eccentricity and end moment on stability of cylinders in compression. *AIAA Journal* **4**, 872–877.

Su, X.Y., Wu, J.K. and Hu, H.C. (987). A large displacement analysis for thin shells of revolution. *Science in China* pp. 400–410.

Teng, J.G. (1994a). Plastic collapse at lap joints in pressurised cylinders under axial load. *Journal of Structural Engineering, ASCE* **120**(1), 23–45.

Teng, J.G. (1994b). Cone-cylinder intersection under internal pressure: axisymmetric failure. *Journal of Engineering Mechanics, ASCE* **120**(9), 1896–1912.

Teng, J.G. (1995). Cone–cylinder intersection under internal pressure: non-symmetric buckling. *Journal of Engineering Mechanics, ASCE* **121**(12), 1298–1305.

Teng, J.G. (1996a). Buckling of thin shells: recent advances and trends. *Applied Mechanics Reviews* **49**(4), 263–274.

Teng, J.G. (1996b). Elastic buckling of cone–cylinder intersection under localised circumferential compression. *Engineering Structures* **18**(1), 41–48.

Teng, J.G. (ed.) (1998). Special Issue on Metal Shell Structures. *Thin-Walled Structures* **31**(1–3).

Teng, J.G. (2000). Intersections in steel shell structures. *Progress in Structural Engineering and Materials* **2**(4), 459–471.

Teng, J.G. and Hong, T. (1998). Nonlinear thin shell theories for numerical buckling predictions. *Thin-Walled Structures* **31**(1–3), 89–115.

Teng, J.G. and Lucas, R.M. (1994). Out-of-plane buckling of restrained rings of general open cross-section. *Journal of Engineering Mechanics, ASCE* **120**(5), 929–948.

Teng, J.G. and Luo, Y.F. (1997). Post-collapse bifurcation analysis of shells of revolution by the accumulated arc-length method. *International Journal for Numerical Methods in Engineering* **40**(13), 2369–2383.

Teng, J.G. and Rotter, J.M. (1988). Buckling of restrained monosymmetric rings. *Journal of Engineering Mechanics, ASCE* **114**(10), 1651–1671.

Teng, J.G. and Rotter, J.M. (1989a). Elastic–plastic large deflection analysis of axisymmetric shells. *Computers and Structures* **31**(2), 211–233.

Teng, J.G. and Rotter, J.M. (1989b). Non-symmetric bifurcation of geometrically nonlinear elastic–plastic axisymmetric shells under combined loads including torsion. *Computers and Structures* **32**(2), 453–475.

Teng, J.G. and Rotter, J.M. (1989c). Non-symmetric buckling of plate-end pressure vessels. *Journal of Pressure Vessel Technology, ASME* **111**(3), 304–311.

Teng, J.G. and Rotter, J.M. (1989d). Buckling of rings in column-supported bins and tanks. *Thin-Walled Structures* **7**(3&4), 251–280.

Teng, J.G. and Rotter, J.M. (1991a). A study of buckling in column-supported cylinders. In *Contact Loading and Local Effects in Thin-Walled Plated and Shell Structures* (eds V. Krupka and M. Drdacky). Springer-Verlag, Berlin, pp. 52–61.

Teng, J.G. and Rotter, J.M. (1991b). Plastic buckling of rings at steel silo transition junctions. *Journal of Constructional Steel Research* **19**, 1–18.

Teng, J.G. and Rotter, J.M. (1991c). Geometrically and materially nonlinear analysis of reinforced concrete shells of revolution. *Computers and Structures* **42**(3), 327–340.

Teng, J.G. and Rotter, J.M. (1992a). On the buckling of imperfect pressurised cylinders under axial compression. *Journal of Engineering Mechanics, ASCE* **118**(2), 229–247.

Teng, J.G. and Rotter, J.M. (1992b). Linear bifurcation of perfect column-supported cylinders: support modelling and boundary conditions. *Thin-Walled Structures* **14**, 241–263.

Teng, J.G. and Rotter, J.M. (1992c). Recent research on the behaviour and design of steel silo hoppers and transition junctions. *Journal of Constructional Steel Research* **23**, 313–343.

Teng, J.G. and Rotter, J.M. (1995). Stability assessment of complex shell structures by numerical analysis. *Australian Civil Engineering Transactions* **CE37**(1), 61–69.

Teng, J.G. and Song, C.Y. (2001). Numerical models for nonlinear analysis of elastic shells with eigenmode-affine imperfections. *International Journal of Solids and Structures* **38**(18), 3263–3280.

Teng, J.G., Zhao, Y. and Lam, L. (2001). Techniques for buckling experiments on steel silo transition junctions. *Thin-Walled Structures* **39**(8), 685–707.

Tennyson, R.C. (1968). The effects of unreinforced cutouts on the buckling of circular cylindrical shells under axial compression. *Journal of Engineering for Industry, ASME* **90**, 541–546.

Thompson, J.M.T. and Hunt, G.W. (eds) (1983). *Collapse: The Buckling of Structures in Theory and Practice*. Cambridge University Press, Cambridge, UK.

Timoshenko, S.P. (1910). Einige stabiats Probleme der Elastizitatstheorie. *Zeitschrift für Angewandte Mathematik und Physik* **58**, 337–357.

Toda, S. (1983). Buckling of cylinders with cutouts under axial compression. *Experimental Mechanics* December, 414–417.

Tooth, A.S. and Susatijo, I. (1983). A study of the stability of horizontal cylindrical vessels filled with varying amounts of liquid: part i and part ii. *Thin-Walled Structures* **1**, 121–138.

Trahair, N.S., Abel, A., Ansourian, P., Irvine, H.M. and Rotter, J.M. (1983a). *Structural Design of Steel Bins for Bulk Solids*. Australian Institute of Steel Construction, Sydney, Australia.

Trahair, N.S., Ansourian, P. and Rotter, J.M. (1983b). Stability problems in the structural design of steel silos. *Proceedings of the International Conference on Bulk Materials Storage, Handling and Transportation*, Newcastle, Institution of Engineers, Australia, August, pp. 312–316.

Tvergaard, V. (1976). Buckling behaviour of plate and shell structures. *Theoretical and Applied Mechanics* (ed. W.T. Koiter). North-Holland, Amsterdam, pp. 233–247.

Uchiyama, K., Uematsu, Y. and Orimo, T. (1987). Experiments on the deflection and buckling behaviour of ring-stiffened cylindrical shells under wind pressure. *Journal of Wind Engineering and Industrial Aerodynamics* **26**(2), 195–211.

Uemura, S. and Morita, M. (1971). Buckling of circular cylindrical shells under linearly varying external pressure. *Transactions of JSME* **37**, 1100–1106 (in Japanese).

Uras, R.A. and Liu, W.K. (1990). Dynamic buckling of liquid-filled shells under horizontal excitation. *Journal of Sound and Vibration* **141**(3), 389–408.

Vandepitte, D., Rathe, J., Verhegghe, B., Paridaens, R. and Verschaeve, C. (1982). Experimental investigation of buckling of hydrostatically loaded, conical shells and practical evaluation of buckling loads. In *Buckling of Shells* (ed. E. Ramm). Springer-Verlag, Berlin, pp. 355–374.

Vandepitte, D., van den Steen, A., van Impe, R., Lagae, G. and Rathé, J. (1988). Elastic and elastic–plastic buckling of liquid-filled conical shells. In *Buckling of Structures, Theory and Experiment*, Josef Singer Anniversary Volume. Elsevier Science Publishers, Amsterdam, pp. 433–449.

Velickov, D. and Schmidt, H. (1998). Beulwiderstand axial gedrückter Metallkreiszylinderschalen mit unversteifter Mantelöffnung auf der Grundlage statistisch ausgewerteter Versuchsergebnisse. *Stahlbau* **67**(5), 459–464.

von Karman, T. and Tsien, H.S. (1941). The buckling of thin cylindrical shells under axial compression. *Journal of Aerospace Sciences* **8**, 303–312.

Wang, D., Zhang, S. and Yang, G. (1993a). Impact torsional buckling of elastic cylindrical shells with arbitrary form imperfection. *Applied Mathematics and Mechanics (English Edition)* **14**(6), 499–505.

Wang, D., Zhang, S. and Yang, G. (1993b). Impact torsional buckling for the plastic cylindrical shell. *Applied Mathematics and Mechanics (English Edition)* **14**(8), 693–698.

Wei, X. and Shun C. (1988). Buckling of locally loaded isotropic, orthotropic, and composite cylindrical shells. *Journal of Applied Mechanics, ASME* **55**(2), 425–429.

Weingarten, V.I. (1962). The buckling of cylindrical shells under longitudinally varying loads. *Journal of Applied Mechanics, ASME* **29**, 81–85.

Weingarten, V.I., Morgan, E.J. and Seide, P. (1965). Elastic stability of thin-walled cylindrical and conical shells under combined internal pressure and axial compression. *AIAA Journal* **3**(6), 1118–1125.

Weller, T. (1978). Combined stiffening and in-plane boundary conditions effects on the buckling of circular cylindrical stiffened-shells. *Computers and Structures* **9**, 1–16.

Weller, T. and Singer, J. (1977). Experimental studies on the buckling under axial compression of integrally stringer-stiffened circular cylindrical shells. *Journal of Applied Mechanics, ASME* **44**, 721–730.

Weller, T., Singer, J. and Batterman, S.C. (1974). Influence of eccentricity of loading on buckling of stringer-stiffened cylindrical shells. In *Thin-Shell Structures, Theory, Experiment and Design* (eds Y.C. Fung and E.E. Sechler). Prentice-Hall, Englewood Cliffs, NJ, pp. 305–324.

Weller, T., Abramovich, H. and Singer, J. (1986). Application of nondestructive vibration correlation techniques for buckling of spot welded and riveted stringer stiffened cylindrical shells. *Zeitschrift für Flugwissenschaften und Weltraumforschung* **10**(3), 183–189.

Wilson, W.M. and Newmark, N.M. (1933). The strength of thin cylindrical shells as columns. Bulletin No. 255, Engineering Experimental Station, University of Illinois.

Winterstetter, Th.A. (2000). Beulen von Kreiszylinderschalen aus Stahl unter kombinierter Beanspruchung. PhD Thesis, Universität Essen, Essen, Germany.

Wunderlich, W. and Albertin, U. (1997). The influence of boundary conditions on the load carrying behaviour and the imperfection sensitivity of conical shells. *Proceedings of the International Conference on Carrying Capacity of Steel Shell Structures*, Brno, 1–3 October, pp. 98–105.

Wunderlich, W., Cramer, H. and Obrecht, H. (1985). Application of ring elements in the non-linear analysis of shells of revolution under non-axisymmetric loading. *Computer Methods in Applied Mechanics and Engineering* **51**, 259–275.

Wunderlich, W., Obrecht, H. and Schnabel, F. (1987). Nonlinear behaviour of externally pressurised toriconical shells-analysis and design criteria. *Stability of Plate and Shell Structures* (eds P. Dubas and D. Vandepitte). *Proceedings of the ECCS Colloquium*, University of Ghent, Belgium, pp. 373–386.

Yamada, S. and Croll, J.G.A. (1993). Buckling and postbuckling characteristics of pressure-loaded cylinders. *Journal of Applied Mechanics, ASME* **60**(2), 290–299.

Yamaki, N. (1984). *Elastic Stability of Circular Cylindrical Shells*. North Holland, Amsterdam.

Yamaki, N. and Kodama, S. (1972). Buckling of circular cylindrical shells under compression – report 3: solutions based on the Donnell type equations considering prebuckling edge rotations. Report of the Institute of High Speed Mechanics, No. 25, Tohoku University, pp. 99–141.

Yamaki, N., Naito, K. and Sato, E. (1979). Buckling of circular cylindrical shells under combined action of a transverse shear edge load and hydrostatic pressure. *Proceedings of the International Conference on Thin-Walled Structures*, University of Strathclyde, Glasgow, UK.

Yang, H.T.Y., Saigal, S. and Liaw, D.G. (1990). Advances of thin shell finite elements and some applications – version I. *Computers and Structures* **35**, 481–504.

Yao, J.C. (1963). Buckling of a truncated hemisphere under axial tension. *AIAA Journal* **1**, 2316–2319.

Yeh, K.Y., Song, W.P. and Rimrott, F.P.J. (1992). Nonlinear instability of corrugated diaphragms. *AIAA Journal* **30**(9), 2325–2331.

Yin, J., Suo, X.Z. and Combescure, A. (1987). About two new efficient nonlinear shell elements. *SMIRT Conference Proceedings*, Lausanne, Paper No. B12/1.

Yun, H. and Kyriakides, S. (1990). On the beam and shell modes of buckling of buried pipelines. *Soil Dynamics and Earthquake Engineering* **9**(4), 179–193.

Zamrik, S.Y. and Dietrich, D. (eds) (1982). *Pressure Vessels and Piping: Design Technology – 1982 – a Decade of Progress*, ASME, New York.

Zhao, Y. and Teng, J.G. (2001). Buckling experiments on cone–cylinder intersections under internal pressure. *Journal of Engineering Mechanics, ASCE* **127**(12), 1231–1239.

Zhou, M., Zheng, S. and Zhang, W. (1992). Study on elephant-foot buckling of broad liquid storage tanks by nonlinear theory of shells. *Computers and Structures* **44**(4), 783–788.

Chapter 2

Cylindrical shells under axial compression

J.M. Rotter

Cylindrical shell structures are often subjected to compressive stresses in the direction of the cylinder axis, which can be either uniform or varying throughout the cylinder. The buckling strength of a thin cylindrical shell under axial compression is particularly sensitive to imperfections in the shell, and the changing patterns of behaviour with changing geometry, loading and boundary conditions make the axially compressed cylinder a classical exemplar for behaviours that may be found in a less marked form in other structures or in shells under other loading conditions. For these reasons, the axially compressed cylinder has probably been the most extensively studied of all shell buckling conditions, giving a wealth of evidence from both experimental and theoretical work.

However, the primary goal of research is to achieve a clear and complete understanding of all the phenomena, and this has been particularly elusive: the axially compressed cylinder may be one of the last classical problems in homogeneous isotropic structural mechanics for which it remains difficult to obtain close agreement between careful experiments and the best predictions from numerical modelling.

This chapter outlines the key features of the buckling behaviour of axially compressed thin cylinders, leading to the latest recommendations for practical design, followed by a brief outline of some key questions that remain unanswered. Very many studies have explored different aspects of the buckling and postbuckling behaviour of these shells (see Timoshenko 1936; Flügge 1973; CRCJ 1971; Fung and Sechler 1974; Brush and Almroth 1976; Arbocz 1983; Calladine 1983; Thompson and Hunt 1983; Yamaki 1984; Bushnell 1985; Samuelson and Eggwertz 1992; Teng 1996). The opening description of elastic buckling phenomena given here naturally follows the very comprehensive work of Yamaki (1984) quite closely.

Specific aspects addressed in this chapter are: buckling and postbuckling of perfect cylinders, buckling of imperfect cylinders, characterisation and control of imperfections and internally pressurised cylinders.

Introduction

Axial compression in cylindrical shells

Structural members subjected to compression are susceptible to Euler buckling, and the Euler buckling stress of a given quantity of material is at its greatest when all the material is placed as far as possible from the axis. This makes the thin cylindrical tube or shell the most efficient form for compression members. However, as the tube wall becomes thinner, other local forms of buckling intervene, and these shell buckling modes control the strength of thin-walled cylinders. Compression members of this form include aircraft, spacecraft and terrestrial vehicles, as well as components of bridges, offshore platforms and other civil engineered structures. Shell structures are also very efficient for containment of fluids and solids, with very thin walled vessels being commonly used for both tanks and silos. This book is chiefly concerned with civil engineering structures, so the assumed geometries and material behaviours relate mostly to them. However, the low resistance to shell buckling of these thin shells means that buckling is a primary design concern.

In many applications, the diameter of the cylinder is controlled by other factors (storage containers, chimneys, aircraft and spacecraft, towers, etc.), so the required thickness may be so small that the buckling stress falls far below the strength of the material. For most thin shells, this is the case.

Axial compression in a cylinder arises from different causes in different structures: in a tower, the weight of the structure may provide a relatively uniform compression. In both towers and chimneys (Fig. 2.1(a)), the transverse loading of wind or earthquake leads to compressive stresses on one side (Fig. 2.2(b)), increasing down the length of the structure, and sometimes in a more complex pattern than that predicted by engineering bending theory. In storage tanks with the cylinder axis vertical, loads on the roof cause axial compression in the shell walls (Fig. 2.2(a)),

Figure 2.1 Examples of axial compression in civil engineering cylindrical shells: (a) tower or chimney; (b) saddle-supported tank; (c) silo.

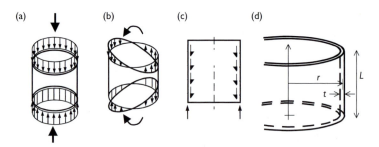

Figure 2.2 Axial compression development in a cylinder: simple cases. (a) Uniform compression; (b) global bending; (c) frictional traction; (d) notation.

but in saddle-supported tanks where the axis is horizontal (Fig. 2.1(b)), the shell must act as a beam in bending, again leading to axial compression stresses that vary around the circumference (Fig. 2.2(b)) and along the length. In silo structures (Fig. 2.1(c)) the stored solid exerts a normal pressure against the wall, but also applies a frictional drag (Fig. 2.2(c)) which accumulates into a substantial axial compression. In addition, unsymmetrical pressures from the solids cause compressions that vary around the cylinder circumference. It is therefore clear that axially compressed thin cylinders occur in many different civil engineering applications, each with its own special features, but with many aspects in common.

There have been many structural failures in thin shells, and the commonest failure mode is probably buckling under axial compression. The reasons for these failures are varied, but a clear statement of the factors known to affect the buckling strength of cylinders is needed before the more complex issues can be addressed. This chapter identifies these factors and reviews current knowledge and design recommendations for buckling of unstiffened thin metal cylinders under axial compression.

Phenomena that affect the buckling strength

The buckling strength of an isotropic axially compressed thin cylindrical shell depends on its geometry (radius r, thickness t and length L), its elastic modulus E and yield stress σ_Y, the amplitudes and forms of minor imperfections in its geometry, the end boundary conditions and the pattern of loading. If the shell has ring stiffeners or changes of plate thickness within it, then these generally act as boundaries between one shell segment and another, separating the shell into zones which can often be considered independent. The strengths of cylinders measured in laboratory tests are very scattered and fall far below the ideal strength (Fig. 2.3). The reasons for both the scatter and the loss of strength long posed a major challenge to researchers, and continue to pose challenges to designers in determining how to ensure that practical shell constructions meet their strength requirements.

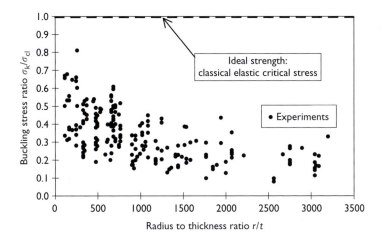

Figure 2.3 Experimental strengths of isotropic axially compressed cylinders (after Harris et al. 1957).

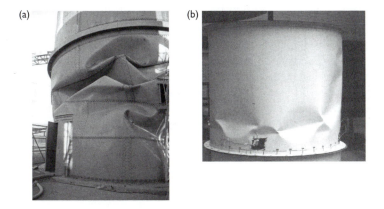

Figure 2.4 Typical appearance of axial compression buckles: (a) failure in service; (b) test in laboratory (Knödel and Schulz 1992).

As it buckles, the surface of a high-quality shell suddenly and dramatically changes from the initial shape into a wave-form which is termed the buckling mode. The mode may be symmetric with respect to the axis, forming a single bulge around the circumference, but this only occurs under special circumstances. Most commonly, the buckling mode involves alternately inward and outward displacements of the shell wall, termed an asymmetric or non-symmetric buckle (Fig. 2.4). The wavelength of the buckles in both the axial and circumferential directions may vary without significantly altering the buckling load. This leads to a multiplicity of possible modes at closely spaced loads, and jumps usually occur from one mode to another as the buckling progresses into the postbuckling region (e.g. Riks *et al.* 1996). This rich potential for many different modes of buckling is also implicated in the sensitivity of the buckling strength to geometric imperfections. Further, the practice, common in other structural forms (e.g. beams and columns), of adding stiffeners to force the structure from one buckling mode into another with a higher strength, is generally rather ineffective for axially compressed cylinders.

Not only is the shell strength seriously affected by the amplitude and form of geometric imperfections in its surface, but the postbuckling path is usually dramatically weakening in character. As a result, buckling is commonly a sudden, dramatic, unpredictable event leading to a substantial loss of load-carrying capacity. As a structural failure mode, it deserves special attention in design and a certain caution if stresses approaching a potential buckling condition may occur.

Typical geometries of thin compressed shells

The thin compressed cylinders described in this chapter have a radius-to-thickness ratio (r/t) in the range 100–2000, and a length such that most buckles occupy only a small part of the distance between boundaries, rings or changes of plate thickness. The thickest of these cylinders $(100 < r/t < 500)$ are generally required in limited zones and for special loading or stress concentration reasons. Most storage containers subject to sensibly uniform axial compression have wall thicknesses in the range $500 < r/t < 2000$.

Most civil engineering cylindrical shells under axial compression fall into the category of medium length cylinders: that is; they are so long that the end boundary conditions do not play a strong role, and so short that the buckling load is far below the Euler load for buckling as a column. The buckling is often well in the elastic range of behaviour, so that the buckling strength is unaffected by the material strength.

Bifurcation and postbuckling in perfect cylinders

Classical elastic buckling

If a geometrically perfect thin elastic cylinder is axially compressed under uniform compression, the load–end shortening relationship is typically as shown in Fig. 2.5. A very linear prebuckling path (with little for the experimentalist to observe) is suddenly terminated as the shell bifurcates into a non-symmetric mode (Fig. 2.6(b)), with several full waves of buckling mode around the circumference, and usually several waves up the height. At buckling, the load falls very rapidly, and the cylinder actually increases in length (Fig. 2.5) as the displacements normal to the surface grow. As the load falls (Fig. 2.5), bifurcation after bifurcation occurs as the mode switches from one circumferential wave number n to another in dynamic jumps, which are particularly difficult to follow with static structural calculations (Riks *et al.* 1996). The bifurcation load is usually slightly lower than

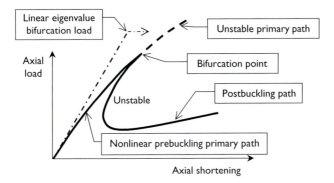

Figure 2.5 Typical load–end shortening relationship for an axially compressed cylinder.

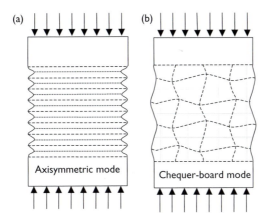

Figure 2.6 Typical buckling modes for axially compressed perfect cylinders: (a) axisymmetric mode; (b) non-symmetric mode.

the load found from a linear eigenvalue calculation, because nonlinearity in the prebuckling path often leads to additional destabilising stresses.

The strength of an axially compressed cylinder under all conditions is commonly compared with the linear bifurcation buckling stress of the perfect shell under these ideal conditions (of medium length, with prebuckling stresses unaffected by the boundary conditions and boundaries that restrain circumferential displacements during buckling). This buckling stress was independently found by Lorenz (1908), Timoshenko (1910) and Southwell (1914) and is known as the 'classical elastic critical stress'

$$\sigma_{cl} = \frac{E}{\sqrt{3(1 - v^2)}} \frac{t}{r} \simeq 0.605 \, E \frac{t}{r} \qquad (1)$$

in which E is Young's modulus, v is Poisson's ratio (typically around $v = 0.3$), t is the shell wall thickness and r the cylinder radius.

At this buckling stress, a very large number of different buckling modes or eigenmodes are all simultaneously critical (sometimes over 100 modes with loads within 1%). Many modes are possible in different 'chequer-board' patterns (Fig. 2.6(b)) whose wavelengths in the circumferential and axial directions are related by the 'Koiter circle' (Koiter 1945; Calladine 1983). The steeply falling postbuckling path (Fig. 2.5) is associated with the proximity of these many modes. It should be noted that Eq. (1) also gives the critical stress for axisymmetric buckling, where the cylinder develops a corrugated appearance, with waves only in the axial direction (Fig. 2.6(a)). This mode has a stable postbuckling path, and is later found to be important when the cylinder is internally pressurised.

For a medium length cylinder, the critical buckling mode (Fig. 2.6(b)) involves square waves, and can be described in terms of the number of full waves around the circumference in this mode, given by

$$n_{cl} = \sqrt[4]{(3/4)(1 - v^2)}\sqrt{r/t} \simeq 0.909\sqrt{r/t} \qquad (2)$$

which indicates that the number of waves falls progressively as the shell becomes thicker. In very thin shells, where the wave number n_{cl} is high, the buckling mode is often very localised, since buckling can occur when a zone of about the size of a single wavelength is critically loaded. However, since a large number of different modes are almost simultaneously critical, Eq. (2) does not have great significance. Further discussion of possible modes may be read in Calladine (1983). It should also be noted that the wide range of possible modes includes an axisymmetric mode (Fig. 2.6(a)) with axial half-wavelength

$$\lambda_{cl} = \frac{\pi}{[12(1 - v^2)]^{1/4}}\sqrt{rt} = \frac{\lambda}{\sqrt{2}} \simeq 1.728\sqrt{rt} \qquad (3)$$

which is half the half-wavelength of the square chequer-board pattern given by Eq. (2).

Although the complex buckling behaviour of a perfect cylinder makes it a popular choice for sophisticated mechanics calculations, Eq. (1) provides a very poor estimate of the experimental strengths of practical cylinders (Fig. 2.3). Despite this, Eq. (1) is almost universally used as the reference buckling stress to which other results are related.

The effect of changing boundary conditions

Real cylinders are restrained against displacements at their ends by boundary conditions that may take a variety of forms. When these practical restraints are taken into account, local bending due to the restraint of prebuckling displacements near the ends can lead to a small loss of bifurcation buckling strength from the ideal

Table 2.1 Boundary condition terminology for cylindrical shells (see Fig. 2.7)

Name	δu	δv	δw	$\delta \beta$	Name	δu	δv	δw	$\delta \beta$
S1	r	r	r	f	C1	r	r	r	r
S2	r	f	r	f	C2	r	f	r	r
S3	f	r	r	f	C3	f	r	r	r
S4	f	f	r	f	C4	f	f	r	r

Notes
f = free to displace during buckling.
r = restrained displacement during buckling.

$\sigma_{cr}/\sigma_{cl} = 1.0$ to between 0.8 and 0.95 (Fig. 2.5). The theoretical prediction of this loss depends slightly on the shell theory that is used (Flügge 1932; Donnell 1934; Sanders 1963), but these differences are minor. The classical elastic critical stress (Eq. 1) is valid for a perfect cylinder if the boundary conditions are chosen in such a way that a purely membrane prebuckling stress state is achieved (e.g. free to expand radially, but rotationally restrained, as in a boundary condition of symmetry).

It is also significant that clamped boundary conditions for medium length shells do not significantly alter the buckling load, which contrasts strongly with their effect in column buckling.

The boundary conditions are best classified according to the system devised by Singer and used extensively by Yamaki (1984). Based on ideas from column buckling the two most obvious classes of boundary condition are pinned (S) and clamped (C) ends, which terms refer to the rotational restraint of the boundary. This rotational restraint only affects the strength of rather short cylinders, but other boundary displacements can affect the buckling load to a much greater extent. The categories of boundary condition are shown in Table 2.1, and relate to the incremental displacements δu, δv, δw and $\delta \beta$ during buckling shown in Fig. 2.7.

Where a pinned boundary condition does not restrain circumferential displacements δv during buckling (S2, S4), the perfect shell buckling load falls to about half the classical value (Fig. 2.7(a)), but loss of restraint of other boundary displacements has a much smaller effect, which is not a particularly obvious outcome. Although the boundary conditions can affect the strengths of axially compressed cylinders, it is clear that the strength losses are insufficient to be the cause of the very low experimental test results shown in Fig. 2.3.

Effect of the length of the cylinder

Cylindrical shells can be divided into three basic length categories according to their buckling response: 'short' cylinders, in which one or two buckle waves form down the cylinder length; 'medium' cylinders in which chequer-board, diamond pattern or outward axisymmetric buckles can form over the surface (Fig. 2.6); and

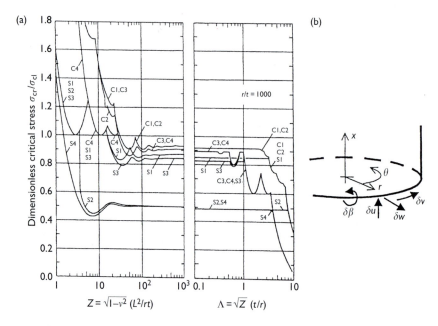

Figure 2.7 Effect of boundary conditions and shell length on perfect shell buckling load. (a) Perfect cylinder buckling strengths; (b) boundary displacements (reproduced from Elastic Stability of Cylindrical Shells, Yamaki, 1984).

'long' cylinders in which Euler buckling as a column occurs without distortion of the cross-section.

In short cylinders, the boundary conditions at the end of the cylinder play a marked role (Fig. 2.7). First, they induce local stresses associated with the restraint of radial expansion during the prebuckling phase, and then they restrain the buckling displacements. In the prebuckling phase, these local stresses occur in a zone roughly limited by the linear half-wavelength λ of meridional bending from each end, where

$$\lambda = \frac{\pi}{[3(1 - v^2)]^{1/4}} \sqrt{rt} \simeq 2.44\sqrt{rt} \tag{4}$$

so it may be expected that shells with lengths below about $L/\lambda = 2$ will have prebuckling stresses that are much affected by end effects and the nature of the end boundary conditions, but that shells longer than this will behave in a manner that is relatively independent of length and boundary conditions. The latter are termed medium length cylinders.

Unfortunately, the shell length is commonly defined in terms of the Batdorf parameter Z, which is related to the above more meaningful bending half-wavelength λ as

$$Z = \sqrt{(1 - v^2)}\frac{L^2}{Rt} \simeq 5.70 \left(\frac{L}{\lambda}\right)^2 \tag{5}$$

Since the Batdorf parameter depends on the square of the cylinder length L between boundaries or rings, it can take very large values: the length $L/\lambda = 2$ becomes $Z \approx 23$. In Fig. 2.7, the short cylinders can be seen as those at low Z where the close boundaries strongly increase the buckling strength. The most commonly assumed boundary condition is S3, which has a plateau buckling strength of approximately 83% of the classical value throughout the range $22 < Z < 0.3(r/t)^2$ and only reliably rises above the classical value when $Z < 2.8$.

The length of the shell also affects the form of the axial load–end shortening relationship, as shown in Fig. 2.8(a). Short cylinders (low Z) have a much more stable postbuckling response, which ultimately leads to a lower sensitivity to geometric imperfections. However, to achieve the higher strengths and more stable behaviour

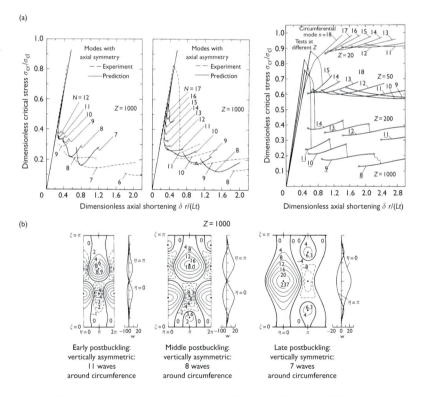

Figure 2.8 (a) Axial load–end shortening relationships and (b) postbuckling modes for axially compressed cylinders (C1 boundaries) (reproduced from Elastic Stability of Cylindrical Shells, Yamaki, 1984).

associated with short lengths, ring stiffeners must be provided at spacings that will reduce each shell segment to around $Z = 20$ ($L/\lambda \approx 1.9$), which for a typical civil engineering shell with $r/t = 500$ means $L/r \approx 0.2$, which is very close. Thus, few practical civil engineering shells can qualify to be classed as short.

The following discussion is therefore principally concerned with medium length cylinders, where the perfect cylinder buckling strength is rather independent of length (Fig. 2.7). It may be noted that the European shells standard (ENV 1993-1-6 1999) classes cylinders as of medium length if they lie in the range $0.70 \leq (L/\lambda) \leq 0.20$ (r/t) or $2.76 \leq Z \leq 0.238$ $(r/t)^2$, which is similar to the above limits for S3 cylinders.

As noted above, the postbuckling behaviour depends on the length of the shell (Fig. 2.8(a)). This leads to differences in the imperfection sensitivity of cylinders of different lengths and also affects the strength of a cylinder with very severe imperfections.

Thicker cylinders and plasticity

The above description relates only to elastic buckling in perfect cylinders. The classical elastic buckling stress (Eq. 1) varies linearly with the thickness t, so in thicker cylinders it rises progressively. In thin cylinders ($r/t \sim 500$), material yield occurs in the postbuckling range, especially in regions of local bending, leading to permanent deformations (Fig. 2.4) that are not necessarily in the form of the initial buckling mode (Fig. 2.8).

In thicker cylinders, first yield occurs in regions of local bending, but this is usually near a boundary condition that provides restraint. As a result, first local yield does not cause a dramatic change to the buckling strength. With increased thickness, the critical mode (Eq. 2) falls and the dimensionless length (L/λ or Z) of the shell falls, leading to a more stable postbuckling behaviour. As a result, the strength of thicker cylinders becomes less variable. In quite thick short cylinders, the axisymmetric buckling mode may become critical, and it has sometimes been thought that axisymmetric buckling was an indication of prebuckling yield, though this is not a reliable conclusion. Design strengths for thicker cylinders are adjusted for the effects of yielding if the critical stress exceeds about 50% of the yield stress.

Buckling of imperfect cylinders

Nature of imperfection sensitivity

Many studies of axially compressed cylindrical shell buckling and postbuckling arose from the difficulty in finding a theoretical treatment that could predict the low strengths and the wide scatter observed in experimental studies (Fig. 2.3). The search for an adequate theoretical prediction of these elastic buckling loads proved to be one of the most formidable tasks in structural mechanics, and the outcome generally one of the least satisfying from the viewpoint of engineering design.

The discrepancy between classical buckling stress predictions and experimental buckling strengths was first shown to be predominantly caused by geometric imperfections in the shell surface by Koiter (1945). His perturbation analysis explored the effect on the buckling stress of minor deviations in geometry in the form of the axisymmetric buckling mode (Fig. 2.6(a)) with wavelength λ_{cl}. This original analysis for a long cylinder with sinusoidal imperfections throughout its length led to the simple asymptotic formula for the buckling stress at bifurcation σ_{cr} as

$$\sigma_{cr} = \sigma_{cl} \left\{ 1 - \psi \left| \frac{w_0}{t} \right| \left[\left(1 + \frac{2}{\psi w_0} \right)^{1/2} - 1 \right] \right\} \tag{6}$$

where the amplitude of the imperfection is w_0 and $\psi = 0.75\sqrt{3(1 - v^2)} \simeq 1.239$. This form of imperfection, the axisymmetric buckling mode for the perfect shell (Fig. 2.6(a)), may be regarded as the most serious for the strength of the shell, leading to a reduction (Fig. 2.9) to only 24% of the classical critical stress for an amplitude of one wall thickness ($w_0/t = 1$). Koiter identified that imperfections in the form of the perfect shell buckling eigenmode (Fig. 2.6) would be very deleterious, and this concept was later widely adopted.

The effect of geometric imperfections on the form of the load–axial shortening curve is shown in Fig. 2.10(a), where it can be seen that the imperfection slightly modifies the shell's passage from the prebuckling path to the postbuckling path, but

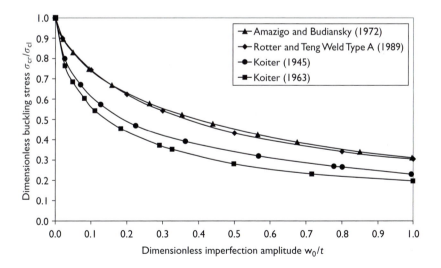

Figure 2.9 Sensitivity of the bifurcation load to the amplitude of axisymmetric geometric imperfections.

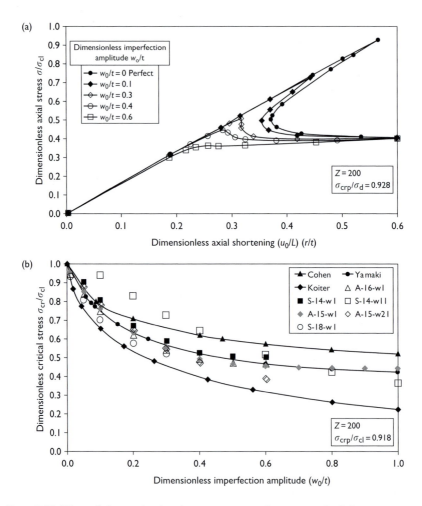

Figure 2.10 Effect of the amplitude of geometric imperfections on the bifurcation stress (after Yamaki 1984). (a) Load–axial shortening relationships for cylinders with circumferentially asymmetric imperfections. (b) Typical strength–imperfection relationships for asymmetric imperfections (reproduced from Elastic Stability of Cylindrical Shells, Yamaki 1984).

the proximity of these paths means that a minor imperfection has a disproportionate effect on the buckling strength (Almroth 1963, 1966). An imperfection with an amplitude of only one wall thickness, which may be only a thousandth part of the radius, has a dramatic effect on the peak load. The ratio (Fig. 2.10(b)) of maximum stress achieved under elastic conditions σ_{cr} to the classical elastic critical stress σ_{cl} is known as the 'knock-down factor' or the 'elastic imperfection factor', and is often denoted by the symbol α ($\alpha = \sigma_{cr}/\sigma_{cl}$).

This sensitivity to geometric imperfections is, however, very dependent on the form of the imperfection, as well as the length of the shell. The peak loads are usually summarised in the form of a strength–imperfection relationship (Figs 2.9 and 2.10(b)), which identifies the peak load achieved, whether this peak occurs by bifurcation into a different mode or by reaching a limit load in a mode that is already present in the geometric imperfection.

The full nonlinear analyses performed by Yamaki (1984) shown in Fig. 2.10(a) relate to imperfections in asymmetric modes and each maximum load occurs as a snap-through limit load. These limit loads are indicated in Fig. 2.10(b) by symbols, and the full curves correspond to asymptotic approximations to these. The latter are valid only for small amplitude imperfections, and are sensitive to the terms that have been retained in the derivation (K corresponds to Koiter's general analysis (Hutchinson and Koiter 1970), C to Cohen's (1971) improved version and Y to Yamaki's (1984) more precise result). However, it is evident that the asymptotic expressions are accurate for small amplitude imperfections if sufficient terms are included.

Careful experiments on cylinders with axisymmetric sinusoidal imperfections (Tennyson and Muggeridge 1969) and others with a single axisymmetric inward imperfection (Hutchinson *et al.* 1971) also showed that the asymptotic analyses described above can represent the strengths very well if the geometric imperfection forms and amplitudes are known.

Once the huge potential fall in the elastic buckling strength due to geometric imperfections had been identified, it was immediately evident that the very low strengths seen in tests (Fig. 2.3) were to be expected when the shell fabrication was not well controlled, and that these test results were also likely to exhibit a high scatter since different specimens might have different amplitudes of imperfections. A very large literature on the effects of geometric imperfections grew up, but few studies have explored the related problems of local irregularities in the boundaries and residual stresses.

Imperfections in different forms

Koiter's seminal work was followed by many extensive studies of the imperfection-sensitivity and postbuckling behaviour of elastic axially compressed shells (e.g. Donnell and Wan 1950; Koiter 1963; Hutchinson and Koiter 1970; Cohen 1971; Arbocz and Sechler 1974; Singer 1982; Yamaki 1984; Rotter and Teng 1989b; Rotter 1997). These studies have all shown that the strengths of elastic shells can be acutely sensitive to the magnitude of initial imperfections in the shell surface (Fig. 2.4), especially in thinner and medium length shells.

However, the buckling strength depends very much on the shape and amplitude of the imperfection, the shell geometry and boundary conditions. Many imperfection forms do not have a very deleterious effect, and the strength reduction caused by an imperfection cannot be easily deduced from the depth of dents in the shell surface. In addition, imperfections of the loading (loss of concentricity, non-uniformity of the applied stress), of the boundary conditions (especially local variations in axial restraint), and residual stresses in the shell due to fabrication processes can all have significant additional impacts on its strength.

Koiter's (1945) original perturbation analysis and subsequent more precise analysis (Fig. 2.9) that included the nonlinear prebuckling effects (Koiter 1963) was applied to sinusoidal imperfections in the form of the axisymmetric eigenmode. The concept that imperfections in the form of the perfect shell linear eigenmode cause the greatest strength losses derives from Koiter's original work, but this concept remains a guide only and is sometimes found to give much higher strengths than the worst possible mode (Schneider *et al.* 2001). Nevertheless, much attention has been given to eigenmode imperfections in recent years.

Imperfections in the shape of the axisymmetric eigenmode (Fig. 2.6(a)) appear to produce the largest reductions in strength (Hutchinson and Koiter 1970) for a given amplitude of imperfection. It may be noted that axisymmetric imperfections lead to a reduced bifurcation load, with the shell bifurcating into an asymmetric buckling mode. By contrast, asymmetric imperfections usually lead to a snap-through buckling or limit load, though bifurcation into a different mode may occur.

An alternative concept for the most severe form of the imperfection is the post-buckling mode for the perfect cylinder that has the smallest postbuckling shortening (Esslinger and Geier 1972). This idea is useful for cases where the postbuckling path is very dramatically softening, but it lacks the generality that is found in a linear eigenmode (which can always be obtained if compressive stresses are present), and for many geometries and load cases, this postbuckling mode cannot be identified.

Yamaki's (1984) extensive study explored many different eigenmode and post-buckling imperfection forms, but all were chosen in modes that were asymmetric with respect to the cylinder axis (i.e. in many waves around the circumference). His results typically (Fig. 2.9(b)) show less imperfection sensitivity than is found with axisymmetric critical mode imperfections. This finding is important in discussions concerning the most damaging forms of imperfections that can occur in practical structures.

The severity of the axisymmetric imperfection form, coupled with its high potential to occur in shells that are fabricated as a series of strakes or 'rings', has led to extensive studies of this imperfection in relation to axial compression (Amazigo 1969; Hutchinson *et al.* 1971; Amazigo and Budiansky 1972; Bornscheuer and Hafner 1983; Bornscheuer *et al.* 1983; Rotter and Teng 1989b; Rotter and Zhang 1990; Teng and Rotter 1992; Knödel *et al.* 1995, 1996; Guggenberger 1996; Rotter 1996, 1997; Berry and Rotter 1996; Holst *et al.* 1996, 2000; Berry *et al.* 2000; Schmidt and Winterstetter 2001). Further discussion of this imperfection form is given below.

Finally, it should be noted that the severity of an imperfection form is not directly and simply discernible from a comparison of curves like those in Figs 2.9 and 2.10(b). The reason is that the imperfection amplitude w_0 may be described in several different ways: the most obvious choices are the amplitude of a defining sine curve, the maximum variation from peak to trough of the imperfection form, the maximum deviation from the intended perfect geometry and the deviation measured using a simple measuring stick of defined length. These four simple alternative methods range over at least a factor of two in the quoted imperfection amplitude, so that comparisons between curves for different imperfection types must be drawn with care. It is later clear that the preferred comparison should be related to the tolerance measurement method to be adopted in design, but further research is certainly needed to improve these methods.

Large amplitude imperfections: limit points and bifurcations

The asymptotic studies of Koiter, Budiansky, Hutchinson, Yamaki and others described above are accurate for small amplitude imperfections only (usually much less than one wall thickness). At these small amplitudes, axisymmetric imperfections lead to bifurcation into circumferentially asymmetric buckling modes, whilst circumferentially asymmetric imperfections either pass a snap-through limit point (Fig. 2.10(a)) or bifurcate into a different asymmetric mode. However, large amplitude imperfections produce different effects. Axisymmetric large amplitude imperfections cause bifurcation into circumferentially unsymmetrical modes with very large buckles. Circumferentially unsymmetrical large amplitude imperfections often produce no maximum load, but a smooth progressive transition from prebuckling to postbuckling (Fig. 2.10(a)).

Thus, at large amplitudes, the circumferentially unsymmetrical imperfection poses problems of interpretation, since no unique buckling load can be identified, and a less clear 'limit of acceptable deformation' must be used. Moreover, the amplitude of the imperfection at which a clear peak load (limit point) is finally lost (leaving only a smoothly rising curve) depends very much on the form of the geometric imperfection. Thus, where large imperfections can occur in practice, unsymmetrical imperfection modes can be difficult to assess.

For large amplitude axisymmetric imperfections, the situation is different. As the imperfection amplitude rises, the size of the critical buckling mode also increases (Fig. 2.11). The size of the critical buckle can easily encompass the whole shell, after which the end boundary conditions may cause the buckling strength to rise. Thus, with very large amplitude imperfections, a shell that was classed as medium length may move towards the short length category and deep imperfections may not cause as much strength loss as smaller ones. This places a burden of care on analysts to ensure that their models always capture the lowest buckling load of a system. A warning on this matter is given in the Eurocode on Shell Strength and Stability (ENV 1993-1-6 1999).

These changes in behaviour are clearly related to the postbuckling curves in Fig. 2.8. In Fig. 2.8(a), repeated bifurcations lead to a steady fall in the circumferential mode, accompanied by progressively larger buckles in the axial direction (Fig. 2.8(b)). Where the shell is short (low Z), and an imperfection transports the

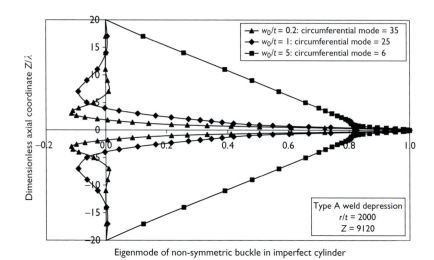

Eigenmode of non-symmetric buckle in imperfect cylinder

Figure 2.11 Changing axial mode of buckling as imperfection amplitude w_0/t rises.

shell rapidly into the postbuckling phase, it is clear that longer buckles will form and the boundary conditions will soon begin to increase the supported load. Current design supposes that medium length cylinders have strengths that are independent of the cylinder length and that the strength reduction for imperfection sensitivity is also independent of cylinder length. However, this is clearly an oversimplification.

Approaches to the design problem

Introduction

From the above, it is clear that geometric imperfections can have a very great effect on the buckling strength of an axially compressed cylinder. However, different forms of imperfection have different effects, so both the form and the amplitude must be characterised if this information is to be given practical application in design.

Three questions arise naturally from the above considerations:

1   What is the worst form the imperfection can take?
2   What is the probability that an identified serious imperfection form will occur in practical structures?
3   How can tolerances be specified for practical construction to ensure that the imperfections assumed in design are not exceeded?

At least four alternative philosophies have developed in attempts to address these questions: (a) simple lower bounds on test results; (b) a search for the worst imperfection pattern for a given structure and loading; (c) a search for theoretical lower bound strengths that make design independent of the form and amplitude of imperfections; and (d) attempts to base design on imperfection patterns found in real structures.

Lower bounds on test results

The simplest and oldest procedure in addressing the scatter shown in Fig. 2.3 has been to draw a lower bound on all available test results. This approach can be traced back at least to Robertson (1928) and Wilson and Newmark (1933), but has been continually used right up to the present time, with the studies of Donnell (1934), Timoshenko (1936), Harris *et al.* (1957), Weingarten *et al.* (1965a), Hoff and Soong (1967), Almroth *et al.* (1970), Steinhardt and Schulz (1971) and Bornscheuer (1982) being steps on this path. The process of empirical rule development may be read in Bornscheuer (1982).

This method has a number of key disadvantages. First, if all tests are included, some have extremely low strengths, and the 'lower bound' or 90 percentile value is found to be very low indeed. Some of the tests are affected by plasticity, and some have poor loading arrangements or poor boundary support, and many are poorly documented so that these questions cannot be resolved. Thus many tests must be eliminated from the database, often with only weak justification, so that the apparently scientific process of devising a lower bound becomes rather dependent on the subjective view of the assessor. Second, there is generally little evidence of the amplitude and pattern of the geometric imperfections in these test shells, so that the huge literature on imperfection sensitivity is useless, and it is not possible to derive tolerance measures in relation to the defined strengths. Third, the method of fabrication of laboratory shells is generally very different from that used for full-scale structures, so the imperfection forms and residual stresses in them are probably unrepresentative of real construction. In addition, the load application and boundary conditions of real construction are rarely represented in laboratory tests, so they may again be quite unrepresentative.

Despite all these disadvantages, all current standards are effectively based on such empirical lower bounds. The most recent change, introduced into the ENV 1993-1-6 standard, attempts to break away from lower bounds by defining different curves with associated amplitudes of imperfections. These curves are still empirically based, but they do attempt to define strengths that can be checked against tolerance measures. Few attempts have been made to use modern calculation methods to relate imperfection sensitivity and tolerance measurements to empirical curves used in design. The only known attempt is that of Rotter (1997).

The 'worst' imperfection pattern

The search for the 'worst' imperfection pattern directly addresses the first question identified above. The mostly widely used interpretation of this concept is that an imperfection in the form of the perfect shell buckling eigenmode is close to the worst form. More recent extensions of the concept have led to automated computer analyses (Deml and Wunderlich 1997; Teng and Song 2001).

This approach seeks to identify the very worst imperfection mode, and has the advantage that the search is formal and mathematical, so it is potentially achievable provided the problem statement and failure criteria have sufficient generality. The procedure is intended to provide a safe lower bound for design. It has the great advantage that, potentially, it can be generalised for all shell forms and load cases, so problems can be studied for which no test data exist (as in most real designs).

However, there are several difficulties with this approach. First, real structures do not generally have severe imperfections in the 'worst' mode, and sometimes the mode is very far from realistic imperfection forms (e.g. for torsional buckling). The amplitude of the imperfection must be chosen for the analysis, and measures must be found that can relate this 'worst' form to tolerances to be achieved in construction.

The second problem is that each load case for a real structure has a different most serious imperfection, so each load case must again have a different set of tolerance measurements. The probability that these imperfection modes will occur in practice is not addressed.

Most serious of all, it is possible that a tolerance measurement system that controls imperfections in the 'worst' mode may miss a much larger amplitude in a very different mode that is only slightly less serious. In other words, a focus on imperfection forms that are potentially serious but unlikely to occur with significant amplitudes in real structures is not good engineering practice.

Theoretical lower bounds on strength

A third approach (Esslinger and Geier 1974; Croll 1984; Croll and Ellinas 1985; Yamada and Croll 1999) has attempted to find lower bounds on the strengths of cylinders, freeing the design process from concerns about the form and amplitude of the imperfections. If this approach is adopted, the problems of tolerance measurement and control are greatly reduced and safety is guaranteed. However, this laudable goal has two key disadvantages.

First, the strengths obtained from this process are often lower than has been assumed in many existing designs, indicating that existing designs are, in some sense, unsafe despite many years of successful use. Moreover, no reward can be given for the use of good quality construction, since the philosophy assumes a priori that design should address the worst of all possible imperfections, both in mode and amplitude. The second disadvantage is that it is difficult to generalise the process: lower bounds are generally either based on or verified against test results, and experimental data exists only for the simplest loading cases and geometries. It should also be noted that the lower bound on experiments is not necessarily a lower bound for real construction because the imperfection patterns found in laboratory specimens may be different from those in full scale construction.

An improved variant on this method is the stochastic reliability assessment based on the ideas of Amazigo (1969, 1974) and extended by Elishakoff (1983). This method attempts to produce a rigorous reliability treatment, but it remains difficult to relate its strength assessments to tolerance measurements in full-scale construction.

Realistic imperfection patterns

The fourth approach has been to use generalised approximations of measured imperfection patterns that attempt to compromise between the known severity of a particular mode and the forms that are observed. This procedure is similar to the computational modelling of tests using experimentally measured imperfections, but it must instead address the patterns observed in real construction. It has the advantage that tolerance measurements can be directly and clearly defined, and that an improved quality of construction can immediately take advantage of the additional strength associated with it. It was probably first proposed by Arbocz (1974), who used measurements of imperfections in aerospace shells (Arbocz and Babcock 1974) to derive key conclusions for shell design. These cylindrical shells were fabricated as an assembly of long cylindrical panels attached at meridional seams, so it was natural that the dominant imperfections were in harmonics of the number of panels used. This study drew attention away from axisymmetric imperfections towards non-symmetric imperfections.

Later research in this area has principally focused on weld depressions and the severe imperfections at fabrication joints. The commonest method of construction in civil engineering is illustrated in Fig. 2.12, which indicates the process of rolling flat sheets into curved panels, joining them into complete strakes and then mounting one strake on another. This fabrication process leads to local imperfections that are close to axisymmetric, at least in certain parts of the circumference, together with local depressions in the axial direction near the vertical seams over each strake height.

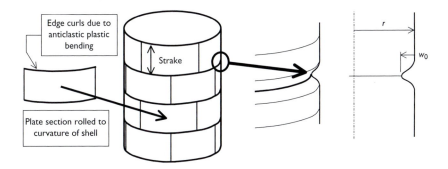

Figure 2.12 Typical fabrication technique for civil engineering cylinders.

The local axisymmetric imperfections associated with these fabricated joints were first studied by Bornscheuer and Hafner (1983), though Amazigo and others (Amazigo 1969; Hutchinson *et al.* 1971; Amazigo and Budiansky 1972) had previously examined a local axisymmetric depression (Fig. 2.9) as a more realistic axisymmetric imperfection that the full eigenmode of Fig. 2.6(a). The first systematic study of the axisymmetric weld depression was produced by Rotter and Teng (1989b), and this was followed by a range of later studies that examined different problems but relied on this imperfection as a realistic form (Teng and Rotter 1992; Rotter and Zhang 1990; Knödel *et al.* 1995, 1996; Rotter 1996, 1997; Berry *et al.* 1997; Berry *et al.* 2000; Holst *et al.* 2000; Schmidt and Winterstetter 2001; Pircher *et al.* 2001). An imperfection–sensitivity curve is shown in Fig. 2.13 for Rotter and Teng's Type A imperfection up to very large amplitudes. These imperfections are defined by

$$w = w_0 \, e^{-(\pi x/\lambda)} \left(\cos \frac{\pi x}{\lambda} + k \sin \frac{\pi x}{\lambda} \right) \tag{7}$$

in which x is the axial distance from the weld centre, λ is the linear bending half-wavelength (Eq. 4), w_0 is the imperfection amplitude and k takes the values 1 for Type A, 0 for Type B, and intermediate values represent alternative rationally based shapes.

One key difference between buckling modes in perfect and imperfect shells is that the circumferential wave number of the buckling mode is far better determined in the imperfect shell than in the perfect shell. As noted above, the perfect shell has many modes that are almost simultaneously critical, so identification of the

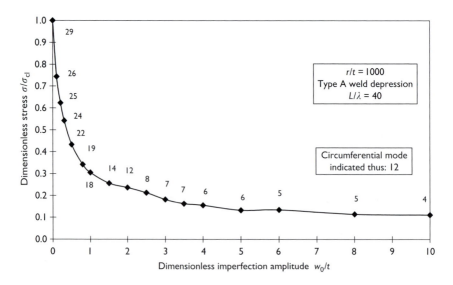

Figure 2.13 Elastic buckling strengths with weld depression imperfections.

'critical' modes is not very meaningful. For imperfect shells, the critical mode is moderately well defined (though a few adjacent modes are often close) for a given imperfection form and amplitude. For example, the Type A weld depression has the critical circumferential mode (Rotter and Zhang 1990)

$$n_{cr} \simeq (0.822 - 0.25\ w_0/t)\sqrt{r/t} \qquad (8)$$

where w_0 is the amplitude of the weld depression imperfection (Fig. 2.12).

The above studies of weld depressions appear to relate only to a fully axisymmetric imperfection, but the high mode number indicated by Eq. (8), and the potential for a buckle to form if the necessary conditions are achieved over a dimension comparable to a critical buckle wavelength means that the imperfection need not be fully axisymmetric. One of the most significant of the above studies is consequently that of Berry and Rotter (1996), which examined 'axisymmetric' imperfections extending around a small part of the circumference. Most of these caused the same dramatic loss of strength (Figs 2.9 and 2.13) as a fully axisymmetric imperfection, and even quite a short length of imperfection in the axisymmetric form, extending only slightly further around the circumference than one full wavelength of the critical buckle, produced buckling strength losses comparable with a fully axisymmetric imperfection. This means that a relatively local imperfection that remains stable in form over a small part of the circumference is as detrimental as a fully axisymmetric imperfection.

The manner in which these weld depressions may be represented has been the subject of considerable research, and can be read in Knödel *et al.* (1995, 1996). Further work on the characterisation of the imperfections measured on a full-scale silo in the large study of Ding (Rotter *et al.* 1992; Ding *et al.* 1996) has recently been produced by Pircher *et al.* (2001) which indicates that this form is a useful representation for real structures.

Imperfection patterns and amplitudes

Both the 'worst imperfection form' strategy and the pursuit of realistic imperfections are usually based on the assumption that the amplitude of the imperfection can be chosen (perhaps with tolerances in mind) after the form has been determined. Unfortunately this is not the case because the shape of the 'worst imperfection form' depends on the amplitude of the imperfection itself. This can be illustrated by calculations of the elastic bifurcation buckling strength of a cylinder with a local axisymmetric imperfection, with a range of different wavelengths used for the imperfection form (Rotter 1997). These calculations were conducted for a fixed amplitude of the imperfection, so that the form of the imperfection is changing but not its amplitude. The result is shown in Fig. 2.14.

For a small imperfection amplitude of $w_0/t = 0.2$, the lowest buckling strength is found to occur at a wavelength just above the axisymmetric eigenmode wavelength (Eq. 3). When the amplitude has risen to $w_0/t = 1.0$, which is perhaps

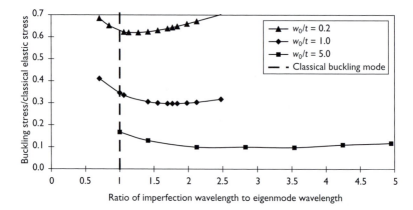

Figure 2.14 Effect of imperfection wavelength on the buckling strength of cylinders with a local weld depression imperfection of fixed amplitude.

typical for large civil engineering shells, the critical mode wavelength is about 1.7 times larger, and use of the eigenmode form overestimates the strength by 16%. At large amplitudes, $w_0/t = 5.0$, the critical mode has a much longer wavelength and the eigenmode buckling strength overestimate has risen to 71%. It is clear that the most critical imperfection form cannot really be specified independently of the amplitude of the imperfection.

Systematic imperfections: plate thickness changes and lap joints

Some forms of geometric imperfection are not accidental but arise as a consequence of the fabrication process. Commonest amongst these are changes of plate thickness and lap joints. Vertical joints do not generally present a significant problem under axial compression because these are orthogonal to the thrust line. Circumferential joints can lead to eccentricities in the line of axial thrust (Fig. 2.15).

Many shell structures (chimneys, silos, tanks) have much higher stress states in one part than another, and the thickness is normally reduced where material is not needed. Most civil engineering shells are made from rolled plates of uniform thickness, so an increase in thickness necessitates a discrete step change. The point at which the shell thickness changes is typically the most highly stressed because the stress resultants rise smoothly but the thickness changes abruptly. This location is therefore important in assessing whether an eccentricity of the thrust line is a serious imperfection.

If the middle surface (centre-line) of the shell is continuous at the step change (no eccentricity), then the shell may be regarded as free of a local eccentricity

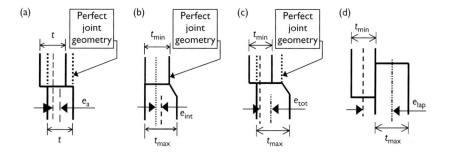

Figure 2.15 Eccentricities at joints. (a) Accidental eccentricity: no change of plate thickness. (b) Intended offset at change of plate thickness. (c) Total eccentricity at change of plate thickness. (d) Lap joint with thickness change.

imperfection. Rotter and Teng's (1989b) study of weld-depression imperfections included a change of plate thickness without eccentricity and showed that the thicker plate always restrains the thinner plate against buckling. Small increases in the buckling stress of the thinner plate result, but these are always ignored in design.

For functional reasons, however, most shells require a smooth surface on one side. This leads to an eccentricity in the line of thrust at the plate thickness change (Fig. 2.15). No study is known to have explored the problem formally. Due to small differences in the geometry of one strake and another (Fig. 2.12), especially near adjacent vertical seams, small errors in alignment also occur (Fig. 2.15(a) and (c)), so the total eccentricity is likely to exceed the intended value in places.

The eccentricity at a circumferential joint has several consequences: local bending leads to axisymmetric deflections, local compressive circumferential membrane stresses, and local bending stresses. The local bending stresses are not very important, but the co-existent axisymmetric inward deflection and circum-ferential compressive membrane stresses are destabilising and can be seriously deleterious.

The best information on the effect of these eccentricities comes from studies of the effect of lap joints (Fig. 2.15(d)). Esslinger and Pieper (1973) appear to have undertaken the first experimental and theoretical study of the problem, from which they concluded (Esslinger and Geier 1977) that systematic imperfections dramatically reduce the uncertainty and scatter in buckling strengths. The range of their experimental and theoretical study was, however, rather limited.

Esslinger's results were generalised by Rotter (1985), who deduced that the elastic buckling strength should be around $\sigma_{cr} = 0.39\sigma_{cl}$, but the first system-atic study of the elastic buckling problem was undertaken by Rotter and Teng (1989a), who showed that bifurcation occurs at an infinitesimally short lap joint

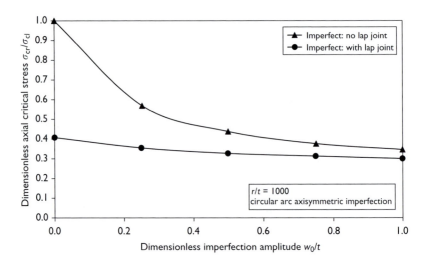

Figure 2.16 Imperfection sensitivity with and without a lap joint.

at $\sigma_{cr} = 0.408\sigma_{cl}$. Realistic lengths and jointing systems, and changes of plate thickness all cause the bifurcation load to increase, so this reference stress is a useful one. However, this buckling stress is well above normal design values for typical imperfect cylinders, so Rotter and Teng considered the simultaneous presence of a dent imperfection just above the lap joint. This exploration (Fig. 2.16) showed that imperfect lap joints are only marginally weaker than perfect lap joints (only 26% strength loss at $w_0/t = 1.0$). Further, if design is based on an imperfection amplitude comparable with realistic accidental dent imperfections (e.g. $w_0/t = 1.0$), the further reduction in strength caused by the lap joint is only some 13%. Since the lap joint is the most extreme version of all the imperfections of alignment at joints (Fig. 2.15), it is evident that most eccentricities at a circumferential joint need not cause concern.

The results of Rotter and Teng (1989a) were later confirmed and extended by Greiner and Yang (1996). Conditions leading to plastic collapse in lap-jointed pressurised cylinders were thoroughly explored by Teng (1994).

Design strengths and tolerances to control geometric imperfections

Introduction

The wide range of laboratory measured strengths for axially compressed cylinders is shown in Fig. 2.3. It has been particularly difficult to achieve reliable correlations between test results and theoretical predictions, even when the geometric imperfections in specimens have been measured. Consequently, the predictions of shell buckling strength have usually been used as an informative guide to key features of shell strength, rather than as a basis for design.

Since the buckling strength is very sensitive to the amplitude and form of geometric imperfections in the surface, its amplitude must be controlled by tolerance measures as part of the design process. However, a major difficulty arises here, as it is not easy for the designer to predict the level of imperfections which may be found in his structure, or for the fabricator to control the magnitude and shape of the imperfections he may produce. Moreover, some of the inevitable imperfection forms which arise in civil engineering practice have received little attention in very thin shells: bolted lap joints, local flats near welds and slight ovalling of the circular planform.

A key aspect of the relationship between shell strengths achieved in practice and tolerances to be achieved in construction is the cause and origin of imperfections. If the mechanics of the causes of imperfections can be understood, then a better estimation can be made of the amplitudes and forms that will occur. Unfortunately, very few studies have explored the causes of imperfections, but recent work on the basic phenomena has been undertaken by Holst *et al.* (1996, 1997, 1999, 2000) and Guggenberger (1996), but these represent only initial basic explorations. This field is likely to prove important in future research in shell buckling.

Since the strengths are highly correlated with the form and amplitude of geometric imperfections, there exists a range of potential imperfections that are associated with the empirical lower bound strength. If the empirical strength designs are to be able to account for the greater strength of accurately fabricated shells, it is vital that fabrication tolerances associated with the design strengths are defined and met in execution. This connection was first suggested by Hoff and Soong (1967), but it has taken many years to be adopted into a design standard.

Strengths and tolerances in standards

As noted above, most design rules have been developed as empirical lower bounds on the results of experiments. Standards that have adopted this approach include API 620 (1978), AWWA D100 (1979), ECCS (1988) and DIN 18800 (1990). An effort to correlate this lower bound with appropriate tolerance measurements was made in the ECCS and DIN standards, but the quantitative basis of the match does not appear to have been published.

The drafting committee for the earlier 1983 version of the ECCS standard recognised that some constructions might fail to achieve the specified tolerances and added, in the commentary, a note to the effect that the tolerances could be relaxed to double their intended values if the assessed buckling strength was halved. Rotter (1985) coupled this idea with the tolerance measure to show that the two rules could be combined to produce an almost unique strength-imperfection sensitivity curve that was not far from that of Koiter (1945). This demonstration led the way to permitting the ECCS rule to be re-cast for the Eurocode (ENV 1993-1-6 1999) in terms of an imperfection sensitivity relation (e.g. like Figs 2.9, 2.10(b) or 2.13) and an explicit imperfection level that could be measured and therefore transformed into a tolerance measurement. This development was described by Rotter (1998).

Measurement and control of imperfections: tolerances

The focus on tolerance measurements led the Eurocode committee for ENV 1993-1-6 (1999) to develop a large group of different measurements (Fig. 2.17). Some of these are intended to control imperfections that relate to buckling under other stress states, but several derive from concerns relating to axial compression.

These tolerance measurements are chiefly conceived in the context of the wavelength of the square eigenmode for the perfect shell (Fig. 2.6(b)), which has a half wavelength in each direction of $2\lambda_{cl}$ or about $3.5\sqrt{rt}$. Since the value to be measured is not so precise, the normal measuring stick has a length of

$$\ell_{gx} = 4\sqrt{rt} \tag{9}$$

to reflect the classical eigenmode, which is used here as the reference worst imperfection. A much shorter length is additionally used across welds

$$\ell_{gw} = 25t \tag{10}$$

in view of the possibility of a local plastic failure if deep local deviations occur.

The measured imperfection is made dimensionless as

$$U_{0x} = \Delta w_{0x}/\lambda_{gx} \tag{11}$$

and similar expressions, and the value of U is limited to fall below the tolerance U_{max}.

These many tolerance measurements may appear to be rather onerous for the fabricator of the shell, but since there is such a strong correlation between strength and imperfection amplitude, it is highly desirable that they should be adopted.

Eurocode hand calculation rule for axial compression buckling

The rules of the draft Eurocode for shell strength and stability (ENV 1993-1-6 1999) are outlined here, since they represent the most recent development and attempt to accommodate more of the above research than previous standards.

Three different fabrication quality classes are defined, as indicated in Table 2.2, allowing highly controlled fabrication to exploit the higher resulting strength, but requiring less controlled construction to assume a lower buckling load.

The characteristic imperfection amplitude Δw_x is found as

$$\Delta w_x = \frac{1}{Q}\sqrt{\frac{r}{t}}t \tag{12}$$

where Q is the meridional compression fabrication quality parameter (Table 2.2).

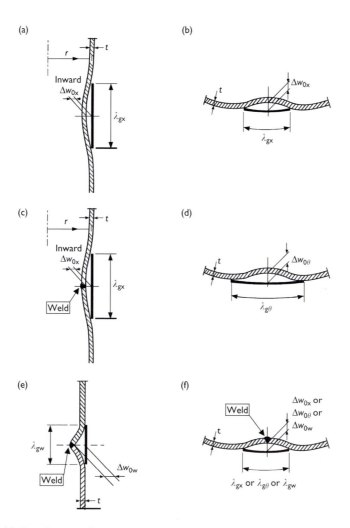

Figure 2.17 Dimple imperfection measurements required by ENV 1993-1-6 (1999). (a) Measurement on a meridian. (b) First measurement on a circumferential circle. (c) First measurement across a weld. (d) Second measurement on circumferential circle. (e) Second measurement across a weld using a special gauge. (f) Measurements on circumferential circle across weld.

The characteristic imperfection is then used in an empirical relationship similar to Figs 2.9 or 2.10(b) for the elastic buckling strength dependency on imperfection amplitude

$$\alpha_{x0} = \frac{0.62}{1 + 1.91\,(\Delta w_x/t)^{1.44}} \qquad (13)$$

Table 2.2 Values for dimple tolerance $U_{0,max}$ and quality parameter Q

Fabrication tolerance quality class	Description	Value of $U_{0,max}$	Value of U_n	Q
Class A	Excellent	0.006	0.01	40
Class B	High	0.01	0.016	25
Class C	Normal	0.016	0.025	16

The characteristic elastic buckling stress for a medium length shell is then found as

$$\sigma_{xcr} = \alpha_{x0}\sigma_{cl} \tag{14}$$

Where elastic–plastic buckling may occur, the characteristic buckling strength is reduced to σ_{xk} for lower values of the relative slenderness of the shell $\bar{\lambda}_x$, given by

$$\bar{\lambda}_x = \sqrt{f_y/\sigma_{cl}} \tag{15}$$

$$\sigma_{xk} = \left\{ 1 - \beta \left(\frac{\bar{\lambda}_x - \bar{\lambda}_0}{\bar{\lambda}_p - \bar{\lambda}_0} \right)^\eta \right\} f_y \qquad \text{for } \bar{\lambda}_0 < \bar{\lambda}_x < \bar{\lambda}_p \tag{16}$$

$$\bar{\lambda}_p = \sqrt{\frac{\alpha}{1 - \beta}} \tag{17}$$

in which β is the plastic range factor ($= 0.6$) and η is the interaction exponent ($= 1.0$), $\bar{\lambda}_0$ is the squash limit relative slenderness ($= 0.2$) and $\alpha = \alpha_{x0}$ for an unpressurised axially compressed cylinder. For shells with $\bar{\lambda}_x < \bar{\lambda}_0$, the characteristic buckling stress σ_{xk} is taken as equal to the characteristic yield stress f_y.

Equations (12)–(17) provide considerable scope for future improvements to the elastic–plastic strength interaction for particular problems (Rotter 2002b) as data on the appropriate values of α, β, $\bar{\lambda}_0$ and η are found in new research.

Eurocode computer buckling predictions and imperfections

All design standards for shells before ENV 1993-1-6 (1999) included only hand calculation methods. A major innovation in this new standard is the regulation of the use of finite element and similar computer calculations in design.

The introduction of these methods is far from simple, since for design purposes the calculations must contrive to capture all the complexity that has been presented in this chapter, but most computer analyses will inevitably be much simpler. Thus a linear eigenvalue calculation cannot accurately capture the sensitivity to geometric imperfections, and a geometrically linear analysis cannot detect snap-through buckling. The regulation of simpler analyses is described by Rotter (2002a), and is beyond the scope of this chapter, but it is appropriate to make note of the highest quality calculation type: geometrically and materially nonlinear with imperfections analysis (termed GMNIA). The nonlinear features concern straightforward issues of mechanics, but the imperfections lead to difficulties for the user of this analysis, especially with reference to tolerances.

The first challenge in using such an analysis is to decide what is meant by failure. The standard defines the following three alternative criteria for identifying the characteristic resistance (strength) R_{GMNIA}, which is defined in terms of a multiplier on the design load set:

1   the maximum load of the load–deformation curve (limit load);
2   the bifurcation load, where this occurs during the loading path before reaching the limit point of the load–deformation curve;
3   the largest tolerable deformation, where this occurs during the loading path before reaching the bifurcation load or the limit load.

Where bifurcation buckling of the type discussed above occurs (Criterion 2), this is not difficult, but where localised high stresses lead to formation of a local dimple (Criterion 2) or a local snap-through buckle (Criterion 1), it can be quite difficult to decide whether the behaviour should be regarded as having reached an ultimate limit state. In particular, where a shell displays a progressively stiffening response under elastic–plastic large displacements (e.g. Teng and Rotter 1989), the criterion of failure (Criterion 3) is rather arbitrary, and more work is needed to decide how the structure's deformations should be limited. It should probably be normally defined in terms of a rotation of the surface, rather than a deflection.

Where an analysis explicitly includes geometric imperfections, the results must be interpreted in a manner that recognises that other imperfections may be present in the real structure (imperfect boundary conditions, non-uniform loading, residual stresses etc.) which may reduce the achieved strength below that which is predicted. It is therefore vital that a GMNIA analysis, which appears to give a realistic strength assessment, is not regarded as an assured lower bound. The standard therefore requires that 'appropriate allowances' be used for the effects of imperfections that cannot be avoided in practice. These are in two groups: first geometric imperfections, such as deviations from the nominal geometry; irregularities at and near welds (minor eccentricities, shrinkage depressions, rolling curvature errors); deviations from the nominal thickness; and lack of evenness of supports. The second group covers material imperfections, such as residual stresses caused by rolling, pressing, welding, straightening; inhomogeneities and anisotropies. Ground settlements or flexibilities of connections or supports are noted as being additional sources of imperfection.

For a GMNIA analysis, it is vital that the amplitude and form of the geometric imperfection is defined. In line with the hand calculations, the amplitude is defined as dependent on the fabrication tolerance quality class, with the maximum deviation from the perfect shape $\Delta w_{0,\text{eff}}$ taken as the larger of $\Delta w_{0,\text{eff},1}$ and $\Delta w_{0,\text{eff},2}$:

$$\Delta w_{0,\text{eff},1} = \lambda_g U_{n1} \tag{18}$$

$$\Delta w_{0,\text{eff},2} = 100\, t\, U_{n2} \tag{19}$$

where λ_g covers all relevant gauge lengths of Fig. 2.17 and U_{n1} and U_{n2} are defined values of the dimensionless imperfection amplitude (Table 2.2).

Despite this effort at amplitude definition, the strength assessment process must recognise that the worst form of imperfection may not be that assumed by the analyst, and that the amplitude of imperfection that causes the greatest strength loss is not necessarily the largest amplitude of the tolerance restriction. To this end, the Eurocode (ENV 1993-1-6 1999) requires that it should also be verified that an imperfection that is 10% smaller than $\Delta w_{0,\text{eff},2}$ does not yield a smaller value for the limit state load R_{GMNIA}. If this occurs, the procedure should be iterated to find the most damaging amplitude.

Finally, since the above procedure may not be sufficiently uniquely and clearly defined to cover all situations that may arise, the requirement is made that the same analysis should be applied to comparable shell buckling cases for which characteristic buckling resistance values $R_{k,\text{known,check}}$ are known (either from calculation or test). The check calculations determine $R_{k,\text{GMNIA,check}}$, and are required to adopt similar imperfection assumptions and concern structural configurations that are similar in their buckling controlling parameters (relative shell slenderness, post-buckling behaviour, imperfection-sensitivity and material behaviour) to the target analysis.

From the reliability check calculation, a calibration factor k_{GMNIA} is evaluated as:

$$k_{\text{GMNIA}} = \frac{R_{k,\text{known,check}}}{R_{k,\text{GMNIA,check}}} \tag{20}$$

The characteristic buckling resistance for the real structure is then obtained from:

$$R_k = k_{\text{GMNIA}} R_{\text{GMNIA}} \tag{21}$$

This procedure may appear rather involved, but the freedom offered by the codification of computer analyses also means that it may be applied to situations that were unforeseen when the rules were drafted, so precautions are needed. Despite the apparent complexity, the formal inclusion of computer analyses in design calculations presents a major opportunity to develop cheaper, safer and more reliable designs than are possible with hand calculations.

Buckling of internally pressurised cylinders

Effect of pressurisation on axially compressed perfect cylinders: postbuckling

Many cylindrical shells that are subject to axial compression also experience a co-existent internal pressure (e.g. silo and tank structures). The internal pressure can substantially increase the buckling strength, so this effect is very important in design.

When a geometrically perfect cylinder is internally pressurised and then axially compressed, the typical axial load–end shortening relationship shown in Fig. 2.5 is progressively modified by the internal pressure, as indicated in Fig. 2.18 (Schnell 1959). The steep postbuckling path of the unpressurised cylinder becomes progressively less severe as the pressure rises, until the axisymmetric buckling mode becomes the critical mode.

To achieve maximum generality, the internal pressure is here made dimensionless in the same manner as in ENV 1993-1-6 (1999) as

$$p' = \frac{pr}{t\sigma_{cl}} \tag{22}$$

The form of Fig. 2.18 suggests that the imperfection sensitivity of the elastic buckling behaviour may decrease as the pressure rises, possibly disappearing altogether when the pressure reaches a value of perhaps $p' = 2$.

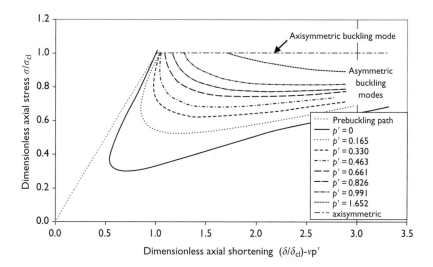

Figure 2.18 Postbuckling paths of perfect internally pressurised elastic cylinders (after Schnell 1959).

Effect of pressurisation on imperfect cylinders

In geometrically imperfect cylinders, the transfer of axial compressive force past an imperfection leads to development of local circumferential compressive membrane stresses in addition to the axial compression. The combination of these two orthogonal compressions gives a useful insight into imperfection sensitivity (Calladine 1983; Rotter and Teng 1989b). When the cylinder is internally pressurised, the local circumferential compression is first reduced and then eliminated, leading to a steady increase in buckling strength with internal pressure (Fig. 2.19(a)). Experimental verifications of this phenomenon include the studies of Harris *et al.* (1957),

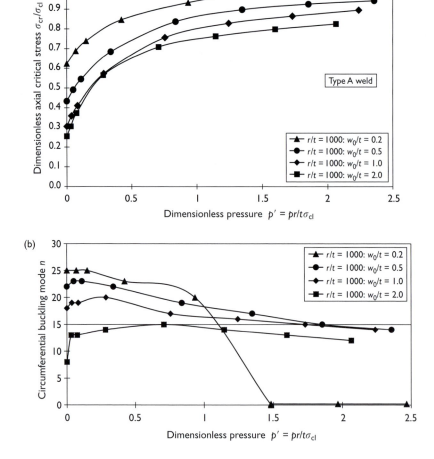

Figure 2.19 Effect of internal pressure on elastic buckling strength. (a) Increases in the elastic buckling stress with internal pressure. (b) Changes in the circumferential mode of buckling with internal pressure.

Weingarten *et al.* (1965b) and Saal *et al.* (1979). In Fig. 2.19(a), the unpressurised buckling strength is affected by the imperfection amplitude, so each of the curves shown begins at a different initial stress. The increases in strength with internal pressure are similar in form for different amplitudes of imperfection, though larger imperfections display faster strength gains. The corresponding circumferential modes are shown in Fig. 2.19(b), where the attainment of axisymmetric buckling for $w_0/t = 0.2$ is seen to be sudden. For very small imperfections, the mode falls as the pressure rises, but for practical imperfection amplitudes, the behaviour is more complicated.

Symmetric and asymmetric imperfections

Hutchinson (1965) made a most useful study of the imperfection-sensitivity of internally pressurised cylinders under axial compression, considering axisymmetric, circumferentially asymmetric and mixed modes of imperfection. The asymptotic analysis was only valid for small imperfections, but it indicated clearly that asymmetric and axisymmetric imperfections are quite differently affected by internal pressure. A sample set of calculations from his study is shown in Fig. 2.20, where several shells with different imperfection forms that all cause the same loss of strength in an unpressurised shell are compared. It is very evident that the asymmetric imperfection gains strength very rapidly when the internal pressure rises, but the axisymmetric imperfection is much less influenced. Moreover, when the two imperfection forms have equal amplitudes (as might occur in practice), the strengths are relatively close to those with axisymmetric imperfections. This

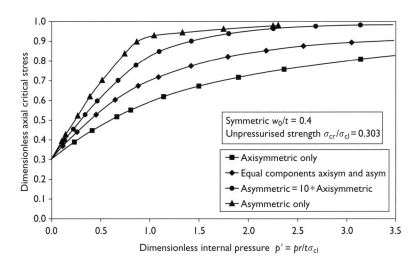

Figure 2.20 Effect of symmetric and asymmetric imperfections on internal pressure elastic buckling strength gain (after Hutchinson 1965).

study has some serious implications for design in view of the above discussion on identification of the most damaging imperfection form.

Laboratory model cylinders are commonly fabricated by wrapping a single sheet of material around a form and making a longitudinal joint. This method leads to geometric imperfections that are predominantly unsymmetrical relative to the axis. As a result, most tests (Weingarten *et al.* 1965b; Saal *et al.* 1979) indicate rather rapid strength gains due to internal pressure. By contrast, the normal full-scale fabrication process (Fig. 2.12) includes circumferential joints that often produce significant axisymmetric imperfection components. Even if these are not the most damaging imperfection for the unpressurised cylinder, they rapidly become the most damaging in the presence of internal pressure. Some earlier tests (Lo *et al.* 1951; Fung and Sechler 1957; Harris *et al.* 1957; Thielemann 1960) show lower strength gains, but the imperfections were not measured and the manner in which the results are presented make them difficult to assess.

Codified rule for pressurisation strength gain

In the context of the above, it is evident that the experimental database for strength gain with internal pressure must be treated with some caution. Earlier proposed expressions for the strength gain (Harris *et al.* 1957; Trahair *et al.* 1983; ECCS 1988) were all derived as empirical lower bounds to test results, and some took little account of variations in the unpressurised strength. These were reviewed by Rotter (1985).

However, following the work of Hutchinson (1965), it is evident that the strength gains for practical construction should be closely associated with the behaviour when axisymmetric imperfections are present (Fig. 2.12), and this is currently only achievable by calculation. The ENV 1993-1-6 (1999) rule is based on such calculation, and permits the elastic buckling strength to rise above the unpressurised value, with a revised value of α in Eqs (14)–(17) as

$$\alpha_{\mathrm{xpe}} = \alpha_{\mathrm{x}0} + (1 - \alpha_{\mathrm{x}0}) \left[\frac{p'}{p' + (0.3/\alpha_{\mathrm{x}0}^{0.5})} \right] \tag{23}$$

where the elastic unpressurised strength reduction $\alpha_{\mathrm{x}0}$ is given by Eq. (13) and p' is defined in Eq. (22). The origin of this expression is given by Rotter (1997). It adjusts the rate of strength gain according to the initial imperfection sensitivity, giving more rapid rises in imperfect cylinders, but the strength gains are generally considerably lower than those associated with asymmetric imperfections.

It should be noted that recent calculations (Walker and Wilson 2001) have revealed that the strength gains due to internal pressure may be substantially lost if the pressure is slightly circumferentially non-uniform within the cylinder. For fluid-filled cylinders, this is only a problem when the cylinder axis is horizontal, but for silos storing soft solids, it may require a careful reassessment of the strength gain provision. Further research is urgently needed on this question.

High internal pressures: elephant's foot failures : boundary conditions, changes of plate thickness and axisymmetric imperfections

The parts of the cylinder that are adjacent to the base, or to ring stiffeners and changes of wall thickness, are subject to local bending as the shell adapts from one membrane deflection state to another. Because of the cantilever bending effect under transverse loads and the cumulative nature of frictional drag loads in silos, these zones are also often points of highest axial stress. In these zones, the local bending deformations caused by both axial stresses and internal pressure are amplified by axial stresses comparable with the buckling stress. Early yielding caused by this amplification can cause early collapse.

In pressurised cylinders the amplification often causes yielding, leading to an elastic–plastic collapse mode known as elephant's foot (Fig. 2.21). Elephant's foot buckling has been investigated elastically, using a first yield test for failure (Rotter 1989) and plastically, using buckling or plastic collapse as the failure criterion (Rotter 1985, 1990). This form of collapse is important at high internal pressures, especially if the stabilising effects of internal pressure are being exploited in the design.

The complete effect of internal pressurisation on the buckling strength is shown schematically in Fig. 2.22, for a typical thin shell ($r/t \sim 500$). The internal pressure strength gains under elastic conditions described in the previous section are abruptly terminated by plasticity developing near a boundary condition or at an imperfection, and elastic-plastic stability failures occur at higher internal pressures. These plastic stability failures occur at load conditions well above first yield, and well below the von Mises yield condition applied to the membrane stresses in the

Figure 2.21 Elephant's foot buckling at the base of a tank.

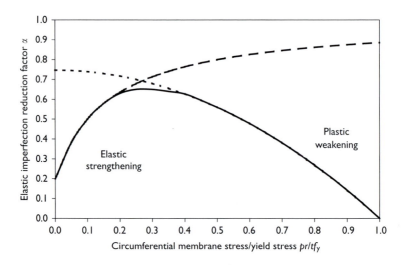

Figure 2.22 Effect of higher internal pressures on elastic–plastic buckling strength.

shell. Although several conditions in the shell (change of plate thickness, local imperfection, boundary condition) lead to elephant's foot failures, the strength loss varies with the condition. The most severe of these is the simply supported boundary condition, so this is currently used as a descriptor of strength.

The phenomenon at fixed and simply supported boundary conditions was first studied by Rotter (1990), who later explored the same effect at weld depressions (1996). Two of the results from these extensive studies are shown in Fig. 2.23. The plastic collapse curves in Fig. 2.23(a) are for a perfect shell at a simply supported boundary condition: the effect of geometric imperfections is not easily included, since these are most damaging in the body of the shell, rather than near the boundary. The corresponding effect at a weld depression axisymmetric imperfection is shown in Fig. 2.23(b), where the weak interaction between the elastic pressurisation strength gain and the plastic instability weakening of elephant's foot is clearly shown. Thicker cylinders show a less dramatic contrast.

The approximate relationship for a simply supported boundary condition developed by Rotter (1990) has now been adopted into several different standards (mostly relating to earthquakes) including ENV 1993-1-6 (1999), where the modified value of α in Eqs (14)–(17) is given by

$$\alpha_{xpp} = \left(1 - \frac{p'^2}{\bar{\lambda}_x^4}\right)\left(1 - \frac{1}{1.12 + s^{1.5}}\right)\left(\frac{s^2 + 1.21\bar{\lambda}_x^2}{s(s+1)}\right) \tag{24}$$

$$s = \frac{1}{400}\frac{r}{t} \tag{25}$$

where the relative slenderness $\bar{\lambda}_x$ is given by Eq. (15) and p' is defined in Eq. (22).

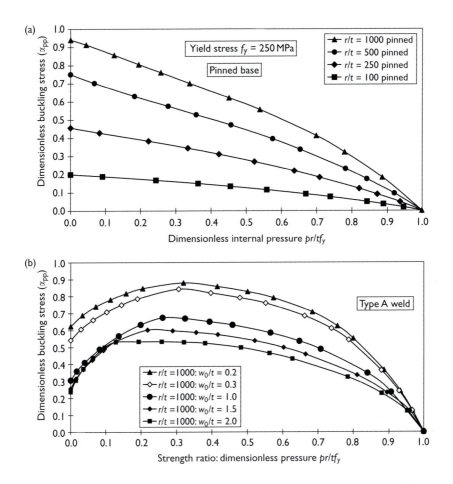

Figure 2.23 Effect of higher internal pressure on elastic–plastic buckling strength: (a) at a simply supported boundary; (b) at a weld depression.

   The above relation was developed from predictions of collapse adjacent to pinned-base details of on-ground silos. Fixed-base details were also explored by Rotter (1990), and were shown to be slightly stronger. At a major change of plate thickness, conditions close to a fixed-base boundary exist. The same remarks may be made concerning zones adjacent to ring stiffeners on the wall. Equation (24) provides a conservative treatment of all such conditions in a pressurised cylinder.

   The process by which a modification of the elastic imperfection factor α is used to treat this elastic–plastic instability is not entirely satisfactory since a plastic phenomenon is being treated as if it were elastic. However, this problem arises because the relative slenderness is defined on the basis of a simple ratio of yield stress to linear eigenvalue stress. If a biaxial yield condition is adopted instead, and this failure mode predicted using the elastic–plastic strength interaction of Eq. (16), the strength is much less precisely predicted and is identified as sensitive to the geometric tolerances, which is not the case. A description of the corresponding effects in lap-jointed pressurised cylinders may be found in Teng (1994).

Finally, caution should be exercised in identifying the load cases that should be covered by this information. The shape of the complete elastic and plastic relationship in Fig. 2.23(b) means that different extreme characteristic values of internal pressure should be taken for each part of the design. The lowest co-existent design internal pressure should be used to assess the elastic buckling strength shown in Figs 2.19(a) and 2.23(b), but the highest coexistent design internal pressure should be used when checking against elephant's foot plastic collapse (Fig. 2.23(a) and (b)).

Conclusions

This chapter has presented a review of current knowledge and design recommendations applicable to the elastic and plastic buckling of axially compressed cylindrical shells in a civil engineering context. It has shown that these shells have many complexities of behaviour and that the translation of these features into design rules is still far from complete.

Many different design methods are in current use and the safety of a given structure as assessed by different rules may vary markedly. Recent changes introduced into the European standard ENV 1993-1-6 (1999) represent a significant move towards integrating current theoretical knowledge into the design process and advancing the use of modern computational procedures. The relationship between tolerances in construction and codified design strengths has also been made much more prominent.

A considerable body of research work is still needed to relate hand calculation design rules to computer predictions that account for geometric imperfections and to appropriate tolerance measures to control and monitor construction. The tolerance measurement systems themselves are also in need of further development. However, the most serious missing element in current research is an understanding of the effects of unsymmetrical axial stress distributions in the cylinder. For very unsymmetrical stresses, additional research is needed to find appropriate criteria of failure.

Notation

E	Young's modulus
f_y	characteristic yield stress
ℓ_{gx}, ℓ_{gw}	lengths of measuring rods used in tolerance measurement
L	length of cylinder between boundaries or rings

n_{cl}	number of full waves around the circumference in classical asymmetric mode
n_{cr}	number of full waves around the circumference in critical mode
k	modifying factor for imperfection form
p	internal pressure within cylinder
p'	dimensionless internal pressure
Q	meridional compression fabrication quality parameter
r	cylinder radius
s	reference value of radius to thickness ratio
t	shell wall thickness
U_{0x}	dimensionless imperfection amplitude
U_{max}	maximum permitted value of U (tolerance measure)
U_{n1}, U_{n2}	values of U for use in computer calculations
w_0	amplitude of imperfection
x	axial distance from centre of weld
Z	Batdorf parameter to characterise shell length
α	ratio of characteristic buckling stress to classical elastic critical stress
α_{x0}	value of α under axial compression alone (no internal pressure)
α_{xpe}	value of α with elastic strength gain due to internal pressure
α_{xpp}	value of α with plastic strength loss due to internal pressure
β	plastic range factor in buckling design
Δw_x	amplitude of imperfection used in axial compression buckling design
$\Delta w_{0,eff,1}$	amplitude of imperfection for computer calculations
η	interaction exponent for buckling design
λ	meridional linear bending half-wavelength
λ_{cl}	axial half-wavelength of axisymmetric buckling mode
$\bar{\lambda}$	relative slenderness of the shell
$\bar{\lambda}_0$	squash limit relative slenderness
ν	Poisson's ratio (typically around $\nu = 0.3$)
σ_{cl}	classical elastic critical stress
σ_{cr}	buckling stress at bifurcation
σ_{xcr}	characteristic value of buckling stress under axial compression

References

Almroth, B.O. (1963). Postbuckling behavior of axially compressed circular cylinders. *AIAA Journal* **1**(3), 630–633.

Almroth, B.O. (1966). Influence of edge conditions on the stability of axially compressed cylindrical shells. *AIAA Journal* **4**(1), 143–140.

Almroth, B.O., Burns, A.B. and Pittner, E.V. (1970). Design criteria for axially loaded cylindrical shells. *Journal of Spacecraft & Rockets* **7**(6), 714–720.

Amazigo, J.C. (1969). Buckling under axial compression of long cylindrical shells with random axisymmetric imperfections. *Quarterly of Applied Mathematics* **26**, 537–566.

Amazigo, J.C. (1974). Buckling of stochastically imperfect structures. In *Buckling of Structures* (ed. B. Budiansky). Springer, New York, pp. 172–182.

Amazigo, J.C. and Budiansky, B. (1972). Asymptotic formulas for the buckling stresses of axially compressed cylinders with localised or random axisymmetric imperfections. *Journal of Applied Mechanics, Transactions of the ASME* **E-39**, 179–184.

API Standard 620 (1978). *Recommended Rules for Design and Construction of Large Welded Low-Pressure Storage Tanks*, 6th edn. American Petroleum Institute, Washington, DC.

Arbocz, J. (1974). The effect of initial imperfections on shell stability. In *Thin Shell Structures*, (eds Y.C. Fung and E.E. Sechler). Prentice Hall, Englewood Cliffs, NJ, pp. 205–246.

Arbocz, J. (1983). Shell stability analysis: theory and practice. In *Collapse* (eds J.M.T. Thompson and G.W. Hunt). Cambridge University Press, Cambridge, UK, pp. 43–74.

Arbocz, J. and Babcock, C.D., Jr. (1974). Prediction of buckling loads based on experimentally measured initial imperfections. In *Buckling of Structures*. (ed. B. Budiansky), Proceedings of the IUTAM Symposium, Harvard University, June, Springer, New York, 1976.

Arbocz, J. and Sechler, E.E. (1974). On the buckling of axially compressed imperfect cylindrical shells. *Journal of Applied Mechanics, Transactions of the ASME* **41**, 737–743.

AWWA D100–79 (1979). *Standard for Welded Steel Tanks for Water Storage*. American Water Works Association, Denver, CO.

Berry, P.A. and Rotter, J.M. (1996). Partial axisymmetric imperfections and their effect on the buckling strength of axially compressed cylinders. *Proceeding of the International Workshop on Imperfections in Metal Silos: Measurement, Characterisation and Strength Analysis*, CA-Silo, Lyon, France, 19 April, pp. 35–48.

Berry, PA., Bridge, R.Q. and Rotter, J.M. (1997). Experiments on the buckling of axially compressed fabricated steel cylinders with axisymmetric imperfections. *Proceedings of the International Conference on Carrying Capacity of Steel Shell Structures*, Brno, 1–3 October, pp. 347–353.

Berry, P.A., Rotter, J.M. and Bridge, R.Q. (2000). Compression tests on cylinders with circumferential weld depressions. *Journal of Engineering Mechanics, ASCE* **126**(4), 405–413.

Bornscheuer, F.W. (1982). To the problem of buckling safety of shells in the plastic range. *Buckling of Shells*, *Proceedings of a State-of-the-Art Colloquium*. (ed. E. Ramm). Springer-Verlag, Berlin, pp. 601–619.

Bornscheuer, F.W. and Hafner, L. (1983). The influence of an imperfect circumferential weld on the buckling strength of axially loaded circular cylindrical shells. Preliminary Report, *3rd International Colloquium on Stability of Metal Structures*, Paris, November, pp. 407–414.

Bornscheuer, F.W., Hafner, L. and Ramm, E. (1983). Zur Stabilität eines Kreiszylinders mit einer Rundschweissnäht unter Axialbelastung. *Der Stahlbau* **52**(10), 313–318.

Brush, D.O. and Almroth, B.O. (1976). *Buckling of Bars, Plates and Shells*. McGraw-Hill, New York.

Bushnell, D. (1985). *Computerized Buckling Analysis of Shells*. Martimus Nijhaff, Dordrecht.

Calladine, C.R. (1983). *Theory of Shell Structures*. Cambridge University Press, Cambridge, UK.

Cohen, G.A. (1971). Computer analysis of imperfection sensitivity of ring-stiffened orthotropic shells of revolution. *Journal of the American Institute of Aeronautics and Astronautics* **9**, 1032–1039.

CRCJ (1971). *Handbook of Structural Stability*. Column Research Council of Japan, Corona, Tokyo.

Croll, J.G.A. (1984). Axisymmetric elastic–plastic buckling of axial and pressure loaded cylinders. *Proceedings of the IMechE* **198C**(16), 243–259.

Croll, J.G.A. and Ellinas, C.P. (1985). A design formulation for axisymmetric collapse of stiffened and unstiffened cylinders. *Transactions of the ASME* **107**, 350–355.

Deml, M. and Wunderlich, W. (1997). Direct evaluation of the 'worst' imperfection shape in shell buckling. *Computer Methods in Applied Mechanics and Engineering* **149**, 201–222.

DIN 18800 (1990). Stahlbauten: stabilitätsfälle, schalenbeulen. DIN 18800 Part 4, Deutsches Institut für Normung, Berlin, November.

Ding, X.L., Coleman, R.D. and Rotter, J.M. (1996). Technique for precise measurement of large-scale silos and tanks. *Journal of Surveying Engineering, ASCE* **122**(1), 14–25.

Donnell, L.H. (1934). A new theory for the buckling of thin cylinders under axial compression and bending. *Transactions of the ASME* **56**, 795–806.

Donnell, L.H. and Wan, C.C. (1950). Effect of imperfections on buckling of thin cylinders and columns under axial compression. *Journal of Applied Mechanics, Transaction of the ASME* **17**(1), 73–83.

ECCS (1988). *Recommendations for Steel Construction: Buckling of Shells*, 3rd edn. European Convention for Constructional Steelwork.

ENV 1993-1-6 (1999). Eurocode 3: Design of steel structures, Part 1.6: General rules – Supplementary rules for the strength and stability of shell structures, Eurocode 3 Part 1.6, CEN, Brussels.

Elishakoff, I. (1983). How to introduce the imperfection-sensitivity concept into design. In *Collapse: The Buckling of Structures in Theory and Practice* (eds J.M.T. Thompson and G.W. Hunt). Cambridge University Press, Cambridge, UK, pp. 345–357.

Esslinger, M. and Geier, B. (1972). Gerechnete Nachbeulasten als untere Grenze der experimentellen axialen Beulasten von Kresiszylindern. *Der Stahlbau* **41**(12), 353–360.

Esslinger, M. and Geier, B. (1974). Calculated postbuckling loads as lower limits for the buckling loads of thin-walled circular cylinders. In *Buckling of Structures* (ed. B. Budiansky). Springer, New York, pp. 274–290.

Esslinger, M. and Geier, B. (1977). Buckling loads of thin-walled circular cylinders with axisymmetric irregularities. *Steel Plated Structures: An International Symposium* (eds P.J. Dowling, J.E. Harding and P.A. Frieze). pp. 865–888.

Esslinger, M. and Pieper, K. (1973). Schnittkrafte und Beullasten von Silos aus uberlappt verschraubten Blechplatten. *Der Stahlbau* **42**(9), 264–268.

Flügge, W. (1932). Die Stabilität der Kreiszylinderschale. *Ing.-Arch.* **3**, 463–506.

Flügge, W. (1973). *Stresses in Shells*, 2nd edn. Springer-Verlag, Berlin.

Fung, Y.C. and Sechler, E.E. (1957). Buckling of thin-walled circular cylinders under axial compression and internal pressure. *Journal of Aeronautical Science* **24**(5), 351–356.

Fung, Y.C. and Sechler, E.E. (eds) (1974). *Thin-Shell Structures: Theory, Experiment and Design*. Prentice Hall, Englewood Cliffs, NJ.

Greiner, R. and Yang, Y. (1996). Effect of imperfections on the buckling strength of cylinders with stepped wall thickness under axial loads. *Proceedings of the International Workshop on Imperfections in Metal Silos: Measurement, Characterisation and Strength Analysis*, CA-Silo, Lyon, France, 19 April, pp. 77–86.

Guggenberger, W. (1996). Effect of geometric imperfections taking into account the fabrication process and consistent residual stress fields of cylinders under local axial loads. *Proceedings of the International Workshop on Imperfections in Metal Silos: Measurement, Characterisation and Strength Analysis*, CA-Silo, Lyon, France, 19 April, pp. 217–228.

Harris, L.A., Suer, H.S., Skene, W.T. and Benjamin, R.J. (1957). The stability of thin-walled unstiffened circular cylinders under axial compression including the effects of internal pressure. *Journal of Aeronautical Science* **24**(8), 587–596.

Hoff, N.J. and Soong, T.C. (1967). Buckling of axially compressed cylindrical shells with non-uniform boundary conditions. In *Symposium on Thin Walled Steel Structures*, University College, Swansea, pp. 61–80.

Holst, J.M.F.G., Rotter, J.M. and Calladine, C.R. (1996). Geometric imperfections and consistent residual stress fields in elastic cylinder buckling under axial compression. *Proceedings of the International Workshop on Imperfections in Metal Silos: Measurement, Characterisation and Strength Analysis*, CA-Silo, Lyon, France, 19 April, pp. 199–216.

Holst, J.M.F.G., Rotter, J.M. and Calladine, C.R. (1997). Characteristic imperfection forms for cylinders derived from misfit calculations. *Proceedings of the International Conference on Carrying Capacity of Steel Shell Structures*, Brno, 1–3 October, pp. 333–339.

Holst, J.M.F.G., Rotter, J.M. and Calladine, C.R. (1999). Imperfections in cylindrical shells resulting from fabrication misfits. *Journal of Engineering Mechanics, ASCE* **125**(4), 410–418.

Holst, J.M.F.G., Rotter, J.M. and Calladine, C.R. (2000). Imperfections and buckling in cylindrical shells with consistent residual stresses. *Journal of Constructional Steel Research* **54**, 265–282.

Hutchinson, J.W. (1965). Axial buckling of pressurised imperfect cylindrical shells. *AIAA Journal* **3**, 1461–1466.

Hutchinson, J.W. and Koiter, W.T. (1970). Postbuckling theory. *Applied Mechanics Reviews* **23**, 1353–1366.

Hutchinson, J.W., Tennyson, R.C. and Muggeridge, D.B. (1971). Effect of a local axisymmetric imperfection on the buckling behaviour of a circular cylindrical shell under axial compression. *AIAA Journal* **9**(1), 48–53.

Knödel, P. and Schulz, U. (1992). Buckling of cylindrical bins – recent results. *Proceedings of the International Conference: 'Silos – Forschung und Praxis'*, Universität Karlsruhe, October, pp. 75–82.

Knödel, P. and Ummenhofer, T. (1996). Substitute imperfections for the prediction of buckling loads in shell design. *Proceedings of the International Workshop on Imperfections in Metal Silos: Measurement, Characterisation and Strength Analysis*, CA-Silo, Lyon, France, 19 April, pp. 87–102.

Knödel, P., Ummenhofer, T. and Schulz, U. (1995). On the modelling of different types of imperfections in silo shells. *Thin-Walled Structures* **23**, 283–293.

Koiter, W.T. (1945). On the stability of elastic equilibrium. PhD Thesis, Delft University (in Dutch).

Koiter, W.T. (1963). The effect of axisymmetric imperfections on the buckling of cylindri-cal shells under axial compression. *Proc. Kon. Ned. Akad. Wet.*, B66, p. 265. (See also *Applied Mechanics Reviews* **18**, Review 3387, 1965.)

Lo, H., Crate, H. and Schwartz, E.B. (1951). Buckling of thin-walled cylinders under axial compression and internal pressure. *Journal of Aeronautical Science* **24**(5), 351–356.

Lorenz, R. (1908). Buckling of a cylindrical shell under axial compression *Zeitschrift des Vereines Deutscher Ingenieure* **52**(1908), 1766 (in German).

Pircher, M., Berry, P.A., Ding, X.L. and Bridge, R.Q. (2001). The shape of circumferential weld-induced imperfections in silos and tanks. *Thin-Walled Structures* **39**, 999–1014.

Riks, E., Rankin, C.C. and Brogan, F.A. (1996). On the solution of mode jumping phe-nomena in thin-walled shell structures. *Computer Methods in Applied Mechanics and Engineering* **136**, 59–92.

Robertson, A. (1928). The strength of tubular struts. *Proceedings of the Royal Society of London, Series A* **121**, 558.

Rotter, J.M. (1985). Buckling of ground-supported cylindrical steel bins under vertical compressive wall loads. *Proceedings of the Metal Structures Conference*, Institution of Engineers Australia, Melbourne, May, pp. 112–127.

Rotter, J.M. (1989). Stress amplification in unstiffened cylindrical steel silos and tanks. *Civil Engineering Transactions, Institution of Engineers, Australia*, **CE31**(3), 142–148.

Rotter, J.M. (1990). Local inelastic collapse of pressurised thin cylindrical steel shells under axial compression. *Journal of Structural Engineering, ASCE* **116**(7), 1955–1970.

Rotter, J.M. (1996). Elastic plastic buckling and collapse in internally pressurised axi-ally compressed silo cylinders with measured axisymmetric imperfections: interactions between imperfections, residual stresses and collapse. *Proceedings of the International Workshop on Imperfections in Metal Silos: Measurement, Characterisation and Strength Analysis*, CA-Silo, Lyon, France, 19 April, pp. 119–140.

Rotter, J.M. (1997). Pressurised axially compressed cylinders. *Proceedings of the Interna-tional Conference on Carrying Capacity of Steel Shell Structures*, Brno, 1–3 October, pp. 354–360.

Rotter, J.M. (1998). Development of proposed European design rules for buckling of axially compressed cylinders. *Advances in Structural Engineering* **1**(4), 273–286.

Rotter, J.M. (2002a). The new European standards for shells and their use in analysis and design. *Thin-Walled Structures* (submitted for publication).

Rotter, J.M. (2002b). Shell buckling and collapse analysis for structural design: the new framework of the European standard. Festschrift Chris Calladine, Celebration volume for the 60th birthday of Prof. C.R. Calladine, University of Cambridge, September.

Rotter, J.M. and Teng, J.G. (1989a). Elastic stability of lap-jointed cylinders. *Journal of Structural Engineering, ASCE* **115**(3), 683–697.

Rotter, J.M. and Teng, J.G. (1989b). Elastic stability of cylindrical shells with weld depressions. *Journal of Structural Engineering, ASCE* **115**(5), 1244–1263.

Rotter, J.M. and Zhang, Q. (1990). Elastic buckling of imperfect cylinders containing granular solids. *Journal of Structural Engineering, ASCE* **116**(8), 2253–2271.

Rotter, J.M., Coleman, R., Ding, X.L. and Teng, J.G. (1992). The measurement of imper-fections in cylindrical silos for buckling strength assessment. *Proceedings of the Fourth*

International Conference on Bulk Materials Storage Handling and Transportation, Institution of Engineers, Australia, Wollongong, June, pp. 473–479.

Saal, H., Kahmer, H. and Reif, A. (1979). Beullasten axial gedruckter Kreiszylinderschalen mit Innendruck – Neue Versuche und Vorshriften. *Der Stahlbau* **48**(9), 262–269.

Samuelson, L.A. and Eggwertz, S. (1992). *Shell Stability Handbook*. Elsevier Applied Science, London.

Sanders, J.L. (1963). Nonlinear theories for thin shells. *Quarterly of Applied Mathematics* **21**, 21–36.

Schmidt, H. and Winterstetter, T.A. (2001). Substitute geometrical imperfections for the numerical buckling assessment of cylindrical shells under combined loading. *Proceedings of the European Mechanics Conference: Euromech 424*, Rolduc-Kerkrade, Germany, 3–5 September, pp. 82–84.

Schneider, W., Höhn, K., Timmel, I. and Thiele, R. (2001). Quasi-collapse-affine imperfections at slender wind-loaded cylindrical steel shells. *Second European Conference on Computational Mechanics*, Cracow, Poland, June, Vol. 2, pp. 1000–1001.

Schnell, W. (1959). Zur Stabilität dunnwandiger Langsgedruckter Kreiszylinderschalen bei Zusatzlichem Innendruck. *Proceedings of the IUTAM Symposium on the Theory of Thin Elastic Shells*, Delft, August, pp. 167–188.

Singer, J. (1982). The status of buckling investigations of shells. In *Buckling of Shells* (ed. E. Ramm). Springer-Verlag, New York.

Southwell, R.V. (1914). *Philosophical Transactions of the Royal Society* **213**, 187.

Steinhardt, O. and Schulz, V. (1971). Zum Beulverhalten von Kreiszylinderschalen. *Schweizerische Bauzeitung* **89**(1), 7.

Teng, J.G. (1994). Plastic collapse at lap-joints in pressurized cylinders under axial load. *Journal of Structural Engineering, ASCE* **120**(1), 23–45.

Teng, J.G. (1996). Buckling of thin shells: recent advances and trends. *Applied Mechanics Reviews, ASME* **49**(4), 263–274.

Teng, J.G. and Rotter, J.M. (1989). Non-symmetric buckling of plate-end pressure vessels. *Journal of Pressure Vessel Technology, ASME* **111**(3), 304–311.

Teng, J.G. and Rotter, J.M. (1992). Buckling of pressurized axisymmetrically imperfect cylinders under axial loads. *Journal of Engineering Mechanics, ASCE* **118**(2), 229–247.

Teng, J.G. and Song, C.Y. (2001). Numerical models for nonlinear analysis of elastic shells with eigenmode-affine imperfections. *International Journal of Solids and Structures* **38**(18), 3263–3280.

Tennyson, R.C. and Muggeridge, D.B. (1969). Buckling of axisymmetric imperfect circular cylindrical shells under axial compression. *AIAA Journal* **17**, 2127–2131.

Thielemann, W.F. (1960). New developments in the non-linear theories of the buckling of thin cylindrical shells. In *Aeronautics and Astronautics*. Pergamon Press, New York, pp. 76–121.

Thompson, J.M.T. and Hunt, G.W. (eds) (1983). *Collapse*. Cambridge University Press, Cambridge, UK.

Timoshenko, S.P. (1910). Einige Stabilitätsprobleme der Elasticitätstheorie. *Zeitschrift für Mathematik und Physik* **58**, 378–385.

Timoshenko, S.P. (1936). *Theory of Elastic Stability*. United Engineering Trustees Inc., New York.

Trahair, N.S., Abel, A., Ansourian, P., Irvine, H.M. and Rotter, J.M. (1983). *Structural Design of Steel Bins for Bulk Solids*. Australian Institute of Steel Construction, November.

Walker, P. and Wilson, D.R. (2001). Buckling of cylinders with local loads normal to the shell. MEng Theses, School of Civil and Environmental Engineering, University of Edinburgh, 150p.

Weingarten, V.I., Morgan, E.J. and Seide, P. (1965a). Elastic stability of thin-walled cylindrical and conical shells under axial compression. *AIAA Journal* **3**(3), 500–505.

Weingarten, V.I., Morgan, E.J. and Seide, P. (1965b). Elastic stability of thin-walled cylindrical and conical shells under combined internal pressure and axial compression. *AIAA Journal* **3**(6), 1118–1125.

Wilson, W.M. and Newmark, N.W. (1933). The strength of thin cylindrical shells as columns. *Bulletin no. 255*, Univ. Illinois Engg Exptl Sta.

Yamada, S. and Croll, J.G.A. (1999). Contributions to understanding the behaviour of axially compressed cylinders. *Journal of Applied Mechanics, Transactions of the ASME* **66**, 299–309.

Yamaki, N. (1984). *Elastic Stability of Circular Cylindrical Shells*, North-Holland, Amsterdam.

Chapter 3

Cylindrical shells above local supports

W. Guggenberger, R. Greiner, and J.M. Rotter

Introduction

Upright cylindrical thin-walled steel containment structures, like silos and tanks, are commonly supported on discrete supports or columns to permit access beneath the hopper for the gravity withdrawal of the contents. In practical silo structures a variety of different support designs are used. The construction details of these designs depend mainly on the magnitude of the local support forces that are to be introduced into the shell wall at its lower edge. The number of supporting columns and the design of the cylinder–cone transition play important roles in determining the local load carrying capacity of the silo shell wall. A simple transition ring may be sufficient in light structures, but in larger silos a skirt may be added, creating a triangular hollow section around the circumference (Fig. 3.1). In large silos, several alternative forms of ring beams may be provided (Fig. 3.1(c)). In addition,

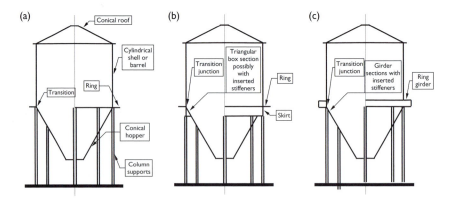

Figure 3.1 Circular planform silos with different constructional forms of the barrel–hopper transition: (a) light silo construction with annular plate ring stiffener; (b) medium construction with integrated transition ring and triangular hollow cross-section; (c) heavy construction with external ring beam.

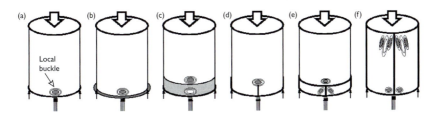

Figure 3.2 Practical stiffening of discretely supported silos and potential buckling mode: (a) unstiffened cylinder; (b) additional edge-ring stiffener at the bottom edge; (c) reinforced wall thickness; (d) partial-height longitudinal stiffeners; (e) additional ring stiffener; (f) full-height longitudinal stiffeners (reprinted from *Thin Walled Structures*, vol 37, Guggenberger, W., Proposal for design rules of axially loaded steel cylinders on local supports, pp. 159–167, with permission from Elsevier Science).

many different stiffening arrangements have been proposed for the cylindrical part (e.g. Greiner 1984; Öry *et al.* 1984; Rotter 1985; Öry and Reimerdes 1987; Galletly 1988), principally with the aim of eliminating potential buckling failures. However, it is difficult for a designer to know whether such stiffening arrangements are needed for a particular structure, as many satisfactory designs have been produced without stiffening.

The initial reference structure to begin evaluating these questions is represented by a locally supported unstiffened cylinder of constant wall thickness (Fig. 3.2(a)). This may then be used to determine the strengthening effect of stiffening measures of different kinds. If the local buckling strength of an unstiffened cylinder can be assessed with confidence, it will be possible to identify which cylinders need stiffening, and to study the changes in buckling behaviour and the increases in buckling strength which stiffening provides.

Stiffening may be provided by integrated edge-ring stiffeners at the bottom edge (Fig. 3.2(b)), an increased wall thickness for the bottom course (Fig. 3.2(c)), longitudinal stiffeners extending over either part of the height (Fig. 3.2(d)) or the full height of the cylinder (Fig. 3.2(f)) and additional ring stiffeners higher above the support (Fig. 3.2(e)). In engineering practice, each of these measures and their combinations have been used. An investigation of the buckling strength of all practically used geometries is, therefore, a very considerable challenge. The idealised arrangements shown in Fig. 3.2 represent the commonest forms of practical support designs. In all these cases, the load-carrying capacity of the shell wall above the supports is closely related to the buckling resistance of the shell wall under local axial compression. The axial compression may be accompanied by circumferential tension due to pressure from the contained solids or liquids.

History of research work

Despite its practical importance, the problem of the axially compressed cylinder on local supports has not been studied widely until recently. The first scientific studies were concerned with the stress distribution above a local support, and these began with supports that applied forces normal to the shell (Bijlaard 1955). Local supports that apply meridional forces were more difficult to analyse, so this development came later (Kildegaard 1969; Gould *et al.* 1976; Rotter 1982; Li 1994; Li and Rotter 1996). These linear stress analyses could define the approximate stress state in the shell, but it was difficult to determine what failure criteria should be used. In addition, there was uncertainty about the effects of geometric nonlinearity and the role of geometric imperfections in causing failure above local supports.

Studies of the buckling strength of axially compressed cylinders were dominated by uniform compression (Yamaki 1984), and only a few studies explored the effects of slight variations from a uniform stress state, such as global bending (Saal 1982) or axial compression along an axial strip (Hoff *et al.* 1964). Design recommendations (NASA SP8007 1968; ECCS 1988) did not attempt to address the local support problem. A few simple empirical treatments were used to propose design concepts (Bodarski *et al.* 1985; Gorenc 1985; Hotala 1986, 1996; Samuelson 1986, 1987, 1990; Samuelson and Eggwertz 1992), but these were neither based on rigorous buckling calculations nor on experiments. Rationally based design proposals have only been developed recently. This chapter outlines these recent developments.

Theoretical studies of the nonlinear buckling behaviour have been undertaken during the past 10 years by several research groups. The buckling behaviour has been found to be quite complicated, with many parameters influencing the buckling strength. Initial work on the subject started in the late 1980s, by performing comprehensive nonlinear numerical parametric studies comprising unstiffened and various forms of stiffened steel cylinders (Fig. 3.2). In the Graz investigations, the cylindrical shell was given fixed dimensions ($R/t = 500$, $H/R = 2$, standard steel grade S 235) and rested on just four local flexible supports whose width was varied (Figs 3.2(a) and 3.3(a)) (Guggenberger 1991, 1992). At the same time, research

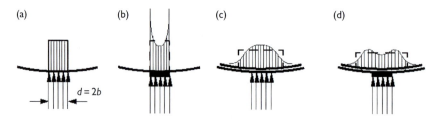

Figure 3.3 Alternative local support boundary conditions: (a) flexible support; (b) rigid support; (c) flexible support with additional ring at bottom edge; (d) rigid support with additional ring at bottom edge.

at Edinburgh involved both experimental and theoretical explorations of similar problems, but used elastic buckling calculations on shells of varying R/t, varying numbers of supports and varying support width, but the supports were treated as rigid, leading to higher strengths than were found in the Graz study. This work was reported by Teng and Rotter (1990, 1992), Rotter *et al.* (1991) and She and Rotter (1993). Both of the above studies used geometrically nonlinear analysis using ABAQUS (Hibbit, Karlsson and Sorensen 1995) of both perfect and geometrically imperfect shells. Finally, a rational design proposal was developed jointly between the two groups for the basic case of the unstiffened cylinder on local supports (Fig. 3.2(a)) (Rotter *et al.* 1993). This was the first time that a rationally based design regulation for this complicated local loading case became available. All previous design rules were concerned only with buckling under uniform compression (NASA SP8007 1968; ECCS 1988; DIN 18800 1990; Beuth 1993; ENV 1993-1-6 1999).

This first design proposal (Rotter *et al.* 1993) was based on numerical calculations with assumed geometric imperfections and idealised elastoplastic material behaviour and was validated against about 20 buckling tests (She and Rotter 1993). Meanwhile, a further project of experiments with matched calculations was undertaken as a collaboration between the University of Gent, where tests were conducted, and TU Graz, where the computational work was undertaken (Guggenberger and Greiner 1994a,b; Greiner and Guggenberger 1996a,b; Guggenberger 1996; Rathé and Greiner 1996). The basic case of investigation was the unstiffened cylinder with constant wall thickness resting on flexible supports (Figs 3.2(a) and 3.3(a)).

In a second group of studies, the strengthening effect of increasing the wall thickness of the bottom course was explored (Fig. 3.2(c)). The shell slenderness ratio R/t was varied between 200 and 750 with dimensionless support widths $0 < d/R \leq 0.2$ (Fig. 3.3). The effect of higher steel grades was also considered (S235, S275 and S355 according to EN 10025). The nonlinear analyses used different levels of sophistication, so the abbreviations of the Eurocode for shell structures (ENV 1993-1-6 1999) are used here for simplicity: MNA, GNA, GMNA and GMNIA. The analyses of this group included geometric and material nonlinearity as well as local geometric imperfections above the supports (GMNIA analyses). The buckling strength predictions were cast into compact and understandable design formulae that allow the designer to make a direct comparison between the buckling capacity of the locally supported cylinder and that of the uniformly loaded cylinder (Guggenberger 1998a; Guggenberger and Greiner 1998).

The test results served a twofold purpose: first, for the validation of the numerical procedure by selective comparison with carefully re-analysed test cases and, second, as an overall check on the new design formulae. In both aspects, a satisfactory agreement was achieved. Additional features of practical importance were also considered, such as rigid support conditions (Fig. 3.3(b)), edge-ring stiffeners at the lower circumference and the strengthening effect of additional elasto-plastically deformable support plates (Figs 3.2(b), 3.3(c) and (d)). The interactive effect of local axial compression with internal pressure was also investigated in two separate studies (Li 1994; Greiner and Guggenberger 1998). Recently, Schulz and Ummenhofer (2000) undertook an experimental study of unstiffened cylinders, with very stiff base rings at the boundary where the local loads are introduced into the cylinder. Following this project, a semi-empirical design proposal was developed by Knödel and Ummenhofer (1998).

Scope of new exploration presented here

The modelling assumptions of the finite element analysis model are described in the next section. In sections on 'Buckling behaviour of an unstiffened cylinder' and 'The effect of internal pressure', descriptions are presented on both the linear and the geometrically and materially nonlinear buckling behaviour of isotropic unstiffened cylinders on discrete supports at the lower edge. The post-buckling behaviour is also outlined. The study of unstiffened cylinders is then extended to explore the effect of internal pressure ('The effect of internal pressure'), and the effect of thickening the lower course. In the final section ('Proposed design rules'), design rules are presented for the basic case of the unstiffened cylinder and for cylinders with increased wall thickness of the bottom course of the cylinder.

Modelling assumptions of the present study

Finite element analysis models

Two equivalent analysis models are shown in Fig. 3.4. Both have been used for the study of the unstiffened cylinder. Throughout this study, the cylinder was supported on four columns of varying width $d = 2b$, so a 45° segment could be used for the finite element model, exploiting symmetry on meridional lines, as indicated by the grey shaded areas in Fig. 3.4. The model in Fig. 3.4(a) represents a more realistic load case and has been used in most of the analyses. However, the descriptions in the section on 'Buckling behaviour of an unstiffened cylinder' are based on the idealised model of Fig. 3.4(b), which assumes symmetry about a horizontal plane at the cylinder half-height. The finite element mesh used is shown in Fig. 3.5.

Teng and Rotter (1990, 1992) and She and Rotter (1993) used a model similar to that of Fig. 3.4(a), but with wall friction loading from the solid for greater realism in modelling the true loading. They showed that the buckling strength at a single discrete support is not sensitive to the number of supports around the circumference, provided that this number is not so large that the nearest edges of adjacent supports are not closer than the support width. Thus, the limitation to only four supports used here is not restrictive.

Values of the radius-to-thickness ratio of $R/t = 200$, 300, 500 and 750 were used in the parametric studies, and a single height-to-radius ratio $H/R = 2.0$ was used throughout. The descriptions in the following sections are performed only

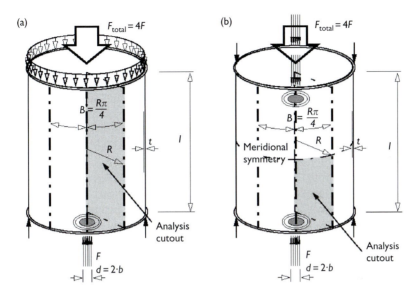

Figure 3.4 Initial models of the locally supported unstiffened cylinder: (a) under uniform roof loading (Edinburgh); (b) with meridional symmetry conditions at mid-span (Graz).

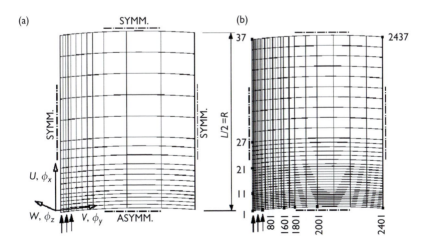

Figure 3.5 Typical finite element mesh and boundary conditions (S8R5 or S4R5 shell elements): (a) displacements and boundary conditions; (b) node and element numbers.

for $R/t = 500$, since their purpose is explanatory. A shell thickness of $t = 10$ mm was used throughout.

Idealisations of boundary, support and loading conditions

In practical silo structures, a conical hopper is often used for discharge of the solids, and it meets the cylinder at the transition, where a ring may be provided at the junction (Fig. 3.1). All the different types of transition shown in Fig. 3.1 apply a stiff radial and circumferential restraint to the cylinder. The analysis of an isolated cylinder with a completely rigid radial restraint at each end thus provides a satisfactory representation of most practical structures. The radial and circumferential displacements at both the lower and upper circumferential edges were, therefore, fixed, but the axial displacement component and the meridional rotations were free to move (i.e. classical hinged boundary conditions at the boundaries). It may be noted that the interaction effects with neighbouring structural components has not been considered here.

The support was treated as axially flexible: the applied support force induces a uniform axial stress across the width of the support, with the support providing no axial warping restraint and no meridional rotational restraint (Fig. 3.3(a)). This is a conservative assumption, leading to lower strengths than are achieved in practical supports. Other studies (e.g. She and Rotter 1993; Guggenberger and Greiner 1994a) have found the same structural behaviour when rigid supports (Fig. 3.3(b)) are used.

The axial stress state in the cylindrical shell vary somewhat according to the geometry and loading applied to it: the dominant forces may arise from pressure on the closure, from wall friction due to stored granular solids or from more uniform loads applied at a distant boundary. Whatever the global loading case, the local stress regime close to the support is similar, and since the buckling behaviour is found to be quite local, the details of the loading which induces the high forces above the supports is not important unless the shell-height-to-diameter ratio is relatively small.

Here, the axial loads are applied uniformly at the upper circumference (Figs 3.4 and 3.6). This corresponds to a reference uniform axial stress of 3.820 MPa (=1200 kN applied to the complete circumference) at the top of the shell. For comparison purposes it may be noted that this reference stress is equivalent to a mean axial membrane stress of 1.53% of the classical elastic critical stress σ_{cl} (Timoshenko and Gere 1961; Yamaki 1984), which is given by

$$\sigma_{cl} = \frac{E}{\sqrt{3(1 - v^2)}} \frac{t}{R} \approx 0.605 \, E \frac{t}{R} \tag{1}$$

If the effect of internal pressure is additionally considered ('The effect of internal pressure'), this may be done in a simplified way (Fig. 3.6(a)). The real pressure distribution that is exerted by the contents is quite complicated (Ooi et al. 1990;

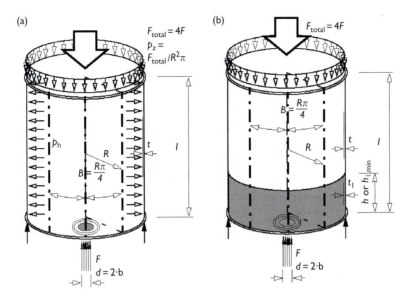

Figure 3.6 Additional features of the locally supported cylinder: (a) with additional internal pressure; (b) with reinforced wall thickness of the bottom course.

Rotter 2001; Zhong *et al.* 2001). However, for an assessment of this effect on the buckling strength, only the local value of pressure in the immediate vicinity of the local buckle above the support is of interest. Therefore, it is justified to assume an uniform internal pressure throughout the shell. The internal (horizontal) pressure magnitude p_h is normally made non-dimensional (Hutchinson 1965; Calladine 1983; Rotter 1997) by dividing the resulting membrane theory circumferential stress $p_h R/t$ by the classical elastic critical stress σ_{cl}, but plasticity dominates in the current work, so this ratio is of less relevance. Instead, the ratio of the membrane theory circumferential stress $p_h R/t$ to the local stress immediately above the flexible support σ_{mm} is used.

$$\alpha = \frac{p_h}{\sigma_{mm}} \frac{R}{t} \tag{2}$$

Support width parameter d/R

The support width can be expressed in terms of either the supported proportion of the circumference, μ, or the ratio of the support width to the radius $\eta = d/R$, which indicates the relative localisation of the support. These two parameters are related through

$$\mu = \frac{b}{B} = \frac{nd}{2\pi R} = \frac{n}{2\pi} \frac{d}{R} = \frac{n}{2\pi} \eta \qquad (3)$$

in which n is the number of supports. Theoretically, the support width can vary from zero to a condition where the entire circumference is supported, which is given by

$$\mu = 1 \quad \text{or} \quad \eta = \frac{d}{R} = \frac{2\pi}{n} \qquad (4)$$

When the silo is supported on four columns ($n = 4$), the value of η becomes $\eta = d/R = \pi/2$. The normalisation relative to the radius (d/R) is probably the more useful representation because the buckling strength of a support of practical size is relatively independent of the number of supports. If the supported circumference is used, different buckling curves are required when the number of supports is changed, making the description much more complicated. Moreover, the proportion of supported circumference draws attention away from the width of the individual support, which determines whether the buckling strength is force-controlled or stress-controlled (She and Rotter 1993; Guggenberger *et al.* 2000).

Geometrical imperfections

In the present study, the basic geometric imperfection form was adopted to be a local bi-cubic spline surface, comparable with the postbuckling mode of the perfect elastic locally supported cylinder. This was chosen because it is often the most detrimental form (imperfection form I2 in Fig. 3.7). This imperfection shape may not necessarily occur in a real shell, since it is clearly not linked to a specific fabrication process, but it is useful in producing buckling strengths that may be lower bound values. This idea is, therefore, generally useful for buckling design purposes if knowledge of real imperfections is uncertain and the method leads to a strength that is relatively independent of imperfection amplitude. The inward amplitude of this imperfection was chosen according to allowable fabrication tolerances specified in design codes, which seems the most reasonable guideline available at present. It amounts to about one wall thickness throughout ($\delta/t = 1.0$) uniformly for all R/t from 200 to 750.

A more refined approach might use the stick tolerance measurement relationship of the ECCS (1988) Recommendations, which would make δ/t explicitly dependent on R/t:

$$\delta/t = 0.01 \times 4\sqrt{R/t} \qquad (5)$$

which would give $\delta/t = 0.65$ for $R/t = 264$, and $\delta/t = 1.0$ for $R/t = 625$.

Here, for calculations with $R/t = 200$ and 300, $\delta/t = 0.65$ was chosen, but for $R/t = 500$ and 750, the larger value of $\delta/t = 1.0$ was used. As a result,

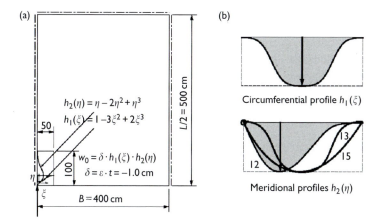

Figure 3.7 Graz local imperfection constructed as bi-cubic Hermite spline surfaces (similar to the elastic postbuckled deformed shape): (a) mesh with local imperfect zone; (b) circumferential and meridional profiles of imperfection geometry (reprinted from *Journal of Constructional Steel Research*, vol. 56, Guggenberger, W., Greiner, R. and Rotter, J.M., The behaviour of locally supported cylindrical shells: unstiffened shells, pp. 175–197, 2000, with permission from Elsevier Science).

slightly higher buckling stresses are found for the shells with lower R/t. However, the results on which the proposed design rules are based and which are presented later, relate to the first approach with constant values of the imperfection amplitude ratio $\delta/t = 1.0$ for all values of R/t.

Material behaviour

Elastic, perfectly plastic material behaviour was assumed with an elastic modulus of $E = 206,000$ MPa and Poisson's ratio $\nu = 0.3$. The reference value of the yield stress was taken as $f_y = 235$ MPa, but values of $f_y = 275$ and 355 MPa were later used to explore the effect of steel grade. To achieve a simple characterisation, the steel grade parameter of Eurocode 3 for steel beams was adopted:

$$\varepsilon = \sqrt{235/f_y} \tag{6}$$

Increased wall thickness of the bottom course

The same computational model was used for cylinders with an increased bottom course wall thickness (Fig. 3.6(b)), with the thickness identified as t_1 and, in this study, assumed to be always 50% thicker than the remainder of the wall. Thus,

$$t_1/t = 1.5 \tag{7}$$

Two shell slendernesses were investigated: $R/t_1 = 200$ with $R/t = 300$ and R/t_1 500 with $R/t = 750$. The study was pursued by seeking the minimal height required of the lower course $h_{1,\min}$ to give the same buckling strength as the value for a uniformly thicker wall (whole shell of thickness t_1). These were geometrically and materially nonlinear analyses (GMNA) without consideration of imperfections. This procedure is relatively quick and yields results that are generally on the safe side (larger $h_{1,\min}$) and independent of uncertain or ambiguous imperfection assumptions.

Buckling behaviour of an unstiffened cylinder

Example problem

An example cylinder is used here to illustrate the phenomena caused by local support forces and the associated shell buckling. A linear stress analysis is shown first to provide a basic understanding of the stress distributions. Thereafter, all the strength predictions are obtained using nonlinear analyses. The load factor Λ used here is applied to the reference loading described in the previous section. The underlying finite element model is shown in Figs 3.3(a) and 3.4(b); the coordinate system and the boundary conditions are shown in Fig. 3.5.

Axial membrane stress distribution in the shell

The pattern of axial compressive membrane stress σ_{xx} above the support is illustrated in Fig. 3.8 for the load factor of $\Lambda = 10$. The stress varies in both the meridional and circumferential directions, with the highest stress immediately above the support, but it declines rapidly as the force is dispersed into the shell. The form of the bell-shaped distribution in the circumferential direction affects the buckling behaviour, since a finite area of the highly stressed shell is needed for a buckle to occur, and the rate at which the axial stress falls with height is closely related to the circumferential width of the highly stressed zone. Because symmetry has been assumed at the cylinder half-height, the axial stress remains quite non-uniform around the circumference at this position, but this does not influence the buckling behaviour markedly because the buckle is very local.

When the support is narrow, very high local axial stresses can occur above it, and buckling may be preceded by yielding. Yielding commences directly above the support (Fig. 3.8), but this position is restrained by the support and the yielded region is small compared with the dimensions of a buckle. Although high bending stresses develop a little higher above the support, and first surface yield occurs at this point, these stresses have little influence on the plastic buckling strength.

The bell-shaped circumferential variation of axial stress plays a key role in the buckling behaviour and strength. The shape and amplitude of the distribution at a height $L/20$ above the support are illustrated in Fig. 3.9(a) and (b) for different support widths.

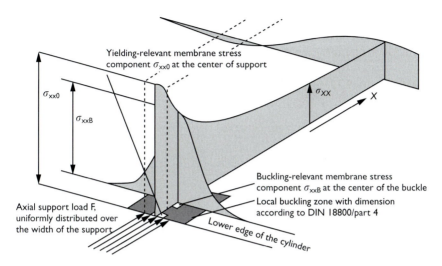

Yielding-relevant membrane stress
component σ_{xx0} at the center of support

σ_{xx0}

σ_{xxB}

σ_{XX}

X

Buckling-relevant membrane stress
component σ_{xxB} at the center of the buckle

Local buckling zone with dimension
according to DIN 18800/part 4

Axial support load F,
uniformly distributed over
the width of the support

Lower edge of the cylinder

Figure 3.8 Axial membrane stress distribution above a local support (critical zones for
local membrane yielding or local buckling are marked) (reprinted from *Journal
of Constructional Steel Research*, vol. 56, Guggenberger, W., Greiner, R. and
Rotter, J.M., The behaviour of locally supported cylindrical shells: unstiffened
shells, pp. 179–197, 2000, with permission from Elsevier Science).

The effect of support width is shown in Fig. 3.9(a) for a fixed mean stress
immediately above the support ($\Lambda = 10$). The integral under each axial stress
distribution is equal to the total applied load at the support, but this load is smaller
for narrow supports than for wide ones because the stress at the support has been
kept constant. Where the support is wide ($d/R = 0.589$, $\mu = 0.375$) there is a
large zone in which the axial stress is still at the value applied to the support, and
the peak stress at this height ($L/20$) is clearly independent of the support width.
By contrast, where the support is narrow (small d/R), the stress pattern at $L/20$ is
quite affected by the small change in level as the load disperses rapidly: the peak
stress magnitude rises almost linearly with the total force at the support (compare
$d/R = 0.033$ and 0.066). Beyond the edge of the support, the rate of decay around
the circumference is similar for both wide and narrow supports, but it is clear that
there is a large zone of reasonably uniform stress above a wide support, which
can easily accommodate a buckle. By contrast, the buckle above a narrow support
must form in a zone of rapidly varying stress. Thus, lower buckling stresses are
found above wide supports, and yielding often precedes buckling above narrow
supports.

The effect of support width is shown in a different manner in Fig. 3.9(b),
where a fixed force is transmitted, irrespective of the width. In the narrow range
($d/R = 0.033$ and 0.066), the stress patterns at $L/20$ for different support widths
are similar (Fig. 3.9(b)). It is reasonable to deduce two simple rules from these

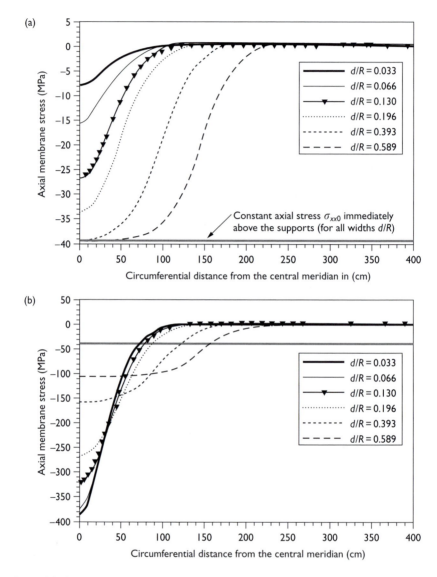

Figure 3.9 Axial membrane stress distributions at $L/20$ above support: (a) fixed mean stress above support; (b) fixed force at support (reprinted from *Journal of Constructional Steel Research*, vol. 56, Guggenberger, W., Greiner, R. and Rotter, J.M., The behaviour of locally supported cylindrical shells: unstiffened shells, pp. 179–197, 2000, with permission from Elsevier Science).

stress distributions: above narrow supports, the *force* at buckling will be more or less independent of the support width, but above wide supports, the *stress* at buckling will be relatively independent of the support width. The transition between these two behaviours may be expected to occur between $d/R = 0.2$ and 0.3. For practical purposes, this corresponds to a rather wide support.

Geometrically nonlinear buckling behaviour of the perfect cylinder

The nonlinear, bifurcation and postbuckling behaviour of elastic shells under uniform axial compression has long been studied (Donnell and Wan 1950; Hoff 1960; Thielemann and Esslinger 1968; Esslinger 1970; Yamaki 1984; Hillmann 1985). In particular, the behaviour of uniformly loaded perfect and near-perfect cylinders immediately after bifurcation is well known to produce a reversed direction of incremental axial deformations. Since significant bending deformations occur above a local support, it was initially unclear whether the same behaviour might be expected in a locally supported cylinder or not.

Load–axial displacement curves are shown in Fig. 3.10(a) for a perfect elastic cylinder on local supports of width $d/R = 0.066$ (refer to Fig. 3.5(b) for node numbering). The two curves show the axial displacements at the centreline and the edge of the support. Whilst the reversal of axial deformations is not marked, it is clear that the buckling phenomenon above a local support may be sudden and imperfection-sensitive, though probably less so than many uniformly compressed

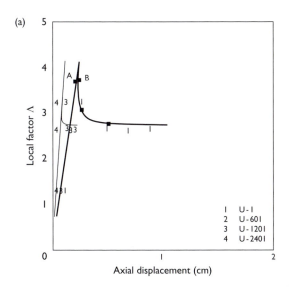

Figure 3.10 (a) Axial displacements of perfect elastic locally supported cylinders on narrow supports ($d/R = 0.066$).

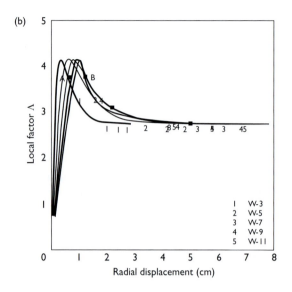

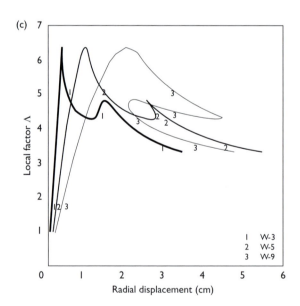

Figure 3.10 (b) Radial displacements of perfect elastic locally supported cylinders on narrow supports ($d/R = 0.066$). (c) Radial displacements of perfect elastic locally supported cylinders on wide supports ($d/R = 0.196$) (reprinted from *Journal of Constructional Steel Research*, vol. 56, Guggenberger, W., Greiner, R. and Rotter, J.M., The behaviour of locally supported cylindrical shells: unstiffened shells, pp. 179–197, 2000, with permission from Elsevier Science).

cylinders. The response well into the postbuckling range is very stable: this is a particularly useful observation when severe imperfections are under consideration. The relationships between load factor and radial displacements for the same cylinder are shown in Fig. 3.10(b). The buckle profile can be identified by comparing curves at different heights, and the rather stable shape of the buckle can be inferred.

Long-wave prebuckling deformations due to inward-bending above the support precede local snap-through buckling. For wider supports (Fig. 3.10(c)), much more complicated curves are obtained, which indicate changes in the buckling mode in the postbuckling region. The smooth changes in radial deformation within the buckle (Fig. 3.10(c)) indicate that this is a snap-through response, not a bifurcation, and that local supports lead, therefore, to a unique path, instead of the multiple clustered bifurcations found for uniformly supported shells. The prebuckling deformations and the development of buckling displacements are shown in Fig. 3.11, which illustrates the transformation from a long-wave bending prebuckling deformation into a local snap-through buckle, which occurs at a load about 10% below the peak value (States A and B in Fig. 3.10(a) and (b)).

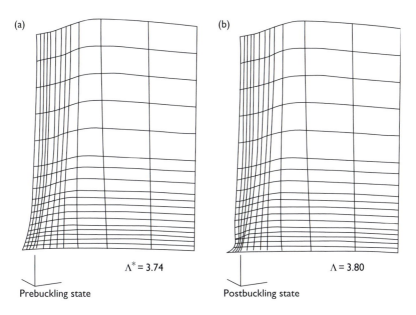

(a)   (b)

$\Lambda^* = 3.74$ $\Lambda = 3.80$

Prebuckling state   Postbuckling state

Figure 3.11 Development of buckling deformations above a narrow support ($d/R = 0.066$), two deformed shapes at the same load level: (a) prebuckling state; (b) postbuckling state (reprinted from *Journal of Constructional Steel Research*, vol. 56, Guggenberger, W., Greiner, R. and Rotter, J.M., The behaviour of locally supported cylindrical shells: unstiffened shells, pp. 179–197, 2000, with permission from Elsevier Science).

Changes in buckling strength with support width for the perfect elastic cylinder

The variation of the elastic buckling strength with support width for all support widths involving four columns is shown in Fig. 3.12. The strengths are displayed in terms of the ratio of the mean axial membrane stress immediately above the support σ_{mm} to the classical elastic critical stress σ_{cl}. The latter is shown as a bold-dashed line for reference. Several alternative methods of calculation have been used, and are marked with simple abbreviations for clarity: GNA–max is the maximum load found using geometrically nonlinear elastic calculations (=buckling strength) and GNA–min is the minimum of the elastic postbuckling response, which has sometimes been used as a lower bound estimate for the strength of imperfect elastic cylinders.

For small support widths, the perfect cylinder buckling strength (GNA–max) falls progressively, as was expected when the prebuckling stress distributions were studied. For impracticably large support widths (e.g. $\mu = 0.5$), an interesting local rise in strength is found as interference occurs between two potential buckles above the support, and finally the strength rises to the value for a uniformly supported cylinder.

For practical supports of practical width ($d/R \leq 0.2$), the changes in the post-buckling minimum (GNA–min) follow those in the buckling strength. A more

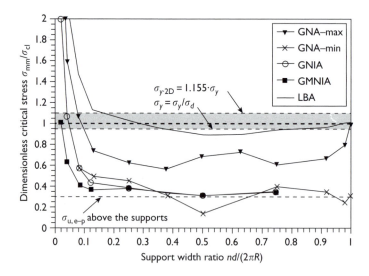

Figure 3.12 Summary of critical stress ratios σ_{mm}/σ_{cl} for perfect elastic and imperfect elastic–plastic cylinders for the full range of support widths ($n = 4$) (reprinted from *Journal of Constructional Steel Research*, vol. 56, Guggenberger, W., Greiner, R. and Rotter, J.M., The behaviour of locally supported cylindrical shells: unstiffened shells, pp. 179–197, 2000, with permission from Elsevier Science).

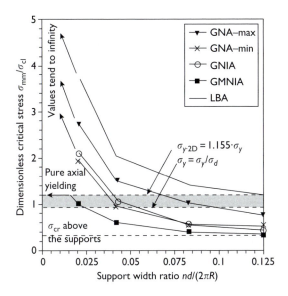

Figure 3.13 Summary of critical stress ratios σ_{mm}/σ_{cl} for perfect elastic and imperfect elastic–plastic cylinders for the full range of support widths with four supports ($n = 4$; $d/R \leq 0.2$) (reprinted from *Journal of Constructional Steel Research*, vol. 56, Guggenberger, W., Greiner, R. and Rotter, J.M., The behaviour of locally supported cylindrical shells: unstiffened shells, pp. 179–197, 2000, with permission from Elsevier Science).

detailed picture is shown in Fig. 3.13, where the progressive rise in the elastic buckling strength is caused by the smaller zone within which the buckle must form. Other studies using slightly different assumptions (She and Rotter 1993) show similar strength predictions for rigid supports and distributed axial loading. The phenomena are, therefore, demonstrably reliable.

Effect of geometric imperfections in elastic cylinders

Although the minimum postbuckling load for a perfect elastic cylinder offers a simple estimate of the worst effects of serious imperfections, there is no formal guarantee of this, and no indication can be obtained of the fabrication tolerances required in design. The next studies examined the effect of a local inward imperfection immediately above the support. The behaviour of shells with a support width of $d/R = 0.13$ is described in detail here to permit comparisons between different analyses. However, similar calculations were performed on other widths with similar outcomes. For this geometry, the perfect elastic shell buckles at a load factor $\Lambda = 5.59$, and the minimum of the postbuckling load is at $\Lambda = 3.04$.

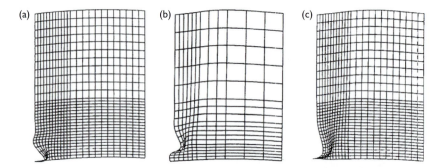

Figure 3.14 Deformed shapes used for specification of geometric imperfections ($d/R = 0.066$): (a) classical buckling eigenmode; (b) snap-through incremental mode; (c) postbuckling mode of the perfect elastic shell (reprinted from *Journal of Constructional Steel Research*, vol. 56, Guggenberger, W., Greiner, R. and Rotter, J.M., The behaviour of locally supported cylindrical shells: unstiffened shells, pp. 179–197, 2000, with permission from Elsevier Science).

Several different forms of imperfection were examined: the first two were in the classical linear buckling eigenmode (Fig. 3.14(a)) and the incremental deformation pattern at the instant of snap-through (Fig. 3.14(b)). A more visual representation of this post-buckled pattern is shown in Fig. 3.15, where it can be compared with a photograph taken from the experimental programme at Gent University. The resulting nonlinear load–axial displacement curves are shown in Fig. 3.16(a) for the classical eigenmode, and in Fig. 3.16(b) for the snap-through incremental deformation pattern (which actually occurs). In the former, an imperfection amplitude of $\delta/t = 0.5$ brings the peak strength down to $\Lambda = 3.28$, though at the larger imperfection ($\delta/t = 1.0$) the strength has risen again to $\Lambda = 3.36$. In Fig. 3.16(b), the snap-through mode shows a similar response, though the strength reductions are not so large, so this imperfection form is less critical in this problem. The classical linear eigenmode has often been recommended in the past as a 'worst' imperfection mode. It may be that it produces low strengths here because it is closely related to the postbuckling shape: this idea needs further investigation.

A further imperfection form was then studied, using an approximation (Fig. 3.7) to the perfect shell elastic postbuckling pattern, which the first author considers as probably the 'worst' imperfection shape (Fig. 3.14(c)). Hermite cubic spline surfaces were used to approximate this pattern using a few dimensional parameters and several different amplitudes, expressed as proportions of the shell wall thickness δ/t. A comparison of the shell response for several alternative imperfection forms is shown in Fig. 3.17, where the imperfection amplitudes and dimensions recommended for nonlinear finite element analysis by DIN 18800 (1990) were used, with both outward and inward amplitudes in the form of Fig. 3.7.

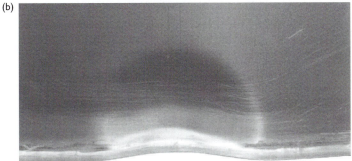

Figure 3.15 Postbuckled deformed shape for $R/t = 500$ and $d/R = 0.118$: (a) GMNIA analysis; (b) test result, Rathé and Greiner (1996).

The inward imperfection of a form similar to the postbuckling mode (DIN 18800 1990) produces the lowest calculated elastic strength in the present study ($\Lambda = 3.10$) and the value is close to the postbuckling minimum ($\Lambda = 3.04$). Similar calculations for much wider supports produced similar results (Guggenberger 1992).

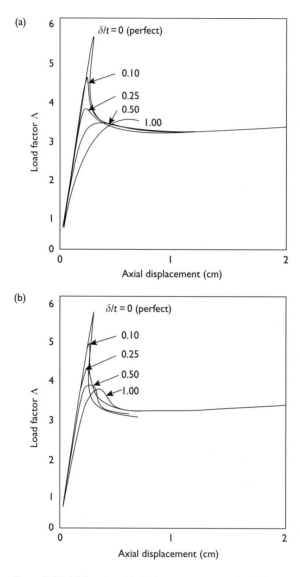

Figure 3.16 (a) Load–axial displacement curves for different imperfection amplitudes in the classical buckling eigenmode (elastic cylinders with $d/R = 0.13$). (b) Load–axial displacement curves for elastic cylinders with various imperfection amplitudes in the snap-through incremental eigenmode ($d/R = 0.13$) (reprinted from *Journal of Constructional Steel Research*, vol. 56, Guggenberger, W., Greiner, R. and Rotter, J.M., The behaviour of locally supported cylindrical shells: unstiffened shells, pp. 179–197, 2000, with permission from Elsevier Science).

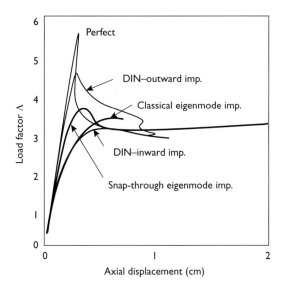

Figure 3.17 Comparison of load–axial displacement curves for cylinders with different imperfection forms but the same amplitude ($\delta/t = 1$ inward, $d/R = 0.13$) (reprinted from *Journal of Constructional Steel Research*, vol. 56, Guggenberger, W., Greiner, R. and Rotter, J.M., The behaviour of locally supported cylindrical shells: unstiffened shells, pp. 179–197, 2000, with permission from Elsevier Science).

Material nonlinearity: elastic–plastic behaviour in perfect and imperfect shells

When the supports are wide and the shell is thin, the elastic buckling load provides an accurate measure of the strength. However, when more practical narrow supports are used, yielding affects the prebuckling and buckling behaviour.

The effect of yielding on both a perfect and an imperfect cylinder with a support width of $d/R = 0.13$ is shown in Fig. 3.18. The form of analysis used to obtain each curve is indicated. The simplest analysis is geometrically linear with material nonlinearity, leading to a classical plastic limit load (marked MNA). The geometrically nonlinear analysis of an elastic perfect cylinder is marked GNA, and that including elasto-plastic yielding is marked GMNA. In this example, the perfect cylinder elastic buckling load is similar to the plastic limit load, placing it in a region of interactive behaviour. The imperfect cylinders are marked GNIA for the elastic calculation and GMNIA for full material and geometric nonlinearity. The adopted imperfection was the approximation to the elastic postbuckling form (Fig. 3.14(c)).

A comparison of these curves shows several interesting things: the prebuckling path is not strongly affected by geometric nonlinearity, though the buckling

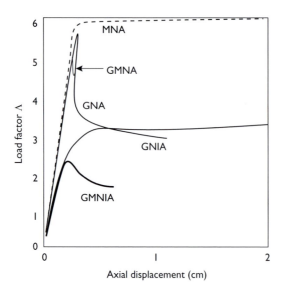

Figure 3.18 Effects of geometric nonlinearity G, plasticity M and imperfections I on the load–axial displacement curves for a cylinder with $d/R = 0.13$.

strength is; the perfect cylinder buckling strength is only slightly altered by yielding (8% reduction), but the imperfect cylinder is much more sensitive (28% reduction); the elastic imperfect cylinder has a maximum strength that is close to the post-buckling minimum of the perfect cylinder curve; and the fully nonlinear analysis GMNIA is far below all other strength predictions.

This example shell has a practical thickness ($R/t = 500$) and rests on four supports of practical width ($d/R = 0.13$). The reduction in strength due to plasticity is very significant. The postpeak behaviour shows that the catastrophically unstable behaviour of the perfect shell becomes much milder in imperfect cylinders, and these smoothing effects are seen in both cases, whether yielding occurs or not.

Elastic–plastic strength predictions for imperfect shells on a wide range of support widths are shown in Fig. 3.19. The buckling strength is presented as the total force carried by the four supports relative to the axial force in the cylinder at uniform axial compression membrane yield. For support widths in excess of $d/R = 0.30$ (a very wide support), the elastic and plastic calculations produce effectively the same result (Fig. 3.19(a)), in spite of some local yielding. For narrower supports (Fig. 3.19(b)), elastic calculations lead to an effectively constant 'force' at buckling, irrespective of the support width. By contrast, the plastic calculation leads to lower strengths, which are ultimately bounded by the condition that the axial membrane stress immediately above the support cannot exceed the 'stress' limit of the von Mises envelope ($\sigma_{mm,max} = 1.155 f_y$). For $R/t = 500$, elastic–plastic interaction occurs in the range $0.03 < d/R < 0.30$. A summary of

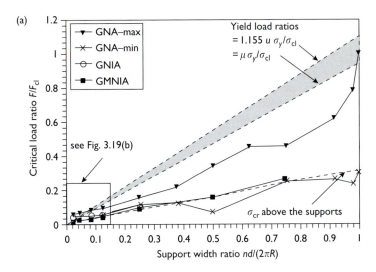

(a)

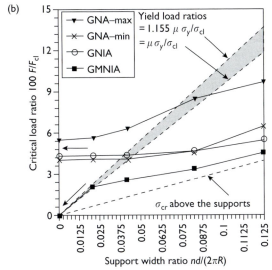

(b)

Figure 3.19 Summary of dimensionless critical loads for perfect elastic and imperfect elastic–plastic cylinders for: (a) the full range of support widths ($R/t = 500; n = 4$); (b) supports of practical width ($d/R < 0.2$) (reprinted from *Journal of Constructional Steel Research*, vol. 56, Guggenberger, W., Greiner, R. and Rotter, J.M., The behaviour of locally supported cylindrical shells: unstiffened shells, pp. 179–197, 2000, with permission from Elsevier Science).

the geometrically and materially nonlinear calculated strengths was also indicated in Fig. 3.12.

The effect of internal pressure

In all silo structures, the stored solid exerts an internal pressure on the wall of the structure, inducing circumferential tension. This tension is well known to increase the buckling strength of uniformly compressed cylinders (Hutchinson 1965; Rotter and Teng 1989; Rotter 1997), so it may be expected to enhance the buckling strength above a local support. The effects of internal pressure on the elastic buckling strength above a support was first explored by Li (1994). At high internal pressures, plasticity may intervene, leading to a dramatic reduction in buckling strength in the elephant's foot mode (Rotter 1990). This applies to both the plastic and buckling limit states.

The present study used the flexible support model, and extends Li's (1994) study into the elastic–plastic domain. The numerical study was extended to very high internal pressures, finally reaching the case where pure circumferential yielding occurs. Because the results are intended to be used directly in design, appropriate geometrical imperfections and a realistic material law for structural steel was adopted (ideal elastic–plastic von Mises behaviour with $f_y = 235$ N/mm^2).

The analysis model ($R/t = 500$) and loading is shown in Fig. 3.6(a). The axial load is applied uniformly at the upper circumference, as described in 'Modelling assumptions of the present study'. Geometrical imperfections were adopted in the form of local inward shapes above the supports, mimicking the postbuckling mode of the perfect elastic structure (Guggenberger 1992), as shown in Fig. 3.7. The boundary condition at the local support was assumed to be 'flexible' (Fig. 3.3(a)), with the support applying a uniform compression throughout the width of the support. Whilst this assumption is not very realistic, it is conservative (Rotter *et al.* 1993). The numerical studies used progressively increased sophistication in the analyses, with the modelling assumptions as follows:

1   Geometrically linear analysis including material nonlinearity (MNA).
2   Geometrically nonlinear analysis of the elastic perfect cylinder (GNA).
3   Geometrically nonlinear analysis of the imperfect elastic cylindrical shell (GNIA).
4   Geometrically and materially nonlinear analysis of the perfect (GMNA) and imperfect cylindrical shell (GMNIA).

A brief representation of some characteristics of the buckling behaviour when the internal pressure magnitude varies is given in Figs. 3.20 and 3.21. It illustrates the buckling behaviour for different levels of internal pressure, expressed by the ratio α of simple circumferential membrane stress (pR/t) to mean axial membrane stress (σ_{mm}) just above the support. This ratio is defined here as α. For zero internal pressure, the buckling mode shows pronounced inward deflections, but high

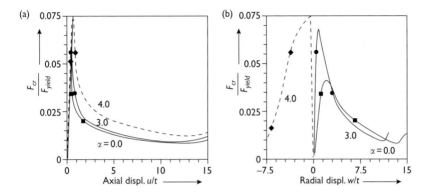

Figure 3.20 Load–displacement diagrams for the point at the centre of the local inward imperfection ($R/t = 500$, $d/R = 0.157$, GMNIA) (reprinted from *Thin Walled Structures*, vol. 31, Greiner, R. and Guggenberger, W., Buckling behaviour of axially loaded steel cylinders on local supports – with and without internal pressure pp. 169–185, 1998, with permission from Elsevier Science).

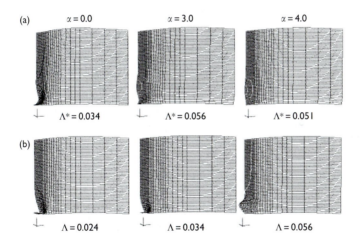

Figure 3.21 Buckling deformations for various levels of internal pressure: (a) prebuckling states; (b) postbuckled states ($R/t = 500$, $d/R = 0.157$, GMNIA with inward imperfection) (reprinted from *Thin Walled Structures*, vol. 31, Greiner, R. and Guggenberger, W., Buckling behaviour of axially loaded steel cylinders on local supports – with and without internal pressure pp. 169–185, 1998, with permission from Elsevier Science).

internal pressure resists them and for values of α in excess of about 3, they become outward deflections. Figures 3.20 and 3.21 relate to an inward imperfection.

The complete interaction diagram is shown in Fig. 3.22 for the support width $d/R = 0.157$, where the load-carrying capacity of the imperfect cylinder

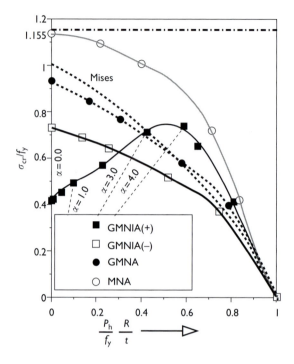

Figure 3.22 Interaction of axial load and internal pressure ($R/t = 500$, $d/R = 0.157$) (reprinted from *Thin Walled Structures*, vol. 31, Greiner, R. and Guggenberger, W., Buckling behaviour of axially loaded steel cylinders on local supports – with and without internal pressure pp. 169–185, 1998, with permission from Elsevier Science).

(GMNIA) rises with internal pressure up to about 50% more than that of the unpressurised cylinder. For higher internal pressures ($pR/t \geq 0.3\,f_y$), the buckling behaviour follows the elephant-foot phenomenon (Rotter 1990) and the capacity drops. The strengths of the inward imperfection model is marked GMNIA(+), whilst that for an outward imperfection is shown as GMNIA(−). At high internal pressures, the buckling resistance of an outward imperfection shape GMNIA(−) is lower, and at very high pressures it approaches the von Mises criterion of the plastic limit state (MNA).

A fuller set of axial compression–internal pressure interaction relationships is shown in Fig. 3.23, where the interaction for the uniformly compressed cylinder (Rotter and Seide 1987; Rotter 1990, 1998) is compared. Locally supported cylinders ($\mu = 0.05$ and 0.1) can be seen to have greater elastic–plastic strengths than uniformly supported cylinders ($\mu = 1.00$), with high internal pressure leading to similar strengths (elephant-foot buckling). When the support is wide, a similar elephant-foot buckle develops at high internal pressure, but it occurs only locally

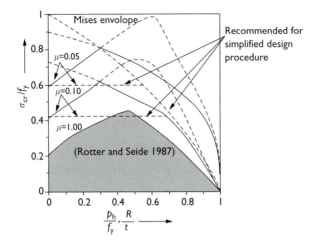

Figure 3.23 Axial load and internal pressure: comparison of locally and uniformly supported cylinders ($R/t = 500$, GMNIA) (reprinted from *Thin Walled Structures*, vol. 31, Greiner, R. and Guggenberger, W., Buckling behaviour of axially loaded steel cylinders on local supports – with and without internal pressure pp. 169–185, 1998, with permission from Elsevier Science).

above the supports. However, the elephant-foot failure is found to have a much smaller effect above a local support, with strengths in excess of the simple membrane von Mises criterion being supportable. This is because the location in which the axial stress is high (immediately above the support) is restrained against radial expansion by the support, so that the circumferential stress does not develop there fully. Although the analyses have been performed only for $R/t = 200$ and 500, it is clear that a simple conservative failure criterion can be made for the elephant-foot buckling by checking the combination of the mean meridional support stress σ_{mm} and the simple membrane circumferential stress pR/t against the von Mises criterion.

If this is the only test used, the strength gains in buckling strength at low internal pressure are ignored (Fig. 3.23), but this loss may not be too severe in many practical designs, and it offers a considerable advantageous simplification.

Proposed design rules

Basic case of the unstiffened cylinder

Comprehensive parametric studies were carried out varying the radius-to-thickness ratio and the steel grade of the cylinders (see 'Modelling assumptions of the present

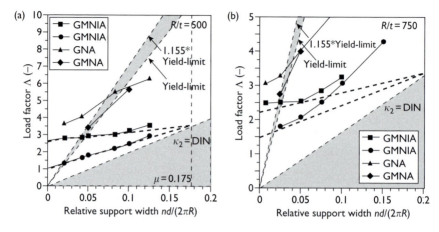

Figure 3.24 Dimensionless load factors $\Lambda = F/F_{\text{ref}}$ for different analyses and shell slenderness ratios: (a) $R/t = 500$; (b) $R/t = 750$ (reprinted from *Thin Walled Structures*, vol. 37, Guggenberger, W., Proposal for design rules of axially loaded steel cylinders on local supports, pp. 159–167, 1998, with permission from Elsevier Science).

study'). The two limiting cases of support rigidity were also investigated: flexible supports and rigid supports (Fig. 3.3(a),(b)). Details of these investigations were described in reports by Guggenberger and Greiner (1994a,b, 1998b), Guggenberger (1998b) and Rathé and Greiner (1996). A small sample of results is shown in Fig. 3.24 for $R/t = 500$ and 750. Similar curves were obtained for the other parameter combinations. Based on the results of these geometrically and materially nonlinear analyses, including imperfections, design rules for the critical load F_{cr} at a support beneath an unstiffened cylinder were developed as follows.

The design rule development began with a 'load-oriented' representation of results, where the critical load Fcr of one support was related to the yield load of the uniformly loaded cylinder divided by the number of columns: $F_{\text{yield}} = (2\pi R t f_y)/4$ (Fig. 3.25). Linear curve fitting was performed for the critical loads in the intermediate range of support widths as indicated by the short dashed lines in Figs 3.24 and 3.25.

For narrow supports, fully plastic conditions are reached (shaded area in Fig. 3.25). For very wide supports, the buckling capacity of the uniformly compressed cylinder is reached, expressed here by the reduction factor κ_2 according to DIN 18800 (1990) (long dashes in Figs 3.24 and 3.25). Thus, these two extreme cases may be described by well-established simple relations.

Finally, the four fitting lines (fitting with respect to the support width ratio; short dashes in Figs 3.24 and 3.25) were joined together to create a continuous surface by performing another nonlinear fit with respect to the slenderness ratio

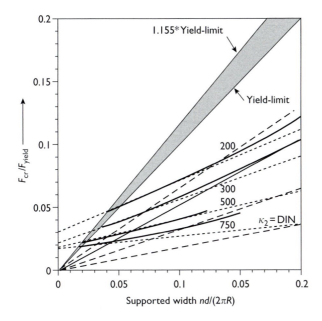

Figure 3.25 Non-dimensional representation of critical support loads for $200 \leq R/t \leq 750$ and steel grade S235 (GMNIA analyses) (reprinted from *Thin Walled Structures*, vol. 37, Guggenberger, W., Proposal for design rules of axially loaded steel cylinders on local supports, pp. 159–167, 1998, with permission from Elsevier Science).

R/t. This is better achieved by using an equivalent stress representation, which yields Fig. 3.26 (solid curves). The dashed curves are the computed critical stress values using a GMNIA analysis, which are closely conservatively approximated by an empirical formula for the practical range of support widths ($\eta < 0.25$). A three-dimensional (3D) representation of the mean buckling stress at the support, related to the yield stress f_y, is shown in Fig. 3.27.

$$\sigma_{cr} = F_{cr}/(dt) = \kappa_{2,\text{local}} \cdot f_y \tag{8}$$

Relating this to the generalised slenderness parameter λ as

$$\lambda = \sqrt{f_y/\sigma_{cl}} \tag{9}$$

where σ_{cl} is the value for uniform axial compression (Eq. 1), leads to the final design diagram for the basic steel grade S235 (Fig. 3.28). This procedure was also applied for the higher steel grades and leads to the final design formula (Eq. 10), specifying the local reduction factor $\kappa_{2,\text{local}}$ for unstiffened cylinders on flexible supports fitting into the calculations procedure used by DIN 18800 (1990),

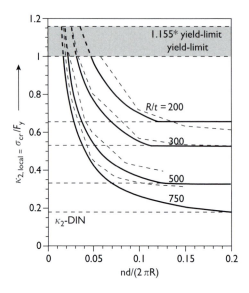

Figure 3.26 Equivalent non-dimensional stress representation of results for $f_y = 235$ MPa (local reduction factor $\kappa_{2,\text{local}}$) (reprinted from *Thin Walled Structures*, vol. 37, Guggenberger, W., Proposal for design rules of axially loaded steel cylinders on local supports, pp. 159–167, 1998, with permission from Elsevier Science).

where the reduction factor κ_2 is for cylinders under uniform axial compression. The critical support force F_{cr} at one support is then calculated according to Eq. (8):

$$\kappa_{2,\text{local}} = \frac{\sigma_{cr}}{f_y} = 0.19 + \left(\frac{0.0283\,c_c}{\eta\lambda^{0.77c_e}}\right) - 1.04\log\lambda \tag{10}$$

but with the additional restrictions

$$\kappa_{2,\text{local}} \leq 1.0 \text{ or } 1.115 \tag{11}$$

$$\kappa_{2,\text{local}} \leq \kappa_2 \tag{12}$$

$$\kappa_{2,\text{local}} \leq \kappa_2/\mu \tag{13}$$

where κ_2 is taken according to DIN 18800 (1990). The modification factors c_c and c_e in Eq. (10) account for the effect of higher steel grades and are given by

$$c_c = -0.43 + 5\varepsilon - 3.57\varepsilon^2 \tag{14}$$

$$c_e = -0.87 + 6.38\varepsilon - 4.51\varepsilon^2 \tag{15}$$

with $\varepsilon = \sqrt{(235/f_y)}$. The condition in Eq. (11) refers to plastic yielding under 1D or 2D stresses at the supports, whilst Eq. (12) states that the buckling stress for

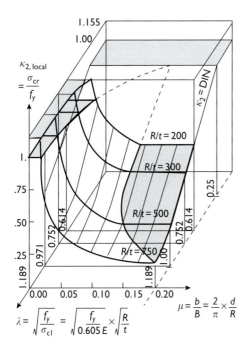

Figure 3.27 Three-dimensional representation of the design formula: dependence of the elasto-plastic critical stress on dimensionless shell slenderness and support width (reprinted from *Thin Walled Structures*, vol. 37, Guggenberger, W., Proposal for design rules of axially loaded steel cylinders on local supports, pp. 159–167, with permission from Elsevier Science).

local loading is always greater than or equal to that for uniform loading. The last condition (Eq. 13) is necessary to prevent the pathologic case in which the cylinder is supported on many narrow, closely spaced supports that might otherwise yield a higher buckling load than the corresponding uniformly loaded cylinder. It is unlikely that this condition will ever be invoked in practical design.

These design formulae have only been verified within a limited range of shell parameters (slenderness ratios $200 \geq R/t \geq 750$, support widths $0 \leq \eta \leq 0.15$ and yield stresses $235 \leq f_y \leq 355$ N/mm^2). The slenderness parameter λ has been chosen for reasons of compatibility with the design rule for uniform axial compression. It depends on the slenderness ratio R/t as well as the steel grade (Fig. 3.28). This mixing could be avoided by substituting Eq. (16) in Eq. (11):

$$\lambda = \lambda_{\text{ref}}/\varepsilon \quad \text{where} \quad \lambda_{\text{ref}} = \sqrt{235/\sigma_{\text{cl}}} \tag{16}$$

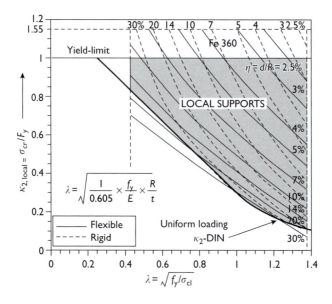

Figure 3.28 Final design diagram for the axially local-loaded unstiffened cylinder (slenderness representation analogous to the German standard DIN 18800/4) (reprinted from *Thin Walled Structures*, vol. 37, Guggenberger, W., Proposal for design rules of axially loaded steel cylinder on local supports, pp. 159–167, 1998, with permission from Elsevier Science).

The final design diagram shown in Fig. 3.28 is valid for S235 (i.e. for $c_c = c_e = 1$) and is an alternative representation of Fig. 3.27. The thin solid curves are valid for flexible supports (Fig. 3.3(a)). The rise in strength caused by rigid supports can immediately be recognised by the thin dashed curves. For each individual support width, represented by $\eta = d/R$ expressed in percent, the strength curve for rigid supports always lies well above that for flexible supports. This is due to the built-in boundary conditions across the width of the local supports (meridional displacement and meridional rotation are fully restrained), which causes a considerable increase of the buckling strength, in particular for narrow supports.

Cylinders with increased wall thickness of the bottom course

Practical silo structures usually have a stepped wall thickness with height, because the axial buckling resistance increases dramatically with depth below the surface of the stored solid. For consideration of the problem of buckling above a support, it is relatively straightforward to increase the buckling resistance by increasing the thickness of the bottom strake or course. Thus, the goal here was to determine the height over which the greater thickness bottom strake should extend to achieve the same strength as if this thickness extended over the whole height of the shell.

To explore this question, it was first decided that the cylindrical shell should consist of only two regions of different wall thickness t and t_1. The thickness of the lower part t_1 was chosen to be always 50% thicker than the upper part. Local support forces were again introduced using flexible supports.

Geometrically and materially nonlinear analyses were performed, both with and without geometric imperfections. The minimum value of h_1/R was determined for each of these cases. It was found that the base strake height requirement was always greater for perfect shells than for imperfect shells, so this critical height was taken as the value for perfect shells. This has the advantage that the outcome is independent of assumptions concerning the form and amplitude of the imperfections.

A qualitative picture of the variation of the critical buckling load with the height of the lower course is shown in Fig. 3.29. The case $h_1/R = 0$ corresponds to the limiting case where no increase has been made in the thickness, giving the smallest buckling loads. As the height of the thicker course is increased, the buckling strength rises as the buckle is driven up the shell, still occurring in the thinner

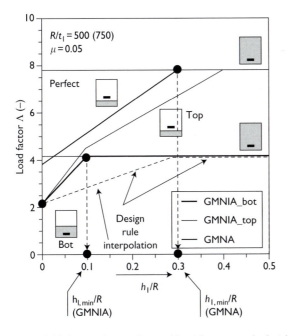

Figure 3.29 Dependence of critical load factors on the height of the thickened course h_1/R (GMNA = perfect, GMNIA_bot/top = local imperfection above the lower or above the upper boundary of the zone of increased wall thickness) (reprinted from *Thin Walled Structures*, vol. 37, Guggenberger, W., Proposal for design rules of axially loaded steel cylinder on local supports, pp. 159–167, 1998, with permission from Elsevier Science).

course. However, at the critical height, the buckling strength for the buckle to form in the thinner course, rather distant from the support, is identical to the buckling strength for the buckle to occur just above the support in the thicker course. At this point, further increases in the height of the thickened lower course do not affect the buckling load at the support, and the critical height h_1 has been found (Fig. 3.29).

The critical height for perfect cylinders is given as a function of the support width parameter η, the steel grade parameter ε and the shell slenderness R/t_1 in design formula Eqs (17)–(19). If the actual height h_1 is smaller than the critical minimum value $h_{1,\min}$ (Fig. 3.29), the strength may be deduced by linear interpolation between the result for the height h_1 and the result for no thickened lower course.

$$h_{1,\min}/R = 0.2 + 0.53(1 - \varepsilon) + 0.2\left(\frac{R/t_1 - 200}{300}\right) \quad \text{for } 0 < \eta < 0.075 \tag{17}$$

$$h_{1,\min}/R = 0.1 + 1.33\eta + 0.53(1 - \varepsilon) + 0.2\left(\frac{R/t_1 - 200}{300}\right)$$

$$\text{for } 0.075 < \eta < 0.15$$

$$h_{1,\min}/R = 4/\sqrt{R/t} \quad \text{for } 0.15 < \eta \tag{18}$$

The problem has only been explored for the range of slendernesses $200 \leq R/t_1 \leq 500$. The last condition (Eq. 19) was added for pragmatic reasons and relates to the width of local buckles. A graphical illustration of the design equation is given in Fig. 3.30.

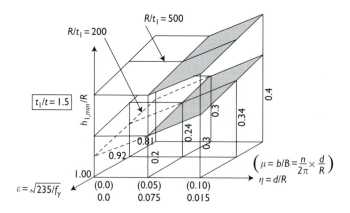

Figure 3.30 Example design diagram for the minimum height of thickened wall $h_{1,\min}/R$ (reprinted from *Thin Walled Structures*, vol. 37, Guggenberger, W., Proposal for design rules of axially loaded steel cylinder on local supports, pp. 159–167, 1998, with permission from Elsevier Science).

Conclusions

This chapter has outlined the historical development of knowledge concerning the buckling and yielding failure of cylindrical shells under axial compression when they are supported on local discrete supports.

These studies are still in an early stage of development, so this chapter has only considered supports beneath the base of the shell. The supports were taken to have a fixed number around the circumference and to act as either flexible (constant stress) supports or rigid (uniform displacements). The strength was shown to vary considerably with the width of the support, and to be affected by plasticity for much of the practical range of support widths. Buckling above narrow supports was shown to be controlled by the force transmitted, whilst buckling above wide supports is controlled by the mean stress. The studies explored a wide range of variables using geometrically and materially nonlinear analyses and including the deleterious effects of geometric imperfections in various forms. This work was summarised in the form of design equations based on the conceptual framework of the DIN 18800 (1990) standard.

Two additional problems were explored in this chapter: the effect of internal pressure on the elastic–plastic buckling strength, and the height over which a thicker base course must extend to achieve the higher buckling strength associated with the additional thickness. The findings from both of these studies were transformed into design equations that can be adopted in design procedures.

Notation

B	circumferential length of the 45° analysis cutout $(= \pi R/4)$
c_c, c_e	modifying factors to account for changed yield stress
d	support width $(= 2b)$
E	Young's modulus
f_y	yield stress (nominal values due to design codes)
F	factored value of vertical reaction force on a support $(= \Lambda F_o)$
F_{ref}, F_o	reference magnitude of vertical reaction force on a support $(= 300$ kN$)$
F_{total}	load applied to the whole cylinder $(= nF)$
F_{yield}	yield load of the uniformly compressed cylinder divided by the number of supports
F_{cl}	classical elastic critical load of the uniformly compressed cylinder divided by the number of supports
F_{max}, F_{cr}	maximum or critical magnitude of vertical reaction force on a single support at the instant of buckling
GNA–min	postbuckling minimum strength from geometrically nonlinear elastic analysis
GNA–max	maximum strength from geometrically nonlinear elastic analysis
LBA	linear buckling eigenvalue analysis

H	cylinder height
L	cylinder length $(= H)$
n	number of discrete supports
R	cylinder radius
t	wall thickness
t_1	thickness of thickened lower course or strake
δ	amplitude of imperfection
ε	steel grade parameter
κ_2	buckling strength reduction factor for uniform axial loading conditions $(= \sigma_{cr}/f_y)$ according to DIN 18 800 (1990)
$\kappa_{2,local}$	buckling strength reduction factor for local axial loading conditions $(= \sigma_{mm}/f_y)$
ν	Poisson's ratio
μ	proportion of the circumference which is supported $(= nd/(2\pi R))$
η	dimensionless support width parameter $(= d/R)$
λ	dimensionless slenderness of the shell
λ_{ref}	reference dimensionless slenderness of the shell (Eq. 16)
Λ	load factor on applied loads
σ_{cl}	classical elastic critical stress for a perfect cylinder under uniform axial compression
σ_{mm}	mean axial membrane stress immediately above a support at buckling
σ_{xx}	axial membrane stress within the shell
σ_{xxo}	axial membrane stress immediately above each support
$\sigma_{cr,GNIA}$	mean axial membrane stress above a local support at elastic buckling in an imperfect structure
$\sigma_{cr}, \sigma_{cr,GMNIA}$	mean axial membrane stress above a local support at elastic–plastic buckling in an imperfect structure

References

Beuth (1993). *Beuth Kommentare. Stahlbauten, Erläuterungen zu DIN 18800/Teil* 1–4, 1st edn, Ernst & Sohn, Berlin, p. 342.

Bijlaard, P.P. (1955). Stresses from local loadings in cylindrical pressure vessels. *Transactions of ASME* **77**, 805–816.

Bodarski, Z., Hotala, E. and Pasternak, H. (1985). Zur Beurteilung der Tragfähigkeit von Metallsilos. *Der Bauingenieur* **60**, 49–52.

Calladine, C.R. (1983). *Theory of Shell Structures*. Cambridge University Press, Cambridge.

DIN 18800 (1990). *Steel Structures: Stability, Buckling of Shells*, DIN 18800/Part 4, Deutsches Institut für Normung, Berlin, November.

Donnell, L.H. and Wan, C.C. (1950). Effect of imperfections on buckling of thin cylinders and columns under axial compression. *Journal of Applied Mechanics, Transactions of ASME* **17**(1), 73–83.

ECCS (1988). *European Recommendations for Steel Construction: Buckling of Shells*, No. 56, 4th edn. European Convention for Constructional Steelwork, Brussels.

ENV 1993-1-6 (1999). Eurocode 3: Design of Steel Structures, Part 1.6: General Rules – Supplementary Rules for the Strength and Stability of Shell Structures. CEN, Brussels.

Esslinger, M. (1970). Hochgeschwindigkeitsaufnahmen vom Beulvorgang dünnwandiger axialbelasteter Zylinder. *Der Stahlbau* **3**, 73–76.

Galletly, G.D. (1988). Buckling of pressure vessels. *Science Progress* **72**, 371–405.

Gorenc, B.E. (1985). Design of supports for steel bins. In *Design of Steel Bins for the Storage of Bulk Solids* (ed. J.M. Rotter). Univ. Sydney, pp. 184–188.

Gould, P.L., Sen, S.K., Wang, R.S.C. and Lowrey, D. (1976). Column-supported cylindrical conical tanks. *Journal of Structural Division, ASCE* **ST2**, 429–447.

Greiner, R. (1984). Zur Längskrafteinleitung in stehende zylindrische Behälter aus Stahl. *Der Stahlbau* **53**(7), 210–215.

Greiner, R. and Guggenberger, W. (1996a). Tragverhalten und Bemessung punktgestützter kreiszylindrischer Silos aus Stahl. In *Silos – Research and Practice Meeting '96, Final Meeting SFB 219 'Silobauwerke und ihre spezifischen Beanspruchungen*, University Fridericiana Karlsruhe, Germany, 29 February – 1 March.

Greiner, R. and Guggenberger, W. (1996b). Stability of column-supported steel cylinders. *Proceedings of the IASS Asia-Pacific Conference on Shell and Spatial Structures*, Beijing, 21–25 May.

Greiner, R. and Guggenberger, W. (1998). Buckling behaviour of axially loaded cylinders on local supports – with and without internal pressure. *Proceedings of the International Conference on Advances in Steel Structures ICASS '96*, Hong Kong, 11–14 December, and *Thin-Walled Structures* **31**, 186–195.

Guggenberger, W. (1991). Buckling of cylindrical shells under local axial loads. In *Buckling of Shell Structures, on Land, in the Sea and in the Air* (ed. J.F. Jullien. International Colloquium, Villeurbanne, Lyon, France, 17–19 September, Elsevier Applied Science, pp. 323–333.

Guggenberger, W. (1992). Nonlinear buckling behaviour of circular cylindrical shells under local axial loads. Doctoral Dissertation (in German), Institute for Steel, Timber and Shell Structures, Report No. 6, Technical University of Graz, Austria.

Guggenberger, W. (1996). Buckling behaviour and design of column-supported cylindrical steel silos. *Proceedings of the 5th International Colloquium on Structural Stability SSRC IC/Brazil '96*, Rio de Janeiro, 5–7 August.

Guggenberger, W. (1998a). Proposal for design rules of axially loaded cylinders on local supports. *Proceedings of the International Conference on Advances in Steel Structures ICASS '96*, The Hong Kong Polytechnic University, Hong Kong, 11–14 December, and *Thin-Walled Structures* **31**, 169–185.

Guggenberger, W. (1998b). Schadensfall, Schadensanalyse und Schadensbehebung eines Silos auf acht Einzelstützen. *Der Stahlbau* **67**(6), 430–433.

Guggenberger, W. and Greiner, R. (1994a). Local buckling of unstiffened steel cylinders – numerical results and comparison with test results. ECSC Research Report G/G No. 1.1, Institute for Steel, Timber and Shell Structures, Technical University of Graz, Austria, February.

Guggenberger, W. and Greiner, R. (1994b). Numerical analysis of unstiffened steel cylinders with artificial imperfections and comparison with test results. ECSC Research Report G/G

No. 2.A, Institute for Steel, Timber and Shell Structures, Technical University of Graz, Austria.

Guggenberger, W. and Greiner, R. (1998). Axialbelastete Kreiszylinderschalen auf Einzelstützen – Numerische Tragverhaltensstudie, Versuchsergebnisse und Bemessungsvorschlag. *Der Stahlbau* **67**(6), 415–424.

Guggenberger, W., Greiner, R. and Rotter, J.M. (2000). The behaviour of locally-supported cylindrical shells: unstiffened shells. *Journal of Constructional Steel Research* **56**(2), 175–197.

Hibbit, Karlsson and Sorensen (1995). *ABAQUS Version 5.4, Theory and Users Manuals*, Pawtucket, Rhode Island.

Hillmann, J. (1985). Grenzlasten und Tragverhalten axial gestauchter Kreiszylinderschalen im Vor- und Nachbeulbereich. Bericht Nr. 85/45, Institut für Statik, Technical University of Braunschweig, Germany.

Hoff, N.J. (1960). The perplexing behaviour of thin circular cylindrical shells in axial compression. *Israel Journal of Technology* **4**(1), 1–28.

Hoff, N.J., Chao, C.C. and Madsen, W.A. (1964). Buckling of a thin-walled circular cylindrical shell heated along an axial strip. *Journal of Applied Mechanics*, **31**, *Transactions of the American Society of Mechanical Engineers*, Series E, June, 253–258.

Hotala, E. (1986). Stability of cylindrical metal shells of silos with circumferential stiffeners (in Polish). Doctoral Dissertation, Technical University of Wroclaw, Poland.

Hotala, E. (1996). Stability of cylindrical steel shells under local axial compression. *Proceedings of the European Workshop on Thin-Walled Steel Structures*, 26–27 September, Krzyzowa, Poland, pp. 61–68.

Hutchinson, J.W. (1965). Axial buckling of pressurised imperfect cylindrical shells. *AIAA Journal*, **3**(8), 1461–1466.

Kildegaard, A. (1969). Bending of a cylindrical shell subject to axial loading. *Proceeding of the Second Symposium on Theory of Thin Shells*, IUTAM, Copenhagen, September. Springer, Berlin, pp. 301–315.

Knödel, P. and Ummenhofer, Th. (1998). Ein einfaches Modell zum Stabilitätsnachweis zylindrischer Schalentragwerke auf Einzelstützen. *Der Stahlbau* **67**(6), 425–429.

Li, H.Y. (1994). Analysis of steel silo structures on discrete supports. PhD Thesis, Department of Civil and Environmental Engineering, University of Edinburgh, Scotland, UK.

Li, H.Y. and Rotter, J.M. (1996). Algebraic analysis of elastic circular cylindrical shells under local loadings (Part 1 and Part 2). *Proceedings of the International Conference on Structural Steelwork*, Hong Kong, December, pp. 801–807, 808–814.

NASA SP8007 (1968). *Buckling of Thin-Walled Circular Cylinders*.

Ooi, J.Y., Rotter, J.M. and Pham, L. (1990). Systematic and random features of measured pressures on full-scale silo walls. *Engineering Structures* **12**(2), 74–87.

Öry, H., Reimerdes, H.G. and Tritsch, W. (1984). Beitrag zur Bemessung der Schalen von Metallsilos. *Der Stahlbau* **53**(8), 243–248. (See also English Translation: A contribution to the layout of large metallic silo shells, internal report, unnumbered, 1984.)

Öry, H. and Reimerdes, H.G. (1987). Stresses in and stability of thin walled shells under non-ideal load distribution. *Proceedings of the International Colloquium on the Stability of Shell and Plate Structures*, Gent, Belgium, 6–8 April, ECCS, pp. 555–561.

Rathé, J. and Greiner, R. (1996). *Local Loads in Cylindrical Structures*, Subproject B of ECSC Contract No. 7210–SA/208: Enhancement of ECCS Design Recommendations and Development of Eurocode 3 Parts Related to Shell Buckling. Final Report, University of Gent, Belgium, May.

Rotter, J.M. (1982). Analysis of ringbeams in column-supported bins. *Proceedings of the Eighth Australasian Conference on the Mechanics of Structures and Materials*, University of Newcastle, August.

Rotter, J.M. (1985). Analysis and design of ringbeams. In *Design of Steel Bins for the Storage of Bulk Solids* (ed. J.M. Rotter). University of Sydney, pp. 164–183.

Rotter, J.M. (1990). Local inelastic collapse of pressurised thin cylindrical steel shells under axial compression. *Journal of Structural Engineering, ASCE* **116**(7), 1955–1970.

Rotter, J.M. (1997). Pressurised axially compressed cylinders. *Proceedings of the International Conference on Carrying Capacity of Steel Shell Structures*, Brno, 1–3 October, pp. 354–360.

Rotter, J.M. (1998). Shell structures: the new European standard and current research needs. *Thin-Walled Structures* **31**, 3–23.

Rotter, J.M. (2001). *Guide for the Economic Design of Circular Metal Silos*. Spon, London.

Rotter, J.M. and Seide, P. (1987). On the design of unstiffened cylindrical shells subject to axial load and internal pressure. *Proceedings of the International Colloquium* 6–8 April, Ghent, Belgium.

Rotter, J.M. and Teng, J.G. (1989). Elastic stability of cylindrical shells with weld depressions. *Journal of Structural Engineering, ASCE* **115**(5), 1244–1263.

Rotter, J.M., Teng, J.G. and Li, H.Y. (1991). buckling in thin elastic cylinders on column supports. In *Buckling of Shell Structures, on Land, in the Sea and in the Air* (ed. J.F. Jullien). Elsevier Applied Science, *International Colloquium*, Villeurbanne, Lyon, France, 17–19 September, pp. 334–343.

Rotter, J.M., Greiner, R., Guggenberger, W., Li, H.Y. and She, K.M. (1993). Proposed design rule for buckling strength assessment of cylindrical shells under local axial loads. *Submitted to ECCS TWG 8.4, 'Buckling of Shells'*, September.

Saal, H. (1982). Buckling of long liquid-filled cylindrical shells. In *Buckling of Shells* (ed. E. Ramm), Springer-Verlag, Berlin.

Samuelson, L.A. (1986). Buckling of cylindrical shells under axial compression and subjected to localized loads. In *Post-Buckling of Elastic Structures* (ed. J. Szabo), *Proceedings of EuroMech Colloquium* No. 200, Matrafüred, Hungary, 5–7 October, Elsevier.

Samuelson, L.A. (1987). Design of cylindrical shells subjected to local loads in combination with axial or radial pressure. *Proceedings of the International Colloquium on the Stability of Plate and Shell Structures*, Gent, Belgium, 6–8 April, ECCS, pp. 589–596.

Samuelson, L.A. (1990). Effect of local loads on the stability of shells subjected to uniform pressure distribution. *Proceedings of the IUTAM Symposium on Contact Loading and Local Effects in Thin-Walled Plated and Shell Structures*, Preliminary Report, Prague, August, pp. 34–38.

Samuelson, L.A. and Eggwertz, S. (eds) (1992). *Shell Stability Handbook*. Elsevier Applied Science, London.

Schulz, U. and Ummenhofer, Th. (2000). *Spannungs- und Stabilitätsverhalten von punktförmig gestützten Schalentragwerken*. Final Research Report, Research Grant No. IV 1-5-813/96, Deutsches Institut für Bautechnik, University Fridericiana Karlsruhe.

She, K.M. and Rotter, J.M. (1993). Nonlinear and Stability Behaviour of Discretely Supported Cylinders. Research Report 93-01, Department of Civil Engineering, University of Edinburgh, March.

Teng, J.G. and Rotter, J.M. (1990). A study of buckling in column-supported cylinders. *Proceedings of the IUTAM Symposium, Preliminary Report 1990*. Also in: *Contact Loading and Local Effects in Thin-Walled Plated and Shell Structures* (eds V. Krupka and M. Drdacky), Academia Press, Prague, pp. 52–61.

Teng, J.G. and Rotter, J.M. (1992). Linear bifurcation of column-supported perfect cylinders: support modelling and boundary conditions. *Journal of Thin-Walled Structures* **14**, 241–263.

Thielemann, F.W. and Esslinger, M.E. (1968). On the postbuckling equilibrium and stability of thin-walled circular cylinders under axial compression. *Proceedings of the 2nd IUTAM Symposium on the Theory of Thin Shells*, Copenhagen, September 1967, Springer, Berlin.

Timoshenko, S.P. and Gere, J.M. (1961). *Theory of Elastic Stability*, 2nd edn. McGraw Hill, New York.

Yamaki, N. (1984). *Elastic Stability of Circular Cylindrical Shells*. North Holland, Elsevier Applied Science Publishers, Amsterdam.

Zhong, Z., Ooi, J.Y. and Rotter, J.M. (2001). The sensitivity of silo flow and wall pressures to filling method. *Engineering Structures* **23**, 756–767.

Chapter 4

Settlement beneath cylindrical shells

J.M.F.G. Holst and J.M. Rotter

Introduction

Storage containers are frequently constructed at ports in coastal regions or on reclaimed land where large settlements arise due to poor soil conditions. Uneven settlement beneath a ground-supported tank may occur, leading to several possible failure modes. The commonest causes are the use of non-uniform fill beneath the vessel and the construction of only part of the vessel on ground that has been consolidated by the earlier presence of a tank (Fig. 4.1). Each failure mode is closely related to a different component of the total settlement. Although the description given in the following outline review is mostly in terms of tank structures, it is equally applicable to all forms of cylindrical metal shells with the axis vertical.

There are several key questions that the designer must face:

- Which components of settlement displacements are critically important?
- How does the structural form of the tank affect the stresses and failure modes?
- Does the settlement lead to loss of serviceability or an ultimate limit state?
- If buckling occurs in the tank wall, is it potentially catastrophic or only unsightly?
- How does the settlement affect the resistance of the tank in other limit states?

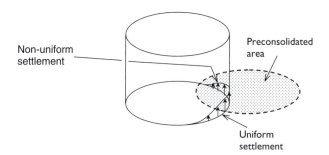

Figure 4.1 Uneven settlement of the base of a shell partially founded on a preconsolidated soil.

Whilst these questions cannot yet be answered comprehensively, this chapter provides an introduction to existing knowledge in the field which gives valuable indicators towards answering these questions.

The ultimate limit state that is most naturally associated with uneven settlement beneath the tank wall is buckling under axial compression. This limit state is normally thought of as catastrophic, and the strengths as highly sensitive to minor geometric imperfections. Although axial compression buckling under uniform stress conditions has been extensively studied (Yamaki 1984; Rotter 2002), relatively few studies have explored the effects of local high axial compression stresses, and even fewer are directly applicable to the problem of uneven settlement. The extent to which the general conclusions on axial compression buckling should be carried over into uneven settlement buckling is therefore a vital part of this description.

This chapter is chiefly concerned with geometrical distortions of the boundary conditions and their consequences for buckling in cylindrical shells. The distortions considered are in the plane of the shell, leading to membrane geometrical misfits. Where such misfits occur at construction joints within the shell, they lead to residual stresses and imperfections of geometry of the shell surface (Rotter 1996; Holst *et al.* 1999, 2000). Where they occur at the boundary, they are usually associated with foundation settlement, but they also lead to both residual stresses and geometric imperfections of the surface.

Settlement components and their consequences

Dishing settlements of a tank floor

Settlement beneath a tank on soft ground is caused by the weight of stored fluid in the tank over an extended period. As a result, the dominant settlement is in the form of a dishing of the base of the tank (Fig. 4.2). The displacements in this mode are so much larger than any other that most publications are concerned with

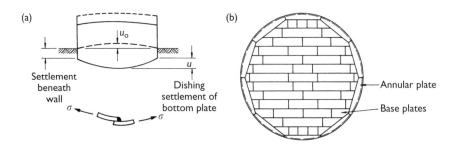

Figure 4.2 Dominant effects of settlement on tank structure: (a) dishing of base of tank due to settlement (b) weld arrangement in base.

them alone. It is also the only settlement mode that is relatively easy to predict in advance using the theories of soil mechanics.

Complete failures of tanks due to settlement are not common, but failure of one tank in Japan (Bell and Iwakiri 1980) and of two tanks in Britain (Green and Hight 1974) were attributed to the fracture of welds in the tank floor. The circular plate which forms the tank floor is commonly assembled by welding a large number of small rectangular plates together (Fig. 4.2). If the welds in the base are stronger than the plate itself, then the plate has an extended hardening and stiffening response up to very large deformations, and rupture is unlikely to occur whilst the settlements are within any reasonable bounds. Recommendations for allowable values of base dishing settlements should therefore relate to weld fracture, as affected by the welding technique and inspection procedure (Crawford 1986).

In their review paper, Marr *et al.* (1982) described a number of proposed relationships to limit the possibility of a failure by weld fracture. They recommended that the non-planar dish-shaped settlement u_{max} of the bottom plate (Fig. 4.2) should be kept below

$$u_{max} = \left(u_0^2 + \frac{3}{8} \frac{\sigma_f}{E} \frac{D^2}{\gamma_M} \right)^{1/2} \tag{1}$$

in which u_0 is the initial upward chamber of the tank floor, σ_f is the ultimate stress of the weld detail used in the floor, E is Young's modulus of elasticity, γ_M is the factor of safety (commonly taken as $\gamma_M = 2$ if rupture is possible), and D is the diameter of the tank. This expression may be derived by assuming the deformed shape of the base to be spherical and the strain in it to be uniform. They examined 30 case histories, including eight cases of bottom plate rupture. They found that their recommended criterion correlated well with the incidence of failure in service.

Settlements beneath the tank wall: settlement components

The second part of the settlement is settlement beneath the wall of the tank. The settlement displacements are always much smaller than base plate settlements, so that many publications have paid relatively little attention to wall settlement. However, it is not the absolute magnitude of displacements that matters, but their effect on the structure: quite small settlements in certain patterns can be very deleterious to the tank. It is difficult to predict the magnitudes of these differential settlements in advance using the theories of soil mechanics, but when a tank is already differentially settling, observations at different times can be most usefully employed to evaluate the likelihood that they will eventually lead to failure. However, a rational assessment of the consequences involves extensive calculation, as described below.

First, the settlements beneath the wall must be resolved into different components, which have different effects. The largest component is normally a uniform settlement, and this rigid body translation does not affect the safety of the structure. The next largest component is normally a uniform tilt of the whole tank wall, and this is only important if it becomes very large. Such tilting is discussed by Greenwood (1974) and Bell and Iwakiri (1980). Their recommended limit is given as 0.5% (i.e. a tilt of 1:200), although Greenwood (1974) suggested that larger tilts could be accepted without causing binding of a tank onto a floating roof. The effect of such a tilt is that the pressure is reduced on one side of the tank and increased on the other, leading to global bending, but not to ovalisation of the tank. However, near the fluid surface the pressure becomes zero around part of the circumference, which does result in ovalisation. The above conclusion is therefore sensitive to the aspect ratio of the tank, and the problem is more serious in very low height to diameter ratio tanks.

The components of settlement remaining after the uniform settlement and the uniform tilt have been removed are usually smaller, but are more important. These can only be determined with reasonable precision if a rather large number of settlement observation stations are set up around the tank (to be useful, the absolute minimum number is eight, but sixteen stations would be preferable, as noted below). These settlement components are referred to here as differential warping settlements.

When examining differential warping settlements, it is important to establish first whether the tank has a fixed roof, a floating roof or no roof. For tanks with a floating roof or no roof, it is necessary to consider the stiffness of the wind girder or ring at the eaves. Where the tank has a fixed roof, the roof plays a major role in the response of the tank to the vertical displacements beneath the wall (differential warping settlements). A fixed roof must be included in any structural model of the tank.

Characterisation of components of wall settlements

In their field study, Sullivan and Nowicki (1974) concluded that tanks with diameters up to 110 m could accommodate differential warping settlements as large as 30 mm without problems, but that values in excess of 45 mm caused problems irrespective of the settlement distribution or tank diameter. Such a finding appears rather sweeping and takes no account of the form of the tank or the effect of differential settlement on it.

A simple crude method of assessing the severity of local differential settlements is to take the difference between the settlements at any two adjacent measurement stations around the tank perimeter, and divide it by the circumferential separation between them. This is sometimes termed the 'angular distortion'. A value of 0.35% was suggested by De Beer (1969) and Belloni et al. (1974). The latter authors improved on this and used the maximum change in slope between three adjacent settlement observation points under the wall. They recommended that this slope be limited to 0.22% as a 'working hypothesis' for satisfactory operation of large tanks with floating roofs. Unfortunately, the procedure leads to quite variable results depending on how many observation stations are used. The criterion is also not rationally founded.

It should be noted that the above recommendations were all made by geotechnical engineers, and seem rather unsatisfactory when viewed from a structural engineering perspective. The distortion which occurs at the tank eaves is strongly dependent on such simple parameters as the tank aspect ratio, the radius to wall thickness ratio, the settlement distribution, and the stiffness of the rings attached to the shell wall. The criteria used to determine a maximum acceptable differential warping settlement should therefore include these parameters.

Greiner (1980) appears to have been the first to propose a rational method, applicable to all tanks, for assessing the form of wall settlements for the purposes of predicting the structural consequences. The same treatment was also used by Rotter (1987), Kamyab and Palmer (1989), Jonaidi and Ansourian (1996) and Hornung and Saal (1996) to describe settlement components. This procedure for determining the wall settlement profile yields the same result when different numbers of settlement measurement stations are adopted, provided the information is sufficiently complete to give an adequate description.

For a circular shell tank, the wall settlements u are described in terms of a truncated Fourier series around the circumference:

$$u = u_0 + \sum_{i=1}^{n}(u_i \cos i\theta + u_i^* \sin i\theta) \qquad (2)$$

The initial term u_0 defines the uniform settlement of the tank. The first harmonic terms u_1 and u_1^* define the tilt of the tank axis, and the orientation of the line of steepest slope. The second harmonic terms u_2 and u_2^* relate to the ovalling of the tank, together with the principal axes about which the ovalling occurs. Higher harmonic terms provide greater detail concerning local abrupt variations in support settlement.

It should be noted that the first term u_0 is normally the largest because the ground is likely to be relatively uniform, even if soft. The first harmonic component is likely to be the next largest, because the stiffness of the tank does not resist settlements in this form. The successive harmonic terms thereafter are then likely to decline progressively, because the shell is progressively stiffer in resisting each component. Soil–structure interaction can play a significant role in the local differential settlement beneath a tank. It should be noted that very rapid local changes in settlement are unlikely to occur because the base of the tank wall will not follow a sharply changing ground profile but will separate from the surface (e.g. it will not displace to the shape of a small hole dug beneath the wall). This problem was explored for unanchored tanks by Hornung and Saal (1996, 1997).

The number of terms that can be included in the series of Eq. (2) is limited by the number of settlement observations available. Five observations are needed to make a first estimate of ovalling, but the tank is not very stiff in ovalling, so the wall stresses induced in the this mode are generally quite small. The stresses arising from higher components induce much higher stresses because the tank is stiff, but the tank stiffness and the relative homogeneity both also tend to make these components small. Less than about eight observation stations around the circumference leads to inadequate evidence of any local variations in settlement. A fairly comprehensive measure, allowing some scope for minor errors, would be provided by 16 observation stations.

Consequences of warping settlements: tanks with flexible eaves

Where the tank has no roof and no eaves ring, vertical differential warping settlements beneath the wall cause small vertical stresses in the wall, but a significant distortion of the tank occurs at the eaves (Fig. 4.3). The largest component of the distortion is likely to be in the form of ovalling, although other components will generally be present, leading to local bulges and depressions. If the tank has a floating roof, binding of the roof against the wall may occur if the wall distortion becomes large.

Malik *et al.* (1977) proposed an inextensional theory to estimate the radial deflections at the top of the shell wall. Their solution may be taken as an upper bound, accurate only for low harmonics, as it neglects the bending stiffness of the shell and any stiffening due to wind girders. The range of these solutions was extended by Kamyab and Palmer (1989, 1991) using the membrane theory of shells and incorporating the stiffening effect of the primary wind girder. The range of harmonics covered was extended, but the analysis is limited to a uniform wall thickness and small settlements. Their prediction for the amplitude of the radial displacement w_{tn} at the top of the shell ($x = H$) in the nth harmonic component

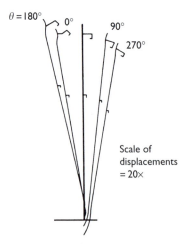

Figure 4.3 Distortion of tank wall and eaves ring caused by settlement at base.

of settlement, is valid for all practical values of n

$$w_{tn} = u_n \left(\frac{6r}{ft}\right)^{0.25} \left(\frac{g^2}{1 + kg^6 + g^8}\right) \tag{3}$$

where r is the shell radius, t is its thickness, f is the ratio of the circumferential bending stiffness of the primary wind girder I_G to that of the complete shell, and $g = n/n_1$ is the ratio of the harmonic n to that at which the inextensional analysis is no longer valid n_1. The parameter k is a dimensionless quantity.

$$f = \frac{12(1 - v^3)I_G}{Ht^3} \tag{4}$$

$$g = \frac{n}{n_1} = n\left\{\left(\frac{f}{36(1 - v^2)}\right)\left(\frac{t}{r}\right)^2 \left(\frac{H}{r}\right)^4\right\}^{0.125} \tag{5}$$

$$k = (2 + v)\left(\frac{3}{2}\frac{t}{r}\right)^{0.5} \left(\frac{r}{H}\right) f^{0.25} \tag{6}$$

The total displacements of the roof may be found by summing the harmonic components w_{tn} arising from the different harmonic components of the settlement u_n.

A more satisfactory procedure is to obtain a good estimate of the complete distribution of settlements beneath the tank perimeter as described above, to perform a shell bending analysis to find the resulting distortion of the eaves (not only ovalling), and to relate the calculated displacements to the tolerances for the floating roof. Appropriate limits on deformations for the top of the wall are set out in EEMUA 1959 (1993) to address the problem of jamming of a floating roof and large distortions that might impair the operation of the tank.

Consequences of warping settlements: tanks with eaves ring or integral roof

Stress analysis

Where the tank has an integral roof or a stiff eaves ring, vertical differential displacements beneath the wall cause only small distortions at the eaves. However, the high ring stiffness leads to high stresses in two locations: large local axial compressive stresses develop at the base of the wall, and large circumferential stresses can develop in the eaves ring.

The axial compressive stresses near the base can lead to axial compression buckling of the wall. The circumferential stresses, associated with circumferential bending of the eaves ring, may lead to yielding in the ring. Buckling of the ring may also occur (Jonaidi and Ansourian 1997, 1998), but has been little studied because the primary action in the ring is bending and not compression, and because the thrusts which develop in the ring are localised, and do not extend all around the ring. Where the ring cross-section cannot distort, buckling is unlikely.

Early attempts to evaluate stresses in the tank wall were described by Marr *et al.* (1982), who concluded that De Beer (1969) had made the only study dealing with overstressing of the shell. However, De Beer assumed that the tank wall is like a horizontal beam in bending, which is a poor representation of the shell structure's response. Gazioglu and Witham (1984) attempted to measure the stresses developing in a tank wall during large non-uniform settlements using strain gauges. Unfortunately, their measurements are insufficiently complete to make any deductions about buckling, because only principal stresses were reported with no indication of tension or compression. They used 0.6 of the yield stress as a failure criterion, which is not very meaningful, and the tank did not fail, so little can be deduced from this test.

The first useful analysis was developed by Greiner (1980) who used the semi-membrane theory of shells to obtain algebraic expressions for the stresses. This line was followed by Kamyab and Palmer (1989, 1991) and Palmer (1994). Finally, Hornung and Saal (1996, 1997) extended the semi-membrane treatment to shells of varying thickness. These studies lead to rather complicated equations that are the most general available solutions for the problem. The solution given by Palmer (1994) for a uniform thickness wall finds the meridional membrane stress resultant N_n in the nth harmonic at the base as

$$N_n = u_n \left(\frac{-3Et}{H} \right) \left(\frac{g^8}{1 + kg^6 + g^8} \right) \tag{7}$$

where f, g and k are given by Eqs (4)–(6).

Since differential settlements beneath the wall are difficult to predict in advance, this problem is more often one of evaluation of an existing structure than prediction of a planned structure. Where a settlement problem is being assessed, much is known about the tank geometry and the pattern and amplitudes of settlement. In these cases, the best method of evaluating the consequences is to obtain a good estimate of the complete distribution of settlements beneath the tank perimeter in the manner described above, to follow this with a full shell bending finite element analysis to determine the deformations and stresses developing in both the shell wall and the ring or roof, and to apply appropriate failure criteria to determine the consequences. Procedures of this kind were employed by Rotter (1987), Jonaidi and Ansourian (1996, 1998) and Hornung and Saal (1996, 1997).

Buckling of the wall under axial compressive stresses

Since differential warping settlements induce axial compressive stresses near the bottom of the tank wall, and catastrophic buckling under axial compression occurs at very low stresses in thin cylindrical shells (Yamaki 1984; Rotter 1985, 2002; ECCS 1988), it is natural to suppose that a buckling failure of the wall should be expected. However, there are reasonable grounds for thinking that these buckles may be stable and not catastrophic, as they are local (taking up only a small part of the circumference) and are induced by displacements, rather than loads. Design limits on these axial stresses are set out in API 653 (1991) and EEMUA 1959 (1993), though these do not appear to have been verified in any scientific study. As noted by Palmer (1994), the API limit to avoid tank overstress is based on the invalid assumption that horizontal stresses are most significant.

The first buckling assessment appears to be that of Rotter (1987), who investigated the effects of large observed settlements on a steel fire water storage tank. He modelled the tank as a shell of revolution, with both the eaves ring and a lower ring as additional shell segments. He used a linear bending theory finite element analysis to predict the radial displacements of the eaves and the stresses throughout the tank wall due to the imposed non-symmetric foundation settlements. He identified the lack of research on buckling under localised axial stresses and adopted the only known design criterion for buckling of imperfect cylinders in limited regions of axial stress (Rotter 1986) which has since been adopted, with modifications, into the European silos standard (ENV 1993-4-1 1999). In this procedure, the local zone of axial compression (Fig. 4.4(a)) is first approximated by the pattern shown in Fig. 4.4(b), and the degree of localisation deduced from it. The effect of non-uniformity of the axial compression stress around the circumference is represented by the parameter ψ, determined from the linear elastic axial compressive stress distribution σ_{xE}.

The design value of the axial compressive stress at the circumferential position where it is highest is defined as σ_{xoEd}, located at the circumferential coordinate θ_0. The design value of the axial compressive stress at a point at the same axial coordinate, but separated from the point in question by the circumferential distance $r\Delta\theta_g$ is found as σ_{xgEd}. The location of this point is best chosen such that the stress σ_{xgEd} is about half the value of the maximum stress σ_{xoEd}. The separation $\Delta\theta_g$ should always be chosen so that:

$$0.2\sigma_{xoEd} < \sigma_{xgEd} < 0.8\sigma_{xoEd} \tag{8}$$

The equivalent harmonic n of the stress distribution is then obtained as:

$$n = \frac{1}{\Delta\theta_g} \cos^{-1}\left(\frac{\sigma_{xgEd}}{\sigma_{xoEd}}\right) \tag{9}$$

where $\Delta\theta_g$ is measured in radians. Following the above approach, the value of n is relatively insensitive to the precise location of the chosen point.

The buckling strength corresponding to this stress distribution is then determined as a proportion of the classical elastic critical stress

$$\sigma_{cl} = 0.605 Et/r \tag{10}$$

The effect of geometric imperfections in reducing the real buckling strength below the value given by Eq. (10) may be found either from Koiter's (1945) asymptotic

(a)

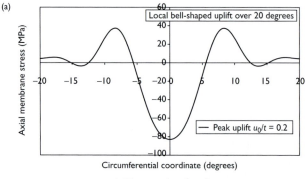

Axial membrane stress (MPa)

Local bell-shaped uplift over 20 degrees

Peak uplift $u_0/t = 0.2$

Circumferential coordinate (degrees)

Axial stresses just above base

(b)

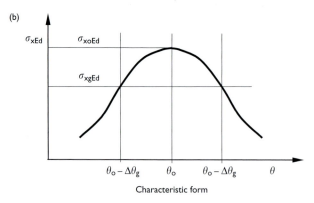

σ_{xEd} σ_{xoEd}

σ_{xgEd}

$\theta_o - \Delta\theta_g$ θ_o $\theta_o - \Delta\theta_g$ θ

Characteristic form

Figure 4.4 Characterisation of local axial stress peak: (a) axial stresses just above base;
(b) characteristic form.

equation, or from codified values (ENV 1993-1-6 1999; ENV 1993-4-1 1999).
The latter give

$$\sigma_{cr,Rk} = \frac{0.62}{1 + 1.91\psi(w_o/t)^{1.44}}\sigma_{cl} \tag{11}$$

where w_o is the design imperfection amplitude and t is the shell wall thickness.
The imperfection modification factor ψ to allow for the localisation of the axial
compression stress state is found from the equivalent harmonic n as:

$$\psi = \frac{(1 - b_1 n)}{(1 + b_2 n)} \tag{12}$$

in which:

$$b_1 = 0.5\sqrt{\frac{t}{r}} \tag{13}$$

$$b_2 = \frac{(1 - b_1)}{\psi_b} - 1 \tag{14}$$

and ψ_b (= 0.40 perhaps) is the value of ψ under conditions of global bending ($n = 1$). The harmonic localisation at which there is no reduction in buckling strength as a result of imperfections is $n_\infty = 1/b_1$.

Rotter (1987) used this procedure to assess the likelihood of failure for a steel water storage tank in Queensland, Australia. The shell was analysed under measured settlement patterns at different times and the calculated maximum compressive axial membrane stress was compared with the buckling strength to obtain the apparent factor of safety $\gamma_M = \sigma_{cr,Rk}/\sigma_{xoEd}$.

The finite element analysis also provided details of the deformed geometry, notably the radial displacements of the eaves. A key feature of the evaluation was a forecasting of the reduction of the safety factor with time, which could be projected into the future for forecasting purposes. It should be noted that it is difficult to forecast the magnitude of warping differential settlements before a tank is constructed, but when a tank is already differentially settling, observations at different times can be most usefully employed to evaluate the likelihood that they will eventually lead to failure.

Yielding of the ring under circumferential stresses

The circumferential stresses that develop in the ring are caused by bending of the ring about a vertical axis, which is known as circumferential bending. These stresses may lead to yielding in the extreme fibres of the ring. When yielding occurs, the ring stiffness declines, leading to a relaxation of these stresses, and slightly larger out-of-round deformations of the tank eaves. The eaves ring may then become less effective in its other roles (e.g. as a wind girder), but first yield of the eaves ring should not be considered a very serious condition.

Buckling of the eaves ring

Very little information is available on the possibility of buckling of the eaves ring due to differential settlements. However, Jonaidi and Ansourian (1997) reported that their calculations of an example tank structure became unstable at a certain amplitude of settlement because the eaves ring experienced distortional buckling (tripping). Moreover, this buckle could not be prevented by the use of stiffeners at 4.3° intervals. This finding matches the expectations from Jumikis and Rotter (1983) and Teng and Rotter (1988) who showed that such buckles in rings should occur with very many waves around the circumference. Further studies of this problem are needed.

Studies of stresses under differential settlement components

Anchored and unanchored tanks

Palmer (1994) showed that local settlement which would give a maximum stress of the order of the yield stress in a fully anchored tank might give a maximum stress of only one-tenth of this value if unanchored. Thus, for unanchored tanks with floating roof he identified tank distortion as the critical design criterion. Hornung and Saal (1997) produced a more extended study of tanks without base anchorage, which showed that the axial stresses are rapidly dissipated when the shell separates from the soil surface (termed uplift in their study). However, both of these studies necessarily used the settlement profile as data, so whether settlements are likely to become so localised that such bridging might occur was not determined.

Tanks with varying thickness walls

Most practical tanks do not have the uniform wall thickness assumed in Palmer's (1994) analysis. Hornung and Saal (1997) extended his analysis to tanks with varying thickness walls, providing more complicated formulae. Their finite element analysis used an example tank to explore the range of validity (Fig. 4.5) when either an eaves ring or a fixed roof was used. The stress induced in each harmonic component of settlement begins as small, rises to a moderate value, and then rises sharply for highly localised settlement effects. The latter are not captured by the membrane theory treatment. In their example, Hornung and Saal (1997) indicated that accounting for variable shell thickness may reduce the axial stresses significantly. Jonaidi and Ansourian (1998) also explored wall thickness variation using an example tank structure.

Nonlinear elastic investigations

The only study to explore the nonlinear response under large amplitude harmonic-form foundation settlements appears to be that of Jonaidi and Ansourian (1996, 1997). The calculations were performed on two laboratory models and one large example tank. No failure condition was obtained, but markedly nonlinear responses (Fig. 4.6) began at settlement amplitudes of the order of two wall thicknesses for harmonic $n = 4$. The calculations on the example tank were undertaken for harmonics $n = 4$, 6 and 8 and showed that the settlement amplitude at which nonlinearity becomes marked falls rapidly with increasing harmonics, and the axial stress associated with it rises. Moreover, it is clear that a snap-through or bifurcation occurs (Fig. 4.6), but this was not identified by the authors. Their chief

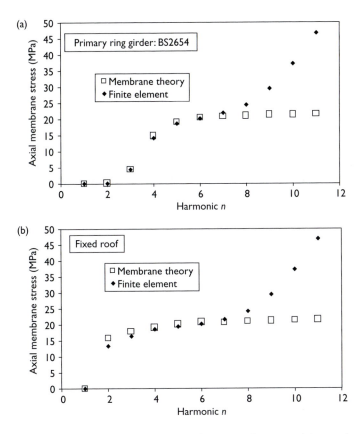

Figure 4.5 Stress development above base as a function of harmonic component (Hornung and Saal 1997): (a) with eaves ring to BS2654; (b) with fixed roof.

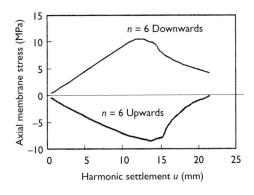

Figure 4.6 Development of axial stresses in nonlinear analysis (Jonaidi and Ansourian 1997).

conclusion was that axial compressive stresses near the base are not critical but that distortion and overstressing of the top ring is important.

Their analyses also show that there is a significant bending effect particularly in the medium range of *n* due to the greater shell bending participation with the heavier wind girder. This leads to circumferential stresses in the wind girder, which are substantially greater than the meridional stresses at the bottom of the shell. For a specific example with a very strong wind girder at a relatively low harmonic, they show that the circumferential stresses are critically high whereas the meridional membrane stresses can easily be tolerated. Thus, they conclude that the circumferential stress at the eaves is critical.

However, studies of this kind are difficult to use. Real uneven settlements do not occur in a single harmonic form, and the chief reason for studying harmonic settlements is that the results can be summed to produce different harmonic profiles that lead to quite different conclusions from those that are obtained from a single harmonic alone. Moreover, the summation of harmonics is only valid for linear behaviour, and one might suppose that it is not too far wrong for mild nonlinearities. However, the calculated response is highly nonlinear (Fig. 4.6), so such a summation is certainly invalid. Given that high axial stresses are induced by small settlements in localised waveforms (corresponding to events like in Fig. 4.1), their conclusion that axial compression stresses are not critical should be treated with caution.

Buckling investigations

Apart from those described below, no calculations are known to have explored snap-through or bifurcation buckling events in cylinders with settlement displacements at the boundaries. Because the severity of a buckling event can only be determined with reference to the postbuckling response, and postbuckling is necessarily highly nonlinear, the study of buckling cannot be usefully conducted by considering the response to individual harmonic components. Instead, it is necessary to explore the response to an individual local settlement pattern to investigate buckling. The design methodology set out in section 'Settlement components and their consequences' suggests that axial compression buckling may be important for any shell that is also subjected to an external axial compression. However, further studies are needed to explore when buckling may occur, the nature of the postbuckling path and the sensitivity to geometric imperfections.

Local settlement studies

Local effects of uneven settlement: uplift displacements

The above studies have attempted to deal with the general case of a settlement pattern that could be in any form by characterising general forms through harmonic analysis. However, the commonest occurrence of uneven settlement arises from the construction of a new tank on soil that has been partly consolidated under a previous tank (Fig. 4.1). This leads to an abrupt kink in the new settlement profile, and when the uniform and tilt terms (harmonics 0 and 1) have been removed, the remaining pattern of settlement is quite well represented by a local uplift over a small part of the shell. In addition, many of the above studies did not address an externally applied axial load, so the proximity of a buckling failure could not be determined. Two recent studies have addressed this situation directly and are outlined here.

Experimental studies

An experimental study (Lancaster *et al.* 1996) investigated the buckling of a cylindrical shell under uniform axial compression in the presence of a local uplift boundary condition defect. Axial displacements of the lower boundary were introduced at zero net axial load by raising a number of local fingers beneath the shell, followed by an uniform axial compression at the upper edge. The effects of the defects on the buckling strength and behaviour were assessed.

The cylindrical shell (Fig. 4.7) was made of Melinex plastic sheet with $E = 5.0 \times 10^3$ MPa and Poisson's ratio $v = 0.3$, with a radius $r = 440$ mm, thickness $t = 0.25$ mm ($r/t = 1760$), and height $H = 690$ mm. Uplift displacements were applied at several locations around the shell circumference. The response of the shell to uplift showed that local dimples could be made to form easily, but that

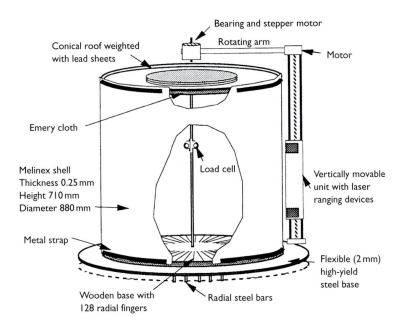

Figure 4.7 Schematic illustration of testing apparatus (from Lancaster *et al.* 1996).

these were stable and did not cause damage to the shell. As the uplift amplitude was increased, the amplitude and size of the dimples grew and the centre of each dimple moved further from the boundary at which the uplift was applied. The buckling tests indicated that the axial compression buckling load was only mildly affected by the amplitude of uplift that had been previously applied (Fig. 4.8), suggesting that local dimple imperfections should not be regarded as seriously degrading to the strength of the shell. The same conclusion can be drawn from the asymptotic analysis of Amazigo and Fraser (1971) for the effect of local dimple imperfections on buckling strength.

Computational studies

Stress analysis

A fuller understanding of the behaviour exhibited by cylindrical shells with local uplift displacements on the lower boundary can be found by detailed examination of the results of computational modelling. Finite element analyses (Holst et al. 1996) were used to model some of the above experiments. This was undertaken as part of a wider study of the effects of misfits in causing imperfections and residual stresses that reduce the buckling strength of cylindrical shells under axial loading (Rotter 1996; Holst et al. 1999, 2000). A brief outline of the chief conclusions found from modelling the tests is given here.

A precise quantitative model of a shell buckling experiment requires precise data on the shell geometry, boundary conditions, material properties, geometric imperfections, residual stresses, uniformity of loading and the pattern of the boundary uplift some of which may show nonlinearities during the test. Because not all this data was available, the calculations described here are more important as descriptive and qualitative explorations of behaviour. All analyses were geometrically nonlinear, performed using ABAQUS with four-noded doubly curved shell elements type S4R, with modified Riks tracking of the pre- and postbuckling paths; and a prescribed incrementation technique to obtain accurate buckling loads (ABAQUS 1997). Multiple symmetry was exploited and a 45° strip analysed to represent four uplift zones as used in some of Lancaster's tests (Fig. 4.8). Locations on the shell are described in terms of dimensionless coordinates $\xi = x/\lambda$ and $\eta = r\theta/\lambda$, where the linear bending half-wavelength λ (= 25.6 mm for this shell) is given by

$$\lambda \cong 2.44\sqrt{Rt} \tag{15}$$

Load path

The analyses followed the process used in the tests, using two stages: first, a sinusoidal uplift with maximum value u_{max} was imposed on the lower boundary

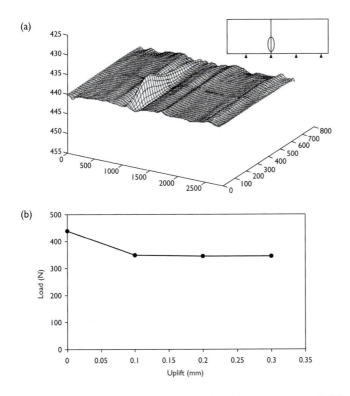

Figure 4.8 Sample experimental results (from Lancaster *et al.* 1996): (a) measured displacements after dimpling; (b) variation of buckling load with uplift.

over an angle θ_u (Stage A).

$$u = \frac{u_{\max}}{2}\left(1 + \cos\left(\frac{\pi\theta}{\theta_u}\right)\right) \tag{16}$$

where $u_{\max}$ is the maximum uplift. In the test described here, the uplift angle was $\theta_u = 11.25°$, corresponding to the edge of one finger. The uplift u is here made dimensionless as $v = u/t$.

When the uplift had attained the chosen value, an external axial load was applied to the upper edge (Stage B) which was kept circular (equivalent to a stiff ring or fixed roof). The mean stress σ_x induced by this load is normalised by the classical elastic critical stress σ_{cl} (Eq. 10) to give the dimensionless externally applied axial stress as $\mu = \sigma_x/\sigma_{cl}$.

Features of the computational and experimental results

The general features of the finite element analyses reproduce those of the experiments. The loading and undeformed geometry used in the numerical investigation are shown in Fig. 4.9. The following observations were seen in both the experiments and the numerical predictions:

- Under uplift alone (zero external axial load, $\mu = 0$), the response was almost linear for small uplift, but marked nonlinearity arose at uplifts in excess of $\upsilon = 0.17$ (Fig. 4.10).

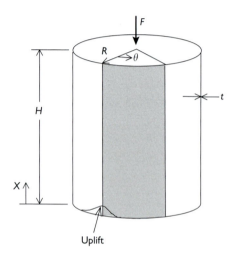

Figure 4.9 Loading and undeformed geometry used in numerical investigation.

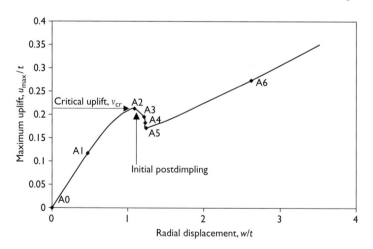

Figure 4.10 Displacements slightly above the base during local uplift.

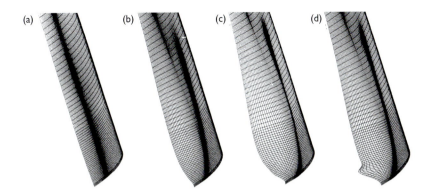

Figure 4.11 Formation of the dimple by snap-through buckling: (a) very small uplift displacements; (b) in predimpling domain; (c) snap-through initiation of dimple;
(d) dimple clearly established.

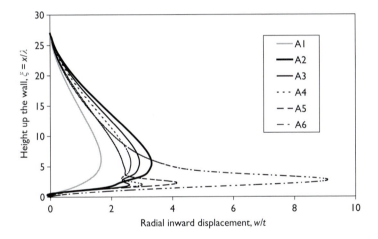

Figure 4.12 Radial displacement up wall through centre of uplift.

- At a critical uplift $v = v_{cr}$ (Fig. 4.10), local snap-through buckling occurred
  and an inward-directed dimple formed above the uplift (Fig. 4.11) ($v_{cr} \approx 0.22$
  here). This stable dimple did not precipitate general buckling.

- At greater uplifts ($v > v_{cr}$), the dimple increased in size and the location
  of the maximum radial displacement moved slightly: first up, and then down
  again (Fig. 4.12). The dimple diameter was typically around 4λ (Fig. 4.11).
  The postdimpling behaviour is markedly nonlinear: further details are given
  by Holst *et al.* (1996).

- The distribution and magnitude of the axial membrane stress near the base
  was very sensitive to the amplitude of the uplift, showing very substantial

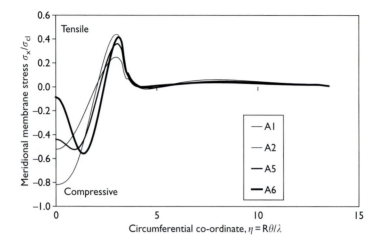

Figure 4.13 Meridional membrane stress around the base.

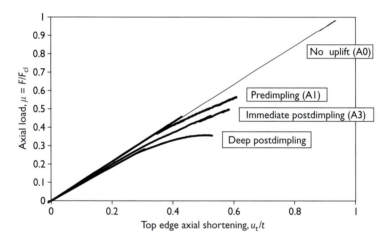

Figure 4.14 Stage B load–displacement curves for different amplitudes of Stage A uplift.

redistributions after dimpling (Fig. 4.13). This indicates that the above linear analyses are only useful for small uplifts (note that $\upsilon_{cr} = u_{cr}/t \approx 0.22$ in this example). Further details are given by Holst *et al.* (1996).

- Under external axial load (Stage B), the load path was increasingly nonlinear as the uplift of Stage A was increased (Fig. 4.14). Small uplifts caused a substantial drop in the bifurcation load, but even very large postdimpling uplifts were not greatly more deleterious (Fig. 4.15).

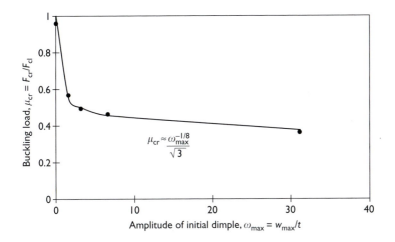

Figure 4.15 Effect of uplift amplitude on global bifurcation load.

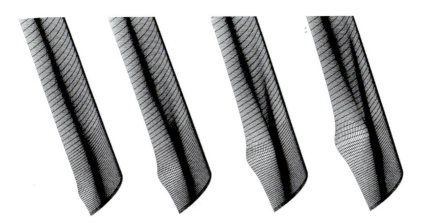

Figure 4.16 Amplification of dimple in Stage B for small initial uplift.

- In Stage B, the dimple moved up the meridian of the shell with increased amplitude and size (Fig. 4.16). Global bifurcation propagated from the dimple (Fig. 4.16) when the dimple width reached about 6λ.
- Where the initial uplift was below the critical value ($\upsilon < \upsilon_{cr}$), the external load of Stage B soon induced dimpling (A1 in Fig. 4.14). The subsequent response was then similar to the case $\upsilon > \upsilon_{cr}$.
- Global bifurcation in the presence of uplift occurred in the load range $0.35 < \mu_{cr} < 0.55$ (Fig. 4.15), which compares with the experimental average $\mu_{cr} = 0.38$. The experiments showed little consistent strength variations with uplift

amplitude (Fig. 4.8(b)): the calculations show the reasons for this, despite the large amplitude of the dimple imperfection (up to $20t$).

• The width of the dimple when global buckling occurred (critical dimple size) appeared to be independent of the uplift and imperfection amplitude. Thus, the critical widths for cases $\upsilon_{max} = 0.203$ and $\upsilon_{max} = 3.7$ were approximately the same ($\approx 5\lambda$).

• One experiment used an uplift of $\upsilon_{max} = 1.0$ at four stations, leading to a Stage B buckling load of $\mu_{cr} = 0.32$. A finite element analysis for $\upsilon_{max} = 1.66$ buckled at $\mu_{cr} = 0.36$, which is very similar. However, other experiments showed a significant experimental scatter, so the close match may not be very meaningful.

Conclusions of local uplift studies

The experiments and calculations of the behaviour of cylinders with local uplift at the base boundary, followed by uniform axial compression loading give a clear insight into the behaviour and severity of local settlement boundary displacements in tanks and silos. The broad conclusions that can be drawn from this study are:

1   Relatively small local uplifts at the shell boundary cause a snap-through dimple buckle that may be unsightly, but is stable. Its amplitude grows with increasing uplift.

2   Axial compression, superposed on the uplift, leads to global bifurcation buckling, generally initiated from the dimple, but at axial compression stress levels that are similar to those associated with random imperfections. The amplitude of the dimple has a minor effect on the bifurcation load.

3   The chief reason why the local dimple does not reduce the buckling strength markedly appears to be its role in redistributing the high axial stresses from its centre to the shell adjacent to it, so that the high local stress is dissipated. This effect is very important in imposed displacement problems (e.g. settlement), but may be less so for load-induced stress concentrations.

Conclusions

This chapter has outlined current knowledge of buckling as a result of settlement beneath a cylindrical shell, usually in the form of a ground-supported tank or silo. The three key features of such knowledge have been explored: characterisation of the settlements and their components, stress analysis and prediction of displacements in the shell structure, and criteria of failure or limit states.

Symmetrical settlements beneath tanks can be predicted using soil mechanical theories, but uneven settlements around the circumference beneath the tank wall are very difficult to predict in advance. As a result, most uses of this knowledge will be in investigations of tanks that already show signs of distress. For such problems, extensive settlement observations at widely spaced times is of critical importance.

The structural studies described here have indicated the nature of the buckling problem: relatively small settlements can lead to nonlinear responses and to snap-through buckling in tanks with fixed roofs or stiff rings. The snap-through buckles produce noticeable dimples in the wall, but these are stable and do not cause a marked reduction in the axial compression buckling load. However, severe distortion of the upper edge can occur in open-topped tanks, and binding onto floating roofs, distortional buckling and yielding of the ring may all be critical.

Acknowledgements

The authors gratefully acknowledge financial support from the EPSRC research grant GR/H 41027, under which the above calculations were performed. They also wish to thank Prof. Richard Greiner and Dr Werner Guggenberger of the Technische Universität, Graz, for their advice and many valuable discussions of shell behaviour.

References

ABAQUS (1997). ABAQUS/Standard User's Manual, Volume 1, Version 5.7. Hibbitt, Karlsson & Sorenson, Inc., Pawtucket, RI.

Amazigo, J.C. and Fraser, W.B. (1971). Buckling under external pressure of cylindrical shells with dimple shaped initial imperfections. *International Journal of Solids and Structures* **7**, 883–900.

API 653 (1991). Tank inspection, repair, alteration and reconstruction. API Standard 653, 1st edn. American Petroleum Institute, Washington.

Bell, R.A. and Iwakiri, J. (1980). Settlement comparison used in tank-failure study. *Journal of the Geotechnical Engineering Division, ASCE* **106**, 153–169.

Belloni, L., Garassino, A. and Jamailkowski, M. (1974). Differential settlements of petroleum steel tanks. *Proceedings of the Conference on Settlement of Structures.* British Geotechnical Society, Cambridge, UK, pp. 323–328.

Crawford, K.N. (1986). Thin-walled steel storage tanks: foundation design considerations. *Proceedings of the Pacific Structural Steel Conference*, Auckland, New Zealand, pp. 209–231.

De Beer, E.E. (1969). Foundation problems of petroleum tanks. *Annales de L'Institute Belge du Petrole* **6**, 25–40.

ECCS (1988). *European Recommendations for Steel Construction: Buckling of Shells*, 3rd edn. European Convention for Constructional Steelwork, Brussels.

EEMUA 1959 (1993). Recommendation for in-service periodic inspection of above-ground vertical cylindrical steel storage tanks, 1st edn. Engineering Equipment and Materials Users Association, London, Publication 1959.

ENV 1993-1-6 (1999). Eurocode 3: Design of Steel Structures; Part 1-6: General Rules: Supplementary Rules for Shell Structures. CEN, Brussels.

ENV 1993-4-1 (1999). Eurocode 3: Design of steel structures, Part 4.1: Silos, Eurocode 3 Part 4.1. CEN, Brussels.

Gazioglu, S.M. and Withiam, J.L. (1984). Evaluation of a differentially settled tank. *Proceedings of the International Conference on Case Histories in Geotechnical Engineering*, Rolla, MO, May, vol. 1, pp. 133–142.

Greiner, R. (1980). Ingenieurmäßige Berechnung dünnwandiger Kreiszylinderschalen. Veröffentlichung des Instituts für Stahlbau, Holzbau und Flächentragwerke der Technischen Universität Graz.

Green, P.A. and Hight, D.W. (1974). The failure of two oil-storage tanks caused by differential settlement. *Proceedings of the Conference on Settlement of Structures*. British Geotechnical Society, Cambridge, UK, pp. 353–360.

Greenwood, D.A. (1974). Differential settlement tolerance of cylindrical steel tanks for bulk liquid storage. *Proceedings of the Conference on Settlement of Structures*. British Geotechnical Society, Cambridge, UK, pp. 35–97.

Holst, J.M.F.G, Rotter, J.M. and Calladine, C.R. (1996). Modelling of experiments on buckling under localised uplift in compressed cylindrical shells. Report No. R96-012, University of Edinburgh, May, 42pp.

Holst, J.M.F.G., Rotter, J.M. and Calladine, C.R. (1999). Imperfections in cylindrical shells resulting from fabrication misfits. *Journal of the Engineering Mechanics Division, Proceedings of the American Society of Civil Engineers* **125**(4), 410–418.

Holst, J.M.F.G., Rotter, J.M. and Calladine, C.R. (2000). Imperfections and buckling in cylindrical shells with consistent residual stresses. *Journal of Constructional Steel Research* **54**, 265–282.

Hornung, U. and Saal, H. (1996). Stresses in tank shells due to settlement taking into account local uplift. *Proceedings of the International Conference on Structural Steelwork*, Hong Kong, December, pp. 827–832.

Hornung, U. and Saal, H. (1997). Stresses in unanchored tank shells due to settlement of the tank foundation. *Proceedings of the International Conference on Carrying Capacity of Steel Shell Structures*, Brno, pp. 157–163.

Jonaidi, M. and Ansourian, P. (1996). Effects of differential settlement on storage tank shells. *Proceedings of the International Conference on Structural Steelwork*, Hong Kong, December, pp. 821–826.

Jonaidi, M. and Ansourian, P. (1997). Non-linear behaviour of storage tank shells under harmonic edge settlement. *Proceedings of the International Conference on Carrying Capacity of Steel Shell Structures*, Brno, pp. 164–170.

Jonaidi, M. and Ansourian, P. (1998). Harmonic settlement effects on uniform and tapered tank shells. *Thin-Walled Structures* **31**, 237–255.

Jumikis, P.T. and Rotter, J.M. (1983). Buckling of simple ringbeams for bins and tanks. *Proceedings of the International Conference on Bulk Materials Storage, Handling and Transportation*, Institution of Engineers Australia, Newcastle, pp. 323–328.

Kamyab, H. and Palmer, S.C. (1989). Analysis of displacements and stresses in oil storage tanks caused by differential settlement. *Journal of Mechanical Engineering Science, Proceedings of the IMechE Part C* **203**, 61–70.

Kamyab, H. and Palmer, S.C. (1991). Displacements in oil storage tanks caused by localized differential settlement. *Journal of Pressure Vessel Technology, Transactions of the ASME* **113**, 71–80.

Koiter, W.T. (1945). On the stability of elastic equilibrium. PhD Thesis, Delft University (in Dutch; see also Translation AFFDL-TR-70-25 Wright Patterson Air Force Base, 1970).

Lancaster E.R., Calladine, C.R. and Palmer, S.C. (1996). Experimental observations of the buckling of thin cylindrical shells subjected to axial compression. Department of Engineering, Cambridge University, D-Struct/TR162, 107p.

Malik, Z., Morton, J. and Ruiz, C. (1977). Ovalisation of cylindrical tanks as a result of foundation differential settlement. *Journal of Strain* **12**, 339–348.

Marr, W.A., Ramos, J.A. and Lambe, T.W. (1982). Criteria for settlement of tanks. *Journal of the Geotechnical Engineering Division, ASCE* **108**, 1017–1039.

Palmer, S.C. (1994). Stresses in storage tanks caused by differential settlement. *Journal of Process Mechanical Engineering, Proceedings of the IMechE Part E* **208**, 5–16.

Rotter, J.M. (1985). Buckling of ground-supported cylindrical steel bins under vertical compressive wall loads. *Proceedings of the Metal Structures Conference*, Institution of Engineers Australia, Melbourne, Australia, May, pp. 112–127.

Rotter, J.M. (1986). The analysis of steel bins subject to eccentric discharge. *Proceedings of the Second International Conference on Bulk Materials Storage, Handling and Transportation*, Institution of Engineers, Wollongong, Australia, July, pp. 264–271.

Rotter, J.M. (1987). Consequences of settlement of a fire water storage tank at Gladstone. Investigation Report S627, School of Civil and Mining Engineering, University of Sydney.

Rotter, J.M. (1996). Elastic plastic buckling and collapse in internally pressurised axially compressed silo cylinders with measured axisymmetric imperfections: interactions between imperfections, residual stresses, and collapse. *Proceedings of the International Workshop on Imperfections in Metal Silos: Measurement, Characterisation, and Strength Analysis*, Lyon, France, pp. 119–140.

Rotter, J.M. (2002). Buckling of cylindrical shells under axial compression. In *Buckling of Thin Metal Shells* (eds J.G. Teng and J.M. Rotter) E&FN Spon, London, pp. 12–25.

Sullivan, R.A. and Nowicki, J.G. (1974). Differential settlements of cylindrical oil tanks. *Proceedings of the Conference on Settlement of Structures*, British Geotechnical Society, Cambridge, UK, pp. 420–424.

Teng, J.G. and Rotter, J.M. (1988). Buckling of restrained monosymmetric rings. *Journal of Engineering Mechanics, ASCE* **114**(EM10), October, pp. 1651–1671.

Yamaki, N. (1984). *Elastic Stability of Circular Cylindrical Shells*. North Holland, Amsterdam, The Netherlands.

Chapter 5

Cylindrical shells under uniform external pressure

R. Greiner

General

This chapter comprises all cases of external pressure uniformly distributed around the circumference. Due to the uniform stress field in the prebuckling state, the theoretical buckling behaviour may consistently and directly be derived from the differential equations. Accordingly, a great number of solutions for different cases are available for shells of constant wall thickness. For variable, that is, stepped wall thickness and for cylinders restrained by stiffening rings, the solutions have been derived numerically on the basis of the linear buckling equations.

R.V. Southwell in 1913 and R.V. Mises in 1914 were the first to start and Flügge (1932) was one of the authors to give a comprehensive solution of the problem. Since the impact of imperfections on cylinders under uniform external pressure is relatively moderate and is normally accounted for by a constant factor for all kinds of applications (geometry, boundary conditions, etc.), the main significance of this buckling problem is to determine the appropriate theoretical buckling pressure.

It is a characteristic that cylinders under uniform external pressure form relatively large buckles extending over the whole length of the shell (Fig. 5.1).

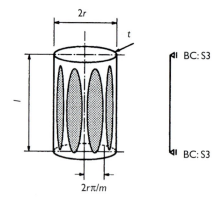

Figure 5.1 Cylindrical shell under uniform external pressure, typical buckling pattern.

Accordingly, they can only resist low pressures compared to the case of axial loading with its small pattern of buckles. However, since the loading usually comes from vacuum-related problems of containments, the range of the external pressure is widely limited by the atmospheric pressure. Different are cases of hydrostatic loads of submarine structures, which, therefore, usually require stiffening of the shell wall.

In accordance with the buckling pattern of large buckles extending over the whole length of the shell, the effects of the cylinder length, of the restraining conditions at the boundaries and of the length of the pressurised zone ('band pressure') have great influence. The same holds for the effects of variable (e.g. stepped) wall thickness, since the thicker courses exercise restraining conditions onto the thinner wall. Amplification of the buckling pressure is most economically reached by ring-stiffening of the shell.

These effects are dealt with in the following with the main emphasis on the essential results. The derivation of the solutions may be very different from case to case and is only discussed in general form. Uniform external pressure is understood as constant along the length of the shell in the following, which means that axially varying pressures like hydrostatic pressure are not considered here (see Weingarten 1962). The theoretical buckling resistance is expressed in the form of the buckling stress σ_{cr} or alternatively in form of the buckling resistance q_{cr}. In general, it holds that

$$\sigma_{cr} = q_{cr} \cdot \frac{r}{t}$$

Cylindrical shells with constant wall thickness

Short, medium and long cylinders

The basic 'classical' case is the cylinder with simply supported ends, that is, radially restrained, axially and rotationally unrestrained. The solutions of the so-called 'linear' buckling theory may be taken from, for example, Flügge (1973, 1981). The buckling pressure can be obtained by variation of the circumferential wave number m only, since in longitudinal direction the wave number is always 1, expressing the shape of a half-sine wave.

The buckling behaviour differs considerably with different 'cylinder lengths'. (The appropriate parameter accounting for the 'cylinder length' is the length parameter $(l/r)\sqrt{r/t}$ or $(l/r)\sqrt{t/r}$ depending on the range of length.) This different behaviour may also be expressed by 'length-specific' buckling formulae derived from the exact Flügge solutions by appropriately neglecting certain terms of minor significance.

Short cylinders are highly influenced by the bending stiffness in axial direction and with shorter length approach the buckling resistance of plates, that is, a plate-strip of the length of the circumference. Medium-length cylinders are practically independent of the longitudinal bending effect. The essential stiffness in axial direction is now the membrane stiffness of the shell. These shells, therefore, may well be approximated by means of the so-called 'semi-membrane theory'. Long cylinders get independent of both the bending and membrane stiffness in axial direction and they then approach the buckling resistance of the ring or the tube. The length-specific formulae for the theoretical buckling resistances are given in the following:

- *Short and medium-long cylinders:*

$$\sigma_{cr} = \frac{E}{m^2} \left[\frac{1}{(1 + (m^2/\pi^2)(l/r)^2)^2} + \frac{(t/r)^2}{12(1 - \mu^2)} \left(m^2 + \pi^2 \left(\frac{r}{l}\right)^2 \right)^2 \right] \quad (1)$$

For very short cylinders, only the second term of the equation is relevant, which – by use of l for the width and $r\pi$ for the length of an equivalent plate-strip – results in

$$\sigma_{cr} = \frac{E\pi^2}{12(1 - \mu^2)} \left(\frac{t}{l}\right)^2 \left[\frac{ml}{r\pi} + \frac{r\pi}{ml} \right]^2 \quad (2)$$

The equivalent formula for the plate is as follows (k is the buckling coefficient of plate):

$$\sigma_{cr,pl} = E \left(\frac{r}{l}\right) \left(\frac{t}{r}\right)^{1.5} \cdot \left[\frac{0.904 \cdot k}{(l/r)\sqrt{r/t}} \right] \quad (3)$$

The results of these formulae are illustrated in the left-hand side of Fig. 5.2, where $k = 2$ for external pressure including the pressure on the end plates.
- *Medium-long and long cylinders:*

$$\sigma_{cr} = E \left[\frac{\pi^4}{m^4(m^2 - 1)} \left(\frac{r}{l}\right)^4 + \frac{(t/r)^2}{12(1 - \mu^2)} (m^2 - 1) \right] \quad (4)$$

$$= E \left(\frac{r}{l}\right) \left(\frac{t}{r}\right)^{1.5} \left[\frac{\pi^4}{m^4(m^2 - 1)} \frac{1}{((l/r)\sqrt{t/r})^3} + \frac{m^2 - 1}{12(1 - \mu^2)} \left(\frac{l}{r}\sqrt{\frac{t}{r}}\right) \right]$$

For medium length, minimising σ_{cr} with respect to m leads to the critical buckling wave number

$$m_{cr}^2 = \pi^4 \sqrt{36(1 - \mu^2)} \cdot \left(\frac{r}{l}\sqrt{\frac{r}{t}}\right) \quad (5)$$

and then to

$$\sigma_{cr} = \frac{0.855}{(1 - \mu^2)^{0.75}} E \cdot \frac{r}{l} \left(\frac{t}{r}\right)^{1.5} \quad (6)$$

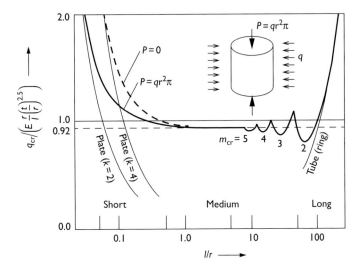

Figure 5.2 Buckling pressure q_{cr} for cylinders with 'classical' boundary conditions and $r/t = 1000$ under uniform external pressure on the shell surface $(P = 0)$ or overall external pressure $(P = qr^2\pi)$.

and further with $\mu = 0.3$ to

$$\sigma_{cr} = 0.92E\frac{r}{l}\left(\frac{t}{r}\right)^{1.5} \tag{7}$$

- *Long cylinders:* Long cylinders approach the tube and the formula may be obtained by setting $m = 2$:

$$\sigma_{cr} = E\left(\frac{t}{r}\right)^2\left[\frac{0.25}{(1-\mu^2)} + \frac{1}{48}\left(\frac{\pi}{(l/r)\sqrt{t/r}}\right)^4\right] \tag{8}$$

$$= E\left(\frac{t}{r}\right)^2\left[0.275 + 2.03\left(\frac{1}{(l/r)\sqrt{t/r}}\right)^4\right] \tag{9}$$

The formula for the tube is

$$\sigma_{cr} = E\left(\frac{t}{r}\right)^2 \cdot 0.275 \tag{10}$$

The results for these formulae are illustrated in the right-hand side of Fig. 5.2.

A formula frequently used in the field of pressure vessels is the one by Windenburg and Trilling (1934):

$$\sigma_{cr} = \frac{r}{t} \cdot \frac{2.60E(t/d)^{2.5}}{(l/d) - 0.45(t/d)^{0.5}} \tag{11}$$

where d is the diameter. This formula gives good results for short- and medium-length cylinders under full external pressure (including the pressure on the end plates). Neglecting the second term in the denominator makes it equivalent with Eq. (7) of Ebner (1952) for medium-length shells.

- *Length-specific effect due to the loading conditions of external pressure with and without pressurised end plates:* Overall loaded shells show differences in their buckling resistance compared with shells pressurised only at their circumferential surface, and this effect depends on the 'lenght parameter' of the shell. While long and medium-long shells are practically unaffected by the end plate pressure, the effect on short shells shows an increase when the length is shorter and the plate behaviour dominates. The plate buckling coefficient k, being $k = 2$ for overall loading, changes to $k = 4$ for pure circumferential loading. The effects are shown in more detail in Fig. 5.4.

Effect of boundary conditions

Boundary conditions of shells should be divided into membrane terms and bending terms. The following five boundary conditions are considered and are written in the usual notation. C_3 and C_4 stand for 'clamped', S_3 and S_4 for 'simply supported' and F for 'free'. Apparently, the first two conditions express the membrane part and the second two the bending part:

$$C_3: n_x = v = w = w' = 0$$

$$C_4: u = v = w = w' = 0$$

$$S_3: n_x = v = w = m_x = 0 \quad \text{(the so-called 'classical' boundary condition)}$$

$$S_4: u = v = w = m_x = 0$$

$$F: n_x = n_{xy} = q_x^* = m_x = 0$$

In accordance with the characteristic length-specific shell behaviour discussed above, the effects of the specific boundary conditions depend on the length of the shell.

In the range of the medium cylinder length only the 'membrane conditions' are governing. For decreasing length the effect of the 'bending conditions' increases gradually until for very short cylinders the plate behaviour is dependent only on the bending terms. For long cylinders, the influence of the boundary conditions ceases to apply.

This behaviour is illustrated in Fig. 5.3 for the case of overall uniform pressure. If the buckling pressure q_{cr} is related to $E(r/l)(t/r)^{2.5}$ according to Eq. (7), the buckling resistance in the medium-length range remains constant for equal membrane boundary conditions. A more detailed illustration for the range of short

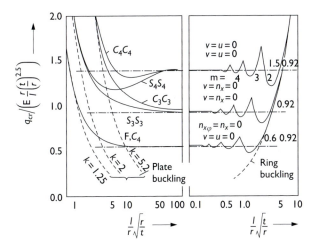

Figure 5.3 Buckling pressure q_{cr} for cylinders with different boundary conditions under overall external pressure.

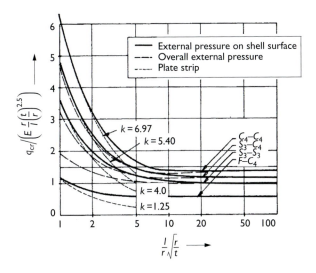

Figure 5.4 Buckling pressure q_{cr} for short cylinders with different boundary conditions under overall external pressure or external pressure on the shell surface only.

cylinders is given in Fig. 5.4, which also shows the differences between the loading conditions with and without pressurised end plates.

The effects of the boundary conditions may be obtained by numerical solution of the differential equations as carried out by Sobel (1964) and Thielemann and

Esslinger (1964). A simplified engineering approach was given by Greiner (1976), leading to a compact formula for the range of medium and long cylinders:

$$q_{cr} = E\frac{r}{l}\left(\frac{t}{r}\right)^{2.5}\left[\frac{C_M}{m^4(m^2-1)}\frac{1}{(l/r\sqrt{t/r})^3} + \frac{m^2-1}{12(1-\mu^2)}\left(\frac{l}{r}\sqrt{\frac{t}{r}}\right)\right]$$

(12)

where C_M is a coefficient representing the effect of the membrane boundary conditions (it is – by the way – equal to the coefficient of the frequency of equivalently supported, fictitious beams).

By setting $(m^2 - 1) \cong m^2$ the range of medium length may be treated analogously to the 'classical case'. The critical buckling wave numbers may be obtained by

$$m_{cr}^2 = C_M\frac{r}{l}\sqrt{\frac{r}{t}}\cdot\sqrt[4]{36(1-\mu^2)}$$

(13)

and

$$q_{cr} = C_M \cdot \frac{0.272}{(1-\mu^2)^{0.75}}\cdot E\frac{r}{l}\left(\frac{t}{r}\right)^{2.5} = C_{BC}\cdot 0.92E\frac{r}{l}\left(\frac{t}{r}\right)^{2.5}$$

(14)

where C_{BC} represents the effect of the boundary conditions in relation to Eq. (7) of the 'classical case' (see Fig. 5.5).

It may be concluded – and physically illustrated by means of the semi-membrane theory – that the medium-length cylindrical shell under uniform external pressure behaves similarly with the vibration behaviour of a fictitious beam as far as the boundary conditions are concerned. Thereby, the meaning of the boundary conditions for the fictitious beam is as follows:

$w = 0$ means lateral support, $u = 0$ means clamped ends,
$n_x = 0$ means free end-rotation, $m_x = q_x = 0$ means free end.

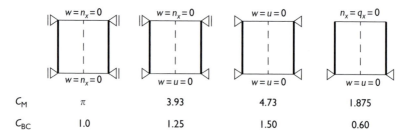

Figure 5.5 Medium-length cylindrical shells under uniform external pressure – effect of boundary conditions.

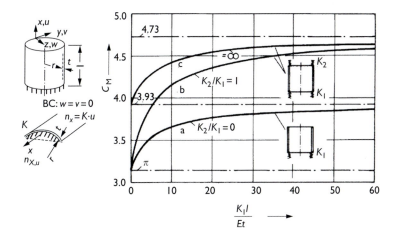

Figure 5.6 Medium-length cylinder under uniform external pressure – effect of elastic axial restraint at the boundaries.

Therefore, also the effect of elastic axial restraint at the boundaries, for example, by axial spring constants K (kN/cm/cm), can easily be accounted for by similar C_M-factors (see Greiner 1976) (Fig. 5.6).

Band pressure loading

The load case of uniform pressure acting in the form of a band around the circumference can be treated similarly as in the previous section, since the effect of the unloaded part of the shell may be considered like a compatibility or boundary condition. The results of such an approach are presented in Fig. 5.7 according to Greiner (1975) for medium-length cylinders.

For b/l approaching unity the buckling resistance accesses the classical buckling pressure of the fully loaded shell. The results are in good accordance with the original work of Almroth and Brush (1961).

Real buckling resistance, effect of imperfections

The real buckling resistance of uniformly pressurised shells is affected by imperfections in a relatively moderate way, if compared with the high imperfection sensitivity of shells under axial compression. This may be explained by the big difference of the dimensions of the buckles in the two cases. The large buckles due to external pressure have a big curvature, which is only moderately impaired by the geometrical predeformations, while the small curvature of the narrow buckling pattern due to axial compression may be of the same size as the pre-deformations.

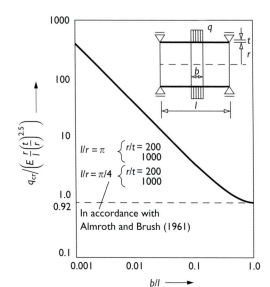

Figure 5.7 Medium-length cylinders with 'classical' boundary conditions under band pressure.

So basically the buckling-relevant membrane stiffness is much more reduced with small buckles than with bigger ones.

The reduction factor α, accounting for the effect of imperfections, is given by different numbers in specific codes. The ECCS Recommendation No. 56 (1988) and the BS 5500 (1985) give a factor of 0.50, while the ASME Code (1989) gives 0.80. The new DIN 18800 (1990) and the EUROCODE 3 (1999) provide an α-factor with 0.65. These different numbers result from different evaluation methods of the buckling tests, which partly do not consider the effects of the specific boundary conditions, of the actual material properties or the difference between overall and pure circumferential pressure.

It is further remarkable that the α-factors are used without relation to the cylinder length. Remembering the fact that short cylinders approach the behaviour of plates, while long cylinders take on the behaviour of rings, one might conclude that in these ranges of length differing imperfection sensitivity should be found. However, specific investigations are not available concerning this aspect.

Nevertheless, based on the evaluation of new test-series, the new part 1–6 of EUROCODE 3 defines α-factors of 0.75/0.65/0.50 for excellent, high or normal fabrication quality, specifying that this factor is related to shells with uniform pressure on the circumferential surface.

The real buckling resistance is also affected by the material law or the plasticity effect of the material. This effect is described by a transition curve in

EUROCODE 3, part 1-6, based on the stress–strain relation of structural steel. Since stainless steel has a different material law the results are not directly transferable. Further investigations for stainless steel are under way (Hautala and Schmidt 1998).

Cylinders with stepped wall thickness

Form of shells

Cylindrical shells with stepped wall thickness are used for vertical tanks and silos, where the increase of internal filling pressure from top to bottom results in an increase in wall thickness. It is usually built up by individual courses (or strakes) of constant wall thickness varying differently over the length of the cylinder in relation to the chosen material strength. In the top region of the shell the wall thickness is controlled by the buckling phenomenon – leading to the specification of a 'minimum wall thickness' in design codes; while in the bottom region the design is determined by the internal pressure – leading frequently to the use of material of higher strength or thicker plates.

The external pressure is normally caused by under-pressure, often controlled by safety valves, or by wind pressure and has a magnitude in the range of 1 kN/m^2 or less. Shell buckling, therefore, is normally an additional check for an already designed shell. It is frequently combined with the axial compression due to the roof load.

The shells under consideration fall into the range of medium-length cylinders; therefore, this chapter is only dealing with this type of solutions. The connections between the courses of the wall may be made by welding or by bolting, the latter resulting in an overlap of the plates. The latter influence may be significant for buckling in axial compression; however, for the given case of external pressure it may be ignored.

In the following, the boundary conditions are restricted to the 'classical' case of both ends radially supported and axially free ($w = n_x = 0$), which is the usual condition in containments.

Theoretical solution

The buckling behaviour of shells with variable wall thickness is dominated by the thin upper courses, while the lower, thicker ones – although weakened in their stiffness by the external pressure – execute a restraining effect on the upper courses. As explained above, the destabilising effect of the thin courses may be analytically treated like the vibration of fictitious beams; and the same idea leads to beams on elastic bedding for the thick, stabilising courses. This follows from the solution of the semi-membrane theory, which results in the following differential equation

(Biezeno and Koch 1938):

$$\frac{E \cdot t \cdot r^2}{m^4} w'''' + \frac{E \cdot t^3 \cdot m^4}{12(1 - \mu^2)r^4} \left(1 - \frac{q}{q_{cr}}\right) \cdot w = 0 \tag{15}$$

where

$$q_{cr} = \frac{E \cdot t^3 \cdot (m^2 - 1)}{12(1 - \mu^2)r^3}$$

is the critical pressure of a free shell segment.

For $q > q_{cr}$, the sign of the second term of Eq. (15) turns into minus and the differential equation of a certain course has the same type as the vibration of a beam: $(EJw'''' - c \cdot w = 0)$. For $q < q_{cr}$, it is analogous to that of a beam with elastic bedding: $(EJw'''' + c \cdot w = 0)$. The problem can, therefore, formally be treated like the vibrational stability of a sequence of coupled beam elements of different behaviour representing the individual courses (Greiner 1972; Resinger and Greiner 1974).

Cylinders with two steps of wall thickness

The effect of a thicker lower course may be demonstrated by Fig. 5.8, where the length and the thickness ratios of the upper course l_o, t_o and the lower course l_u, t_u are varied. The buckling pressure q_{cr} is related to a fictitous shell with the thickness t_o and the overall ('total') length $l = l_o + l_u$, which is called $q_{o,tot}$.

It is illustrated that up to a certain ratio t_u/t_o the buckling behaviour is well described by the buckling pressure q_{tm} of a shell with mean thickness $t_m = (t_o + t_u)/2$. However, for higher thickness ratios the behaviour changes significantly, since the buckling pressure remains nearly constant – smoothly approaching the limit of the case, where the lower boundary condition is radially and axially fixed through the restraint of the lower course.

A more appropriate presentation of this behaviour is given in Fig. 5.9 for the buckling pressure q_{cr} and in Fig. 5.10 for the buckling wave number m_{cr}. Both values are related to the analogous terms of a cylinder with the length l_o and the thickness t_o of the upper course assuming 'classical' boundary conditions at both ends:

$$q_o = 0.92E \frac{r}{l_o} \left(\frac{t_o}{r}\right)^{2.5} \tag{16}$$

$$m_o^2 = 7.52 \cdot \frac{r}{l_o} \sqrt{\frac{r}{t_o}} \tag{17}$$

Thus, the increasing thickness of the lower course can at most result in an amplification factor of 1.25, which would represent full restraint at the bottom end of the upper course.

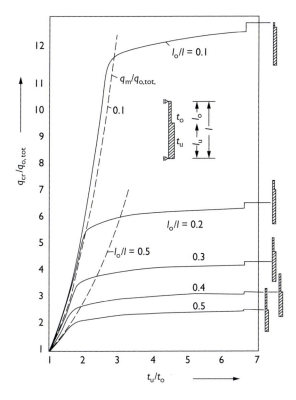

Figure 5.8 Medium-length cylinders with two steps of wall thickness.

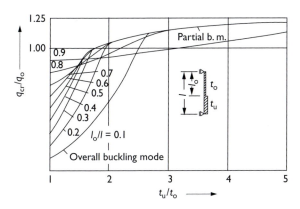

Figure 5.9 Critical buckling pressure of medium-length cylinders with two steps of
wall thickness.

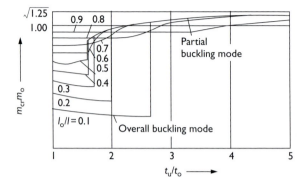

Figure 5.10 Critical buckling wave number of medium-length cylinders with two steps of wall thickness.

The two ranges, where 'overall' buckling appears (mean thickness) or where 'partial' buckling of the upper course alone is caused, are clearly shown.

Cylinders with linearly stepped wall thickness

The behaviour of cylinders with two, three and six steps of linearly varying wall thickness is compared to similar examples discussed earlier (Figs 5.11 and 5.12).

The buckling pressures – related to that of the upper course q_0 – are presented in Fig. 5.11, indicating the transition points from overall buckling to partial buckling. The sixfold stepped wall results in the highest buckling resistance; however, due to the relation to the higher value q_0 it formally appears as lowest curve in the diagram.

The presentation in Fig. 5.12 is given to show that partial buckling is initiated at a wall thickness ratio t_2/t_0 of about 1.5. This means that any method using the mean thickness of adjacent courses should be limited at ratios of $t_2/t_0 \leq 1.5$.

Practical design, different code regulations

Practical cylindrical shells frequently behave in the way illustrated by the examples of Fig. 5.13. In the figure, the variation of the wall thickness results in buckling of the upper courses, while the lower ones provide full restraint, both in radial as well as in axial directions. The axial shape of the buckling deformation resembles the vibrational deformation of a fictitious beam restrained by beams on elastic bedding continuously connected with it. All three cylinders have the same buckling resistance and this makes it clear that the use of the mean wall thickness as the design basis can be totally wrong in such cases (see Resinger and Greiner 1976).

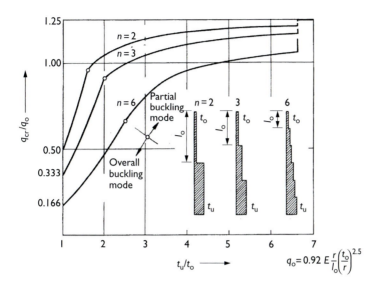

Figure 5.11 Critical buckling pressures of medium-length cylinders with linearly stepped wall thickness.

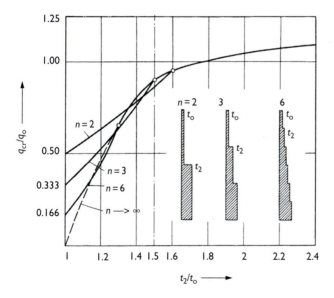

Figure 5.12 Critical buckling pressures of medium-length cylinders with linearly stepped wall thickness related to t_2/t_o.

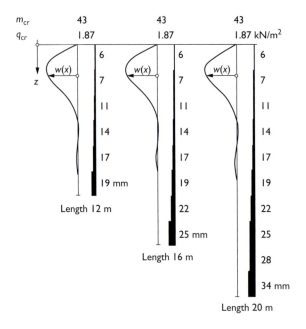

Figure 5.13 Buckling behaviour of cylindrical tanks of 40 m in diameter and different heights (steel grade S 235).

Design regulations given by the present codes tackle the buckling problem by quite different procedures, which will be discussed briefly below. This is because these procedures are to be found also in the new EUROPEAN Standards of CEN/TC 265 for storage tanks and of CEN/TC 250 for general shell buckling design. The first one is related to the procedure of the British Standard BS 2654 (1989), the latter to that of the German Standard DIN 18800.

The general approach of DIN 18800 replaces a specifically stepped cylinder wall by a fictitious cylinder with three courses of mean wall thickness each and then gives diagrams to calculate the buckling pressure for this substitute shell structure. According to Fig. 5.12, only sections of the wall may be transferred into a mean thickness that does not exceed the limit of $t_i/t_o = 1.5$. The diagrams for the three-stepped cylinders were elaborated on the basis of the semi-membrane theory as presented by Resinger and Greiner (1976).

The general approach of BS 2654 replaces a specifically stepped cylinder wall by a fictitious cylinder with the constant wall thickness t_o of the upper course and a reduced length. The length reduction of each course is carried out by a factor $(t_o/t_i)^{2.5}$, where t_i is the actual thickness of the course (i).

For cylinders of constant thickness the two procedures yield equivalent critical buckling pressures; however, the BS 2654 results in a different design due to omitting any further reduction for the effect of imperfections or material plasticity.

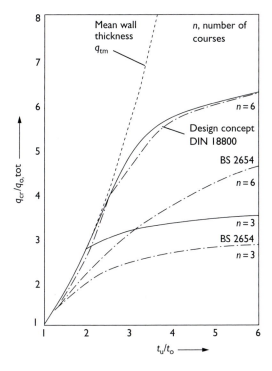

Figure 5.14 Comparison of critical buckling pressures with code results for cylinders built up by three and six linearly stepped courses of equal length.

Independently of this fact, Fig. 5.14 presents a comparison of the two procedures for the case of linearly stepped cylinders of different ratios t_u/t_o, consisting of three or of six courses.

The results are related to $q_{0,tot}$, that is, the buckling pressure of a fictitious cylinder with the constant wall thickness t_o and the same total length as the real cylinder. The diagram shows a considerable discrepancy of the BS 2654 results for higher ratios t_u/t_o. Other cases of thickness variation may result in a different response. In general, it may be said that the method of length reduction is an intuitively correct approach to the right side; however, it is not able to quantitatively account for the specific restraining effects of the various types of thickness distribution of practical cylinders (Saal and Hornung 1999).

Effect of additional stiffening rings

For many cylinders with the upper wall thickness according to the minimum thickness of codes, stiffening by a ring is necessary in order to increase the buckling resistance of this region to carry both circumferential and axial compressions.

The buckling resistance of the cylinder wall is then increased by the ring in analogy to the effect of an elastic spring action on a vibrating beam. The stiffness of this elastic spring may be expressed by

$$C_R = \frac{(m^2 - 1)^2 E I_R}{r^4} \left(1 - \frac{q}{q_{R,cr}} \right) \tag{18}$$

where

$$q_{R,cr} = \frac{(m^2 - 1) E I_R}{r^3 \cdot l_R} \tag{19}$$

and I_R, l_R and $q_{R,cr}$ are the moment of inertia, the effective length and the buckling resistance of the ring stiffener, respectively. The analysis of the cylinder with the stepped wall described above is only supplemented by the term of the ring stiffener.

The result of a practical cylinder 56 m in diameter and 26 m in length stiffened by a ring at a level of 8 m below the top is demonstrated in Fig. 5.15 (see Greiner 1975). The buckling resistance is increased by about 50% related to the unstiffened case by a ring of very small dimension (annular plate: 100×12 mm). This is due to the fact that the high buckling wave number of $m_{cr} = 41$ reduces the effective span of the ring to about 2.30 m only. The buckling deformation illustrates the effect of the ring stiffener. The figure indicates further that part of the courses stabilise, if the buckling pressure q_{cr} is lower than $q_{cr,i}$ of the individiual courses, and part of the courses destabilise, if $q_{cr} > q_{cr,i}$. This makes the physical behaviour of the shell more apparent.

Another example of a cylinder 25 m in diameter and 21.55 m in length stiffened by three rings is presented in Fig. 5.16. Two buckling modes may occur, the first one for $m_{cr} = 21$ and the second one for $m_{cr} = 33$, both on the approximately same level of buckling pressure (Greiner 1975). The critical buckling pressure is

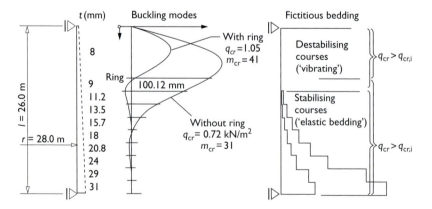

Figure 5.15 Buckling behaviour of cylinder with stepped wall thickness and an additional ring-stiffener.

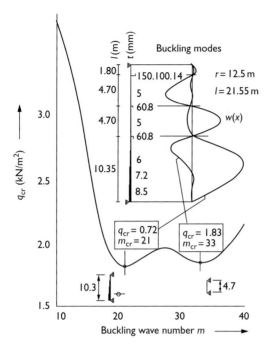

Figure 5.16 Buckling behaviour of a cylinder with stepped wall thickness and three additional ring-stiffeners.

plotted against the buckling wave number. The buckling deformations illustrate the two different buckling modes.

Ring-stiffened cylinders

Thin-walled cylindrical shells may effectively be strengthened by stiffening rings. For design purposes a concept may be recommended, which considers the two failure modes of local interring buckling, that is, the buckling of the unstiffened shell between the rings, and of general instability, that is, the overall resistance of the rings.

A common design principle of ring-stiffened cylinders is the use of the 'limiting ring-stiffness' (in German: 'Mindeststeifigkeit'). It may be derived from the linear buckling theory as the ring-stiffness, which shows that the load-carrying capacity of the unstiffened shell between the rings is equivalently provided by the stiffeners.

The limiting stiffness γ_R^* as derived by Resinger and Greiner (1982), Greiner (1987) may be related to the dimensionless factors

$$\gamma_R = \frac{I_R}{(lt^3/12(1-\mu^2))} \tag{20}$$

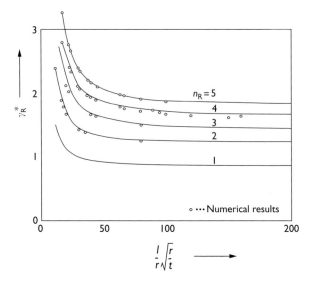

Figure 5.17 Limiting stiffness γ_R^* of ring-stiffened cylinders with one to five equidistant rings.

and

$$\delta_R = \frac{A_R}{lt} \tag{21}$$

where A_R is the cross-section area of the ring stiffener.

A parametric study results in the limiting stiffness γ_R^* as given in Fig. 5.17 for one to six ring stiffeners and for overall external pressure (n_R, number of stiffening rings).

The coefficients γ_R^* are influenced by the ratios l/r and r/t, which could be eliminated by the use of the parameter $(l/r)\sqrt{r/t}$. It is also influenced by the extensional deformation of the stiffener, which is expressed by the parameter δ_R, but due to the small size of this effect, δ_R-values up to 0.02 were covered on the conservative side in the diagram.

An analoguous investigation was performed by Blackler (1986), resulting in the formula for the limiting value I_R^*:

$$I_R^* = 0.077t^3l \cdot n_R^{0.45} \tag{22}$$

This result is in good accordance with the data in Fig. 5.17, if higher values of $(l/r)\sqrt{r/t}$ are considered.

Design rules for the stability of ring-stiffened cylinders are given in ECCS Recommendation No. 56 and in the German Recommendations DASt-Richtlinie 017 (1992), where the effect of imperfections is accounted for by a second-order analysis of the pre-deformed ring stiffener.

References

Almroth, B.O. and Brush, D.O. (1961). Buckling of a finite-length cylindrical shell under a circumferential band of pressure. *Journal of the Aerospace Sciences* **28**, 573–579.

ASME-Code (1989). ASME Boiler and Pressure Vessel Code. American Society of Mechanical Engineers.

Biezeno, C.B. and Koch, J.J. (1938). The buckling of a cylindrical tank of variable thickness under external pressure. *Proceedings of the 5th International Congress on Applied Mechanics*, Cambridge, MA.

Blackler, M.J. (1986). Stability of silos and tanks under internal and external pressure. PhD Thesis, University of Sydney.

BS 5500 (1985). Specification for unfired fusion welded pressure vessels. British Standards Institution.

BS 2654 (1989). Specification for manufacture of vertical steel welded non-refrigerated storage tanks with butt-welded shells for the pretroleum industry. British Standards Institution.

DASt-Richtlinie 017 (1992). Beulsicherheitsnachweise für Schalen – spezielle Fälle. Deutscher Ausschuss für Stahlbau (DASt), Stahlbau-Verlagsgesellschaft.

DIN 18800 (1990). Part 4, Steel Structures, Stability, Buckling of Shells. Deutsches Institut für Normung.

Ebner, H. (1952). Theoretische und experimentelle Untersuchung über das Einbeulen zylindrischer Tanks durch Unterdruck. *Stahlbau* **21**, 153–159.

ECCS-Recommendations No. 56 (1988). Buckling of Steel Shells. European Recommendations, ECCS – Technical Committee 8.

EUROCODE 3 (1999). Part 1–6, Design of Steel Structures, Supplementary Rules for the Strength and Stability of Shell Structures. European Committee for Standardisation.

Flügge, W. (1932). Die Stabilität der Kreiszylinderschale. *Ing.-Archiv* **III**, 463–506.

Flügge, W. (1973). Stresses in Shells. Springer Verlag, Berlin.

Flügge, W. (1981). Statik und Dynamik der Schalen. Springer Verlag, Berlin.

Greiner, R. (1972). Ein baustatisches Lösungsverfahren zur Beulberechnung dünnwandiger Kreiszylinderschalen unter Manteldruck. Bauingenieur-Praxis, Heft 17 Verlag Ernst & Sohn.

Greiner, R. (1975). Beulberechnung von Kreiszylinderschalen unter Außendruck. Sonderheft der DFVLR, Schalenbeultagung in Braunschweig 1975, pp. 106–133.

Greiner, R. (1976). Zur Klärung des Tragverhaltens von Zylindern unter Aussendruck mit besonderer Berücksichtigung des Randeinflusses. Sonderheft des DFVLR, Schalenbeultagung in Meersburg 1976, pp. 182–194.

Greiner, R. (1987). Ring-stiffened cylindrical shells under external pressure – an extended proposal for design recommendations. *Proceedings of International Colloquium on Stability of Plate and Shell Structures*, Ghent University, 6–8 April 1987, pp. 457–466.

Hautala, K. and Schmidt, H. (1998). Buckling tests on axially compressed cylindrical shells made of various austenitic stainless steels at ambient and elevated temperatures. Research Report No. 76, University Essen.

Resinger, F. and Greiner, R. (1974). Zum Beulverhalten von Kreiszylinderschalen mit abge stufter Wanddicke unter Manteldruck. *Stahlbau* **43**, 182–187.

Resinger, F. and Greiner, R. (1976). Praktische Beulberechnung oberirdischer zylindrischer Tankbauwerke für Unterdruck. *Stahlbau* **45**, 10–15.

Resinger, F. and Greiner, R. (1982). Buckling of wind loaded cylindrical shells – application to unstiffened and ring-stiffened tanks. *Proceedings of the State of the Art Colloquium*, University of Stuttgart, Germany, 6–7 May.

Saal, H. and Hornung, U. (1999). Design Rules for Tank Structures – Different Approaches; Light-weight Steel and Aluminium Structures, Elsevier Science, pp. 399–407.

Sobel, L.H. (1964). Effects of boundary conditions on the stability of cylinders subject to lateral and axial pressures. *AIAA Journal* **2**(8), 1437–1440.

Thielemann, W. and Esslinger, M. (1964). Einfluß der Randbedingungen auf die Beullast von Kreiszylinderschalen. *Stahlbau* **33**, 353–361.

Weingarten, V.I. (1962). The buckling of cylindrical shells under longitudinal varying loads. *Journal of Applied Mechanics* **29**(1), 81–85.

Windenburg, D.F. and Trilling, Ch. (1934). Collapse by instability of thin cylindrical shells under external pressure. *U.S. Transactions of ASME* **56**, 819–825.

Chapter 6

Cylindrical shells under non-uniform external pressure

P. Ansourian

Early studies

In Chapter 5, cylindrical shells under uniform external pressure with particular reference to short, medium-long and long cylinders, boundary conditions, stepped wall thicknesses and ring stiffeners were discussed. This chapter deals with buckling under non-uniform external pressure, examples of which abound in civil/mechanical engineering, marine, ballistic and aerospace structures.

Earlier work on the stability of cylinders under external pressure was based on the Donnell stability equations, modified by Batdorf (1947). While it is more accurate to apply Flügge's (1973) equations in coupled form to the buckling problem for either uniform or variable pressure, the former solutions remain useful especially in the 'medium' range of thin cylinders. Almroth (1962) gave solutions based on the minimum potential energy principle for a centrally located band of pressure (Fig. 6.1) varying circumferentially as in Eqs (4) and (5); the band varied in width from 10% of the cylinder length to 100% ($b/L = 1$).

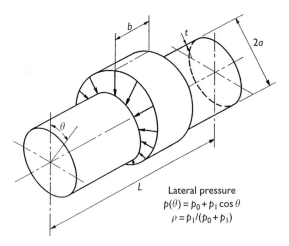

Lateral pressure
$$p(\theta) = p_0 + p_1 \cos \theta$$
$$\rho = p_1/(p_0 + p_1)$$

Figure 6.1 Cylinder with band of varying external pressure (Almroth 1962).

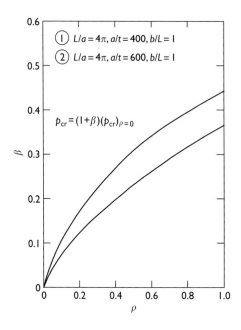

Figure 6.2 Gain in buckling strength over uniform pressure case (Almroth 1962).

For the latter configuration, results are reproduced in Fig. 6.2. As the 'skew' pressure factor ρ rises from zero (uniform pressure) to 1 (cosine variation), the buckling pressure is seen to rise monotonically from that calculated for a uniform pressure distribution. The rise is most pronounced and is in the order of 40% in long cylinders ($L/a = 4\pi R/t = 400\text{--}600$) for which the number of half waves is small; it reaches 60% in thicker cylinders ($R/t = 200$, $\rho = 1$). In shorter cylinders, the gain is less pronounced.

Weingarten (1962) solved the Donnell equation modified by Batdorf for the buckling of cylinders subjected to a lateral pressure varying linearly in the meridional direction, but uniform circumferentially. Using a harmonic series formulation and the Galerkin method, he established the homogeneous system of equations, the determinant of which, when set equal to zero, yielded the critical pressure. The pressure-variation parameter α (Fig. 6.3) is so defined that when it is less than 1, the pressure is external over the entire length of the cylinder, but reversal of pressure occurs when α is greater than 1. Figure 6.3 shows the critical value of the lateral pressure coefficient based on peak pressure $K_p = p_0 R L^2 / \pi^2 D$; for $Z = (L^2/Rt)(1 - v^2)^{1/2}$ greater than 100 (i.e. not very short), all the α-curves become straight lines proportional to $Z^{1/2}$. For the same range of Z, if L is replaced by the length of the positive pressure zone, and the pressure by a *uniform* pressure equal to the mean positive pressure, then the maximum variation of lateral pressure coefficient compared with the uniform case is less than 5%; this effectively

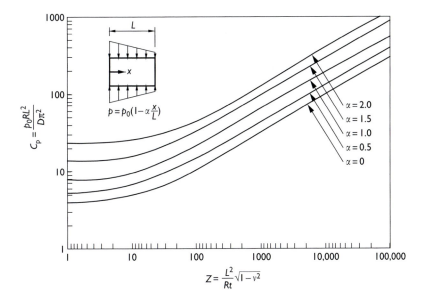

Figure 6.3 Critical buckling coefficient for meridionally linear variation of lateral pressure (Weingarten 1962).

replaces the complex non-uniform problem by the simpler case of uniform pressure buckling.

Theory

Many theoretical analyses in the literature of the buckling of circular cylinders under external pressure have been performed within the framework of the Donnell linear stability theory with simple supports (e.g. Batdorf 1947). It is, however, more accurate to apply Flügge's equations (1973) in coupled form to the buckling problem for uniform or variable pressure. In this more refined theory, the in-plane displacements are not neglected when compared with transverse deflections of the surface. The differences in the two solutions are exacerbated as the length of the cylinder increases. A further advantage of the method is that solutions can be obtained for non-classical edge conditions. The usual assumption of a homogeneous and isotropic material holds, while the Fourier series representation of the displacement functions includes a sufficient number of harmonics for results of high accuracy.

Consider a thin, elastic circular cylindrical shell of length L, radius R and thickness h (Fig. 6.4), compressed by a non-uniform lateral pressure $p(\theta)$ (Figs 6.6 and 6.7). The external load and the stress resultants on a shell element are given in Fig. 6.5 with their positive directions.

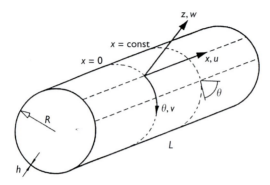

Figure 6.4 Cylinder coordinate system.

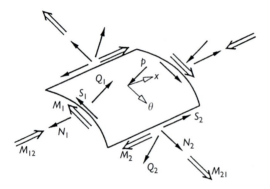

Figure 6.5 Stress resultants.

As a membrane prebuckling state is assumed, the stability problem is reduced to solving Flügge's partial differential equations that relate the small incremental non-dimensional displacements $\bar{u}$, $\bar{v}$, $\bar{w}$ to $\bar{x}$ ($\bar{u} = u/R$, $\bar{v} = v/R$, $\bar{w} = w/R$ and $\bar{x} = x/R$) and θ:

$$L_{11}(\bar{u}) + L_{12}(\bar{v}) + L_{13}(\bar{w}) = 0$$
$$L_{21}(\bar{u}) + L_{22}(\bar{v}) + L_{23}(\bar{w}) = 0 \qquad (1)$$
$$L_{31}(\bar{u}) + L_{32}(\bar{v}) + L_{33}(\bar{w}) = 0$$

where the linear partial differential operators are given in the Appendix (Eqs A1); the stress resultants are also given in Eqs (A2) and (A3).

The longitudinal coordinate is non-dimensionalised as $\bar{x} = x/R$; the incremental displacements are $\bar{u}$, $\bar{v}$ and $\bar{w}$. In a complete cylinder, the incremental displacements must satisfy circumferential periodicity, and are assumed as series:

$$\bar{u}(\bar{x}, \theta) = \sum_{n=0}^{\infty} \bar{U}_n \cos(\lambda\bar{x}) \cos(n\theta)$$

$$\bar{v}(\bar{x}, \theta) = \sum_{n=0}^{\infty} \bar{V}_n \sin(\lambda\bar{x}) \sin(n\theta) \tag{2}$$

$$\bar{w}(\bar{x}, \theta) = \sum_{n=0}^{\infty} \bar{W}_n \sin(\lambda\bar{x}) \cos(n\theta)$$

where n is the circumferential wave number and $\lambda = m\pi/l, l = L/R$. Only one half-sine function is taken meridionally ($m = 1$). Thus, Eqs (1) produce an infinite system of linear homogeneous algebraic equations in $\bar{U}_n$, $\bar{V}_n$ and $\bar{W}_n$ with variable coefficients:

$$\sum_{n=0}^{n=\infty} \cos(n\theta) \left\{ \bar{U}_n \left[-\lambda^2 - n^2 \left[\frac{1-\mu}{2} \left(1 + \frac{h^2}{12R^2}\right) - \frac{p(\theta)R}{D}\frac{h^2}{12} \right] \right] \right.$$

$$+ \frac{1+\mu}{2} n\lambda \bar{V}_n + \lambda \bar{W}_n \left(\mu - \frac{h^2}{12R^2}\frac{1-\mu}{2}n^2 \right.$$

$$\left. \left. + \frac{p(\theta)R}{D}\frac{h^2}{12} + \frac{h^2}{12R^2}\lambda^2 \right) \right\} = 0 \tag{3a}$$

$$\sum_{n=0}^{n=\infty} \sin(n\theta) \left\{ \frac{1+\mu}{2}\lambda n\bar{U}_n + \bar{V}_n \left[-n^2 \left(1 - \frac{p(\theta)R}{D}\frac{h^2}{12}\right) - \frac{1-\mu}{2} \right. \right.$$

$$\left. \times \left(1 + \frac{h^2}{4R^2}\right)\lambda^2 \right] + n\bar{W}_n \left(-1 - \frac{3-\mu}{2}\frac{h^2}{12R^2}\lambda^2 + \frac{p(\theta)Rh^2}{12D} \right) \right\} = 0 \tag{3b}$$

and

$$\sum_{n=0}^{n=\infty} \cos(n\theta) \left\{ \bar{U}_n\lambda \left[-\lambda^2 - \frac{12R^2}{h^2}\mu + \frac{1-\mu}{2}n^2 - \frac{R^3}{D}p(\theta) \right] \right.$$

$$+ \bar{V}_n n \left[\frac{3-\mu}{2}\lambda^2 + \frac{12R^2}{h^2} - \frac{R^3}{D}p(\theta) \right] + \bar{W}_n \left[(\lambda^2 + n^2)^2 \right.$$

$$\left. \left. -2n^2 + 1 + \frac{12R^2}{h^2} - \frac{n^2p(\theta)R^3}{D} \right] \right\} = 0 \tag{3c}$$

Two cases of circumferential variation are considered: hydrostatic, defined in a horizontal cylinder as a linear variation from crown to invert; and partial (patch), where a sinusoidal pressure acts over a limited part of the circumference. In all these cases, since there is circumferential variation in pressure, all harmonics are coupled, and the algebraic equations have variable coefficients, whereas in the case of uniform pressure (Vodenitcharova and Ansourian 1994, 1996), the harmonics are uncoupled and the coefficients constant. Thus, the stability problem is reduced to the determination of the minimum value of pressure at $\theta = 0$ that results in a non-trivial solution of the infinite algebraic homogeneous system (Eqs 3).

Hydrostatic pressure on horizontal cylinder

In this configuration, the cylinder is horizontal and the fluid pressure linearly distributed with height (Fig. 6.3); the skew is expressed by the skew coefficient or pressure factor ρ as

$$\rho = \frac{p_1}{p_0 + p_1} \tag{4}$$

where p_0 is the pressure at mid-height of the cylinder. The pressure is linearly distributed with height and is given by

$$p = p_0 + \frac{\rho}{1-\rho} p_0 \cos\theta \tag{5}$$

$\rho = 0$ corresponds to uniform pressure; $\rho = 0.5$ to zero pressure at the top and $\rho = 1$ to $p = -p_1$ at the top and $+p_1$ at the bottom (Fig. 6.6). The hydrostatic pressure can also be expressed as

$$p = \sum_{r=0}^{r=1} a_r \cos(r\theta) \quad \text{where } a_0 = p_0, \quad a_1 = \frac{\rho}{1-\rho} p_0 \tag{6}$$

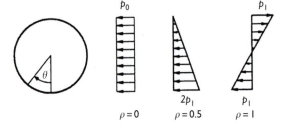

Figure 6.6 Fluid pressure distribution and skew factor ρ.

Partial or patch pressure

The partial pressure, normal to the surface and linearly distributed with height above the invert of the cylinder (Fig. 6.7), is expressed in terms of the maximum pressure p_{st} as

$$p(\theta) = \frac{\cos\theta - \cos\alpha}{1 - \cos\alpha} p_{st}, \quad \theta \in (0, \pm\alpha) \tag{7}$$

Since $p(\theta)$ must be a continuous symmetric function of θ, it is expanded as an even series:

$$p(\theta) = \frac{a_0}{2} + \sum_{k=1}^{\infty} a_k \cos(k\theta) \tag{8}$$

The coefficients in Eq. (8) are:

$$a_0 = \frac{2}{\pi} \int_0^{\pi} p(\theta)\, d\theta = \frac{2p_{st}}{\pi(1 - \cos\alpha)} (\sin\alpha - \alpha\cos\alpha) \tag{9}$$

$$a_1 = \frac{p_{st}}{\pi(1 - \cos\alpha)} \left[-\frac{1}{2}\sin 2\alpha + \alpha \right], \quad \lim_{k \to 1} \frac{\sin(k-1)\alpha}{k-1} = \alpha \tag{10}$$

$$a_k = \frac{2}{\pi} \int_0^{\alpha} p(\theta)\cos(k\theta)\, d\theta = \frac{2p_{st}}{\pi(1 - \cos\alpha)}$$
$$\times \left\{ -\frac{1}{k}\cos\alpha\sin(k\alpha) + \frac{1}{2}\left[\frac{1}{k+1}\sin(k+1)\alpha + \frac{\sin(k-1)\alpha}{k-1} \right] \right\} \tag{11}$$

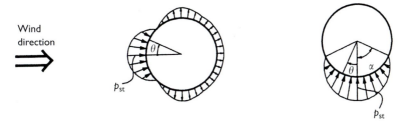

Wind direction

Figure 6.7 Wind and partial (patch) pressure distribution.

Finally

$$p(\theta) = \frac{p_{st}}{\pi(1 - \cos\alpha)}(\sin\alpha - \alpha\cos\alpha)$$

$$+ \frac{p_{st}}{\pi(1 - \cos\alpha)}(-0.5\sin 2\alpha + \alpha)\cos\theta + \frac{2p_{st}}{\pi(1 - \cos\alpha)}$$

$$\times \sum_{k=2}^{k=\infty}\left\{-\frac{1}{k}\cos\alpha\sin k\alpha + 0.5\left[\frac{1}{k+1}\sin(k+1)\alpha\right.\right.$$

$$\left.\left. + \frac{1}{k-1}\sin(k-1)\alpha\right]\cos k\theta\right\} \tag{12}$$

where α is the half subtended angle of the patch loading.

Approximate solution of Eq. (3)

Galerkin's method is employed to orthogonalise the error made with the introduction of the finite series Eq. (2) in Eq. (1). If the left part of the first equation in Eqs (3) is denoted by $X1$, of the second by $X2$ and of the third by $X3$, then:

$$\int_0^{2\pi} X1\cos(q\theta)\,d\theta = 0, \quad \int_0^{2\pi} X2\sin(q\theta)\,d\theta = 0, \quad \text{and}$$

$$\int_0^{2\pi} X3\cos(q\theta)\,d\theta = 0 \tag{13}$$

for $q = 0, 1, 2, \ldots, \infty$, leading to the following infinite algebraic system:

$$\sum_{n=0}^{n=\infty}\bar{U}_n\left\{\left[-\lambda^2 - n^2\frac{1-\mu}{2}\left(1 + \frac{h^2}{12R^2}\right)\right]J_1 + \frac{R^3}{D}\frac{h^2}{12R^2}n^2\sum_{r=0}^{r=\infty}a_r J_2\right\}$$

$$+ \frac{1+\mu}{2}n\lambda\bar{V}_n J_1 + \lambda\bar{W}_n\left\{\left(\mu - \frac{h^2}{12R^2}\frac{1-\mu}{2}n^2 + \frac{h^2}{12R^2}\lambda^2\right)J_1\right.$$

$$\left. + \frac{h^2}{12R^2}\frac{R^3}{D}\lambda\sum_{r=0}^{r=\infty}a_r J_2\right\} = 0 \tag{14}$$

$$\sum_{n=0}^{n=\infty} \frac{1+\mu}{2} \lambda n \bar{U}_n J_1 + \bar{V}_n \left\{ \left[-n^2 - \frac{1-\mu}{2} \left(1 + \frac{h^2}{4R^2} \right) \lambda^2 \right] J_1 \right.$$

$$+ \left(n^2 \frac{R^3}{D} \frac{h^2}{12R^2} \right) \sum_{r=0}^{r=\infty} a_r J_2 \right\} + n\bar{W}_n \left\{ \left(-1 - \frac{3-\mu}{2} \frac{h^2}{12R^2} \lambda^2 \right) J_1 \right.$$

$$\left. + n \frac{R^3}{D} \frac{h^2}{12R^2} \sum_{r=0}^{r=\infty} J_2 \right\} = 0 \tag{15}$$

$$\sum_{n=0}^{n=\infty} \bar{U}_n \lambda \left\{ \left(-\lambda^2 - \frac{12R^2}{h^2} \mu + \frac{1-\mu}{2} n^2 \right) J_1 - \frac{R^3}{D} \sum_{r=0}^{r=\infty} a_r J_2 \right\}$$

$$+ \bar{V}_n n \left\{ \left(\frac{3-\mu}{2} \lambda^2 + \frac{12R^2}{h^2} \right) J_1 - \frac{R^3}{D} \sum_{r=0}^{r=\infty} a_r J_2 \right\}$$

$$+ \bar{W}_n \left\{ \left[(\lambda^2 + n^2)^2 - 2n^2 + 1 + \frac{12R^2}{h^2} \right] J_1 - \frac{n^2 R^3}{D} \sum_{r=0}^{r=\infty} a_r J_2 \right\} = 0 \tag{16}$$

where

$$J_1 = \int_0^{2\pi} \cos(n\theta) \cos(q\theta) \, d\theta, \qquad J_2 = \int_0^{2\pi} \cos(n\theta) \cos(q\theta) \cos(r\theta) \, d\theta \tag{17}$$

The same equations may also be independently derived from the product of the trigonometric functions in Eqs (3) after substitution of the series for $p(\theta)$. Neglect of the small terms containing the pressure in Eqs (14) and (15) results in a large saving in computation time with almost no loss of accuracy, as these terms are multiplied by the small factor $h^2/12R^2$; the present solution includes these small terms. The linear system is

$$[A]\{\bar{U} \, \bar{V} \, \bar{W}\}^t = 0$$
$$\{\bar{U}\} = \{\overline{U_1} \, \overline{U_2} \dots \overline{U_N}\}^t \tag{18}$$
$$\{\bar{V}\} = \{\overline{V_1} \, \overline{V_2} \dots \overline{V_N}\}^t$$

and the stability determinant is of order $3N$ by $3N$. To decrease its order, we express the sets of coefficients $\bar{U}_n$ and $\bar{V}_n$ in terms of $\bar{W}_n$:

$$\left[\begin{array}{c|c} M_1 & M_2 \\ \hline M_3 & M_3 \end{array} \right] \left\{ \begin{array}{c} \bar{U} \\ \bar{V} \end{array} \right\} = - \left[\begin{array}{c} B_1 \\ B_2 \end{array} \right] \{\bar{W}\} \tag{19}$$

Eq. (16) becomes:

$$[R_1 \quad R_2 \quad R_3]\{\bar{U} \mid \bar{V} \mid \bar{W}\} = 0 \tag{20}$$

Then, the $2N$ calculated coefficients $\bar{U}_n$ and $\bar{V}_n$ are substituted into Eq. (20):

$$[Q]\{\bar{W}\} = 0$$
$$[Q] = [-R_1(\bar{M}_1 B_1 + \bar{M}_2 B_2) - R_2(\bar{M}_3 B_1 + \bar{M}_4 B_2) + R_3] \tag{21}$$

where

$$\begin{bmatrix} \bar{M}_1 & \bar{M}_2 \\ \bar{M}_3 & \bar{M}_4 \end{bmatrix} = \begin{bmatrix} M_1 & M_2 \\ M_3 & M_4 \end{bmatrix}^{-1} \tag{22}$$

The stability equation becomes

$$\det[Q] = 0 \tag{23}$$

Increasing the value of p_{st} in Eq. (23), starting at zero, a value is reached that makes det[Q] vanish and this is the critical stagnation pressure, denoted by p_{st} in Tables 6.1 and 6.2 and Figs 6.8 and 6.9. The iterative process converges to a highly accurate value of p_{cr} by interval halving. To obtain a good representation of $p(\theta)$ in the case of partial loading, the Fourier series was truncated at $N = 25(\alpha = 90°)$ and $N = 100(\alpha = 60°, 36.7°$ and $18.5°)$; this made the discretised function effectively indistinguishable from the theoretical. For an accurate (converged) critical pressure p_{st}, the number of linear equations and therefore the order of the stability determinant was taken as $N = 33$ in the cases of hydrostatic pressure, $N = 29$ for partial load with $\alpha = 90°$, and $N = 40$ with $\alpha = 60°, 36.5°$ and $18.5°$. With hydrostatic pressure, the pressure increase in the iterative process was effected by factoring the fluid density, keeping the height of fluid constant.

Numerical results

Cylinder buckling pressures are given for the geometric range $L/R = \pi/4 - 4\pi$, $R/h = 100$–1000, and Poisson's ratio $\mu = 0.3$; p_{st} was non-dimensionalised by division by Young's modulus, and is listed in Tables 6.1 and 6.2. Further details may be found in Vodenitcharora and Ansourcan (1998).

Hydrostatic pressure

Values of $(p_{st}/p_{cr} - 1)$ in the case of hydrostatic pressure are plotted against R/h in Fig. 6.8 for $L/R = \pi$ and $\rho = 0.1, 0.3$ and 0.55; more complete results are given in Table 6.1, which also contains buckling pressures for the case of uniform pressure, for comparison pressure (Vodenitcharova and Ansourian 1994, 1996). A specific

Table 6.1 Uniform pressure (p_{cr}) and hydrostatic pressure at buckling (p_{st}) (non-dimensionalised by Young's modulus, four significant digits)

L/R	R/h	Uniform			Hydrostatic					
		p_{cr}/E $\times 10^{-9}$	n_{cr}	ρ	p_{st}/E $\times 10^{-9}$	ρ	p_{st}/E $\times 10^{-9}$	ρ	p_{st}/E $\times 10^{-9}$
$\pi/4$	100	13,960	9	0.1	14,540	0.3	15,070	0.5	15,440
	200	2333	11		2425		2502		2557
	300	832.0	13		857.4		882.5		900.3
	400	399.7	13		411.2		422.6		430.5
	500	225.7	14		232.9		239.0		243.3
	600	141.8	15		146.5		150.2		152.8
	700	96.29	16		99.01		101.4		103.1
	800	68.47	16		70.56		72.23		73.39
	900	50.93	17		52.34		53.54		54.39
	1000	38.94	17		40.09		40.97		41.60
π	100	3001	4	0.1	3258	0.3	3565	0.5	3772
	200	528.2	4		568.4		610.1		639.6
	300	195.2	5		205.0		217.7		227.3
	400	92.4	5		98.84		105.0		109.2
	500	53.31	5		56.4		59.61		61.91
	600	33.97	5		35.69		37.57		38.96
	700	22.73	6		24.16		25.43		26.33
	800	16.33	6		17.26		18.14		18.76
	900	12.34	6		12.84		13.47		13.91
	1000	9.37	6		9.86		10.31		10.66
2π	100	1577.0	4	0.1	1695	0.3	1854	0.5	1995
	200	256.4	4		279.6		312.4		336.5
	300	96.71	5		103.9		112.6		119.5
	400	45.27	5		49.21		54.07		57.36
	500	26.11	5		28.20		30.67		32.44
	600	17.20	5		18.06		19.34		20.39
	700	11.43	6		12.23		13.10		13.77
	800	8.04	6		8.67		9.34		9.80
	900	5.99	6		6.43		6.93		7.26
	1000	4.61	6		4.94		5.30		5.56
4π	100	803.4	4	0.1	881.5	0.3	1011	0.5	1114
	200	122.7	4		135.3		159.7		180.5
	300	47.39	5		51.69		58.20		63.71
	400	24.17	5		26.07		28.60		30.79
	500	13.10	5		14.33		16.14		17.43
	600	8.10	5		8.87		10.07		10.90
	700	5.49	6		6.00		6.77		7.33
	800	3.97	6		4.33		4.83		5.21
	900	3.03	6		3.27		3.60		3.86
	1000	2.40	6		2.54		2.76		2.94

Table 6.2 Critical pressure (p_{st}) for partial (patch) pressure (non-dimensionalised by Young's modulus)

L/R	R/h	Wind pressure, $p_{st}/E \times 10^{-9}$	Partial (patch) loading, $p_{st}/E \times 10^{-9}$			
			$\alpha = 90°$	$\alpha = 60°$	$\alpha = 36.7°$	$\alpha = 18.5°$
$\pi/4$	100	19,110	16,150	17,220	19,510	26,170
	200	3081				
	300	1069				
	400	505.7				
	500	283.6	251.4	263.3	288.9	357.7
	600	176.9				
	700	118.8				
	800	84.2				
	900	62.2				
	1000	47.4	42.8	44.5	48.2	58.6
π	100	6098	4167	4743	6036	8098
	200	962.7	697.0			
	300	329.5	245.8			
	400	154.4	117.5			
	500	85.93	66.34	72.9	86.33	126.9
	600	53.28	41.61			
	700	35.60	28.07			
	800	25.11	19.96			
	900	18.47	14.77			
	1000	14.04	11.30	12.27	14.31	20.23
2π	100	3953	2288	2706	3628	4288
	200	601.7				
	300	202.4				
	400	93.79				
	500	51.73	35.77	40.64	51.09	70.90
	600	31.86				
	700	21.17				
	800	14.86				
	900	10.86				
	1000	8.24	6.04	6.80	8.19	11.99
4π	100	2826	1314	1660	1989	2393
	200	410.7				
	300	134.9				
	400	61.34				
	500	33.37	19.91	23.33	31.37	37.66
	600	20.34				
	700	13.41				
	800	9.36				
	900	6.81				
	1000	5.13	3.31	3.83	5.06	6.44

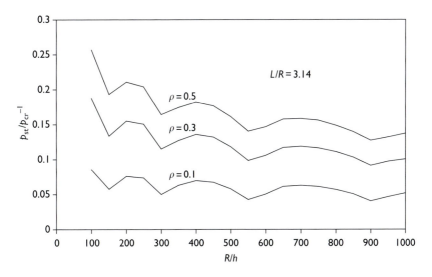

Figure 6.8 Variation of hydrostatic stagnation pressure $(p_{st}/p_{cr}-1)$ versus $R/h(L/R=\pi)$.

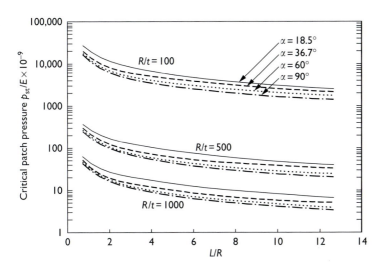

Figure 6.9 Buckling under partial (patch) loading.

solution may be compared with Almroth (1962), whose study is based on Donnell's equations and prebuckling deformations: in the case $L/R=2\pi$, $R/h=400$ and $\rho=0.5$, the value of $(10^5 R/h)p_{st}/E=2.25318$ is given, to be compared with the present value of 2.294.

A study of Table 6.1 shows that for the case $\rho = 0.5$ corresponding to the maximum skew in the table, the greater percentage rise in buckling pressure calculated relative to the uniform buckling pressure occurs in the longer and thicker cylinders. For the case $L/R = 2\pi$, the rise is 26.5% at $R/h = 100$, and 20.6% at $R/h = 1000$. The rise is only 6.8% at $L/R = \pi/4$, $R/h = 1000$, and naturally decreases with reducing skew.

Partial loading

Table 6.2 contains results for partial loading ($\alpha = 18.5$–$90°$, Fig. 6.7). Figure 6.9 gives a selection of graphs for this load case. With partial loading, the subtended angle of $\alpha = \pm36.7°$ corresponds approximately to a positive wind pressure zone. On the other hand, $\alpha = \pm60°$ is closer to the extent of positive wind pressure on a closely spaced group of cylindrical tanks or silos, for which the pressure distribution is not unlike that on a block-like envelope of the group; a consequent substantial fall in buckling strength occurs, by an average 18% and a maximum 26%.

Experimental behaviour

Considerable experimental effort has been exerted around the world to gain a clear understanding of the response of cylinders under external pressure. In this section, thin-walled cylinders are considered in the slenderness regime close to $R/h = 1000$, common in silos and tanks. It is clear that the response of these cylinders depends strongly on the restraint conditions at the two ends of the cylinder. These restraints are radial, tangential, axial and rotational. A detailed study of this effect is given in Vodenitcharova and Ansourian (1994, 1996) and Ansourian *et al.* (1995) in the context of the Flügge formulation, and includes numerical data for many combinations of restraint.

It is clear that the buckling strength of thin cylinders under uniform or non-uniform pressure is very low when the two ends are free (wave number $n = 2$), and reaches a maximum when fully clamped at both ends; while a wide range of strengths is encompassed within the two extremes, the most significant rise occurs in a 'free' cylinder with the addition of radial restraint at each end; for example, with $L/R = 1$, $R/h = 1000$, an 82-fold increase in buckling pressure occurs, with n rising from 2 to 15. The addition of tangential restraint has an almost negligible effect ($<1\%$ except in very short and thick cylinders where the increase is 4%).

Of particular significance is the case of non-symmetric boundary conditions in which one end of the cylinder has radial and axial restraint, while the other end is entirely free. This is typically the case of a tank bolted to the foundations but totally free at the top. Here, the buckling strength does not reach the 'full' restraint value, but is nevertheless high with a 48-fold increase over the 'free' condition. The presence of meridional restraint at the foundation is however crucial, as a significant loss of axial restraint stiffness at the base would almost eliminate resistance of the shell to external pressure. At least one major silo failure is attributed to such a loss of stiffness by failure of the anchorage of the vertical stiffeners into the foundation.

The addition of axial restraint to one end of a cylinder with classical restraint, causes a very significant further rise in buckling strength (Blackler and Ansourian 1986a; Ansourian *et al.* 1995) by an average of 20% in practical cases; when applied at both ends, the rise becomes about 40%. A detailed study of this effect is given in Vodenitcharova and Ansourian (1994, 1996). It is, however, difficult to provide effective axial restraint in silo and tank structures, except when the lower end is firmly bolted to foundations, or a stiff 'skirt' is provided. The required stiffness of such a skirt may be deduced from long wavelength bending theory (Blackler and Ansourian 1986b).

Early experiments on cylinders and curved plates (e.g. Windenburg and Trilling 1934; Sturm 1941), led Gerard and Baker (1957) to the conclusion that small imperfections had a negligible effect on buckling strength under lateral pressure. It is, however, likely that the strength gain due to additional restraint (e.g. meridional) over and above the 'classical' restraints assumed in calculations may have masked the significant imperfection sensitivity that is now generally accepted. From experimental studies of full-scale tanks by Blackler and Ansourian (1986a), from model specimens of Ansourian *et al.* (1995), tests by Thielemann and Esslinger (1967) and by many others, it is now clear that the effect of initial imperfections is significant to the onset of buckling, whether the external pressure is uniform or not, and especially in relation to the premature appearance of one or two initial buckles prior to the full development of the buckling pattern. In this context, Schweizhof and Ramm (1985) also concluded that imperfection sensitivity occurred in cylinders of medium length, causing a reduction of up to 50–60%, under uniform or non-uniform pressure.

In the case of slender shells, the experimentally observed response to increasing external pressure may be described as prebuckling initial buckling, general buckling, postbuckling and collapse. In the prebuckling phase, displacements remain small although some non-linearity may exist due to amplification of initial deformations; there are no plastic deformations and recovery upon pressure release is complete. At initial buckling, one or more buckles form in the area of greatest imperfection, accompanied by large deflections and some plasticity; a deep rumble caused by vibration of the wall may be heard in the larger specimens. Buckling is of the snap-through type and gives no warning; upon release and reload, the same buckles will reappear at slightly reduced pressure. As pressure rises, the global buckling pattern develops, normally with a lesser number of lobes than predicted by perfect cylinder bifurcation theory, for example, 12 lobes versus 15; this effect is caused by imperfections that govern the formation of the initial and final buckling pattern, ensuring that several buckles have longer wavelength than the 'perfect' prediction.

With maximum imperfections in the range $5t$–$10t$, initial buckling is observed in the range $0.4p_{cr}$–$0.7p_{cr}$, where p_{cr} is the bifurcation pressure of the perfect shell, computed taking account of the actual edge conditions, while the general buckling

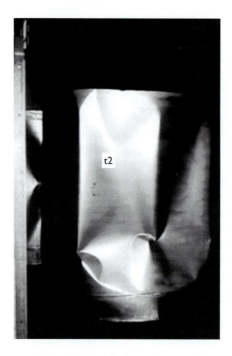

Figure 6.10 Ultimate state of cylinder under external pressure [showing torsional failure of 'ridge' ($D = 1$ m, $R/t = 910$, $L/R = 2$)].

pattern occurs later in the range $0.6p_{cr}$–$0.9p_{cr}$. The observed wave number is invariably smaller than perfect shell predictions.

Formation of the general buckling pattern does not normally announce imminent collapse. The postbuckling regime in which the buckling pattern is strongly amplified and pronounced plasticity develops ends by a local plastic torsional failure of the ridge or V-shaped 'curved beam' bent in the meridional direction, at the nodal line of the circumferential buckles (Fig. 6.10, $D = 1$ m, $R/t = 910$, $L/R = 2$). Unless fractures of welds or other connections occur in the neighbourhood of the ridge under the high distortions caused by buckling, then the postbuckling range becomes significant, and can raise the collapse load of the shell to a maximum of $1.4p_{cr}$ (a minimum of 10% is normally available). However, use of this reserve in design is inadvisable because of the danger of buckles appearing on the shell surface at working loads.

Tests on horizontal cylinders (Ansourian *et al.* 1995; Sengupta 1997) subjected to hydrostatic external pressure by submersion in a water tank and additional internal vacuum have shown that the response of the shell is not unlike that under uniform external pressure in terms of buckling behaviour and ultimate strength. The combination can be expressed in terms of the skew factor ρ (Fig. 6.6). In contrast to

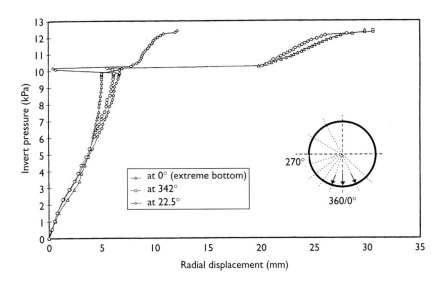

Figure 6.11 Horizontal tank immersion. Load history for non-uniform pressure at midspan ($R/h = 670, L/R = 1.55$).

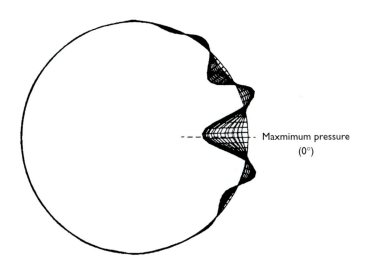

Figure 6.12 Horizontal tank immersion – decaying buckling mode.

uniform pressure testing, a resultant force now acts on the cylinder which therefore must be restrained; the restraint is normally applied at the end supports where the 'classical' boundary restraints are also applied. Further, the resultant force applies global bending to the cylinder and the resulting meridional compressive stresses

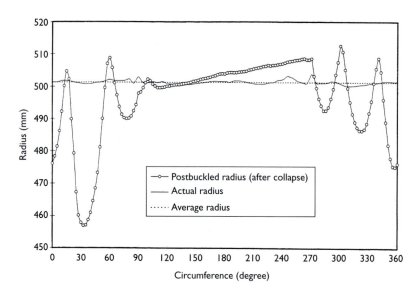

Figure 6.13 Horizontal tank immersion – initial and final geometry at midspan.

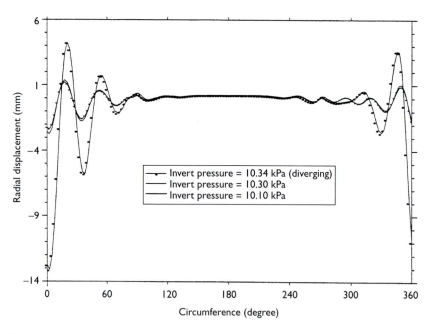

Figure 6.14 Horizontal tank immersion – nonlinear analysis prediction.

may, if of sufficient magnitude, reduce the buckling pressure. Tests were carried out on 1-m diameter cylinders in the slenderness range $R/h = 500–900$ with $L/R = 1.55–2.38$.

A typical response when the meridional stresses are minor is shown in Fig. 6.11. The level of imperfection in the zone of peak buckling deformations is of the order of t. The full buckling pattern is of decaying sinusoidal form (Fig. 6.12) and is developed at 90% of the linear buckling prediction, but the collapse pressure is slightly higher. The wave number is reduced by 1 from the predicted value of 10. The geometry at the midspan section near collapse is shown in Fig. 6.13, while non-linear analysis predictions based on the measured initial imperfections is shown in Fig. 6.14.

Appendix

The linear partial differential operators are:

$$L_{11}(u) = \frac{\partial^2 u}{\partial x^2} + \left[\frac{1-\mu}{2R^2} \left(1 + \frac{h^2}{12R^2} \right) - \frac{p}{DR} \frac{h^2}{12} \right] \frac{\partial^2 u}{\partial \theta^2}$$

$$L_{12}(v) = \frac{1+\mu}{2R} \frac{\partial^2 v}{\partial x \, \partial \theta}$$

$$L_{13}(w) = \frac{\partial w}{\partial x} \left(\frac{\mu}{R} + \frac{p}{D} \frac{h^2}{12} \right) - \frac{h^2}{12R} \frac{\partial^3 w}{\partial x^3} + \frac{1-\mu}{2R} \frac{h^2}{12R^2} \frac{\partial^3 w}{\partial x \, \partial \theta^2}$$

$$L_{21}(u) = \frac{1+\mu}{2R} \frac{\partial^2 u}{\partial x \, \partial \theta}$$

$$L_{22}(v) = \frac{\partial^2 v}{\partial \theta^2} \left(\frac{1}{R^2} - \frac{pR}{D} \frac{h^2}{12R^2} \right) + \frac{1-\mu}{2} \frac{\partial^2 v}{\partial x^2} \left(1 + \frac{h^2}{4R^2} \right) \tag{A1}$$

$$L_{23}(w) = \frac{\partial w}{\partial \theta} \left(\frac{1}{R^2} - \frac{pR}{D} \frac{h^2}{12R^2} \right) - \frac{3-\mu}{2} \frac{h^2}{12R^2} \frac{\partial^3 w}{\partial x^2 \, \partial \theta}$$

$$L_{31}(u) = \frac{1-\mu}{2R^3} \frac{\partial^3 u}{\partial x \, \partial \theta^2} - \frac{1}{R} \frac{\partial^3 u}{\partial x^3} + \frac{\partial u}{\partial x} \left(\frac{12}{h^2 R} \mu + \frac{p}{D} \right)$$

$$L_{32}(v) = -\frac{3-\mu}{2R^2} \frac{\partial^3 v}{\partial x^2 \partial \theta} + \frac{\partial v}{\partial \theta} \left(\frac{12}{h^2 R^2} - \frac{p}{DR} \right)$$

$$L_{33}(w) = \Delta^4 w + \frac{\partial^2 w}{\partial \theta^2} \left(\frac{2}{R^4} + \frac{p}{DR} \right) + \frac{1}{R^2} w \left(\frac{1}{R^2} + \frac{12}{h^2} \right)$$

The stress resultants are assumed as in Flügge (1973):

$$N_1 = \frac{Eh}{(1-\mu^2)} \left(\frac{\partial u}{\partial x} + \frac{\mu}{R} \frac{\partial v}{\partial \theta} + \frac{\mu}{R} w - \frac{h^2}{12R} \frac{\partial^2 w}{\partial x^2} \right)$$

$$N_2 = \frac{Eh}{(1 - \mu^2)} \left[\mu \frac{\partial u}{\partial x} + \frac{1}{R} \frac{\partial v}{\partial \theta} + \frac{1}{R} w \left(1 + \frac{h^2}{12R^2} \right) + \frac{h^2}{12R^3} \frac{\partial^2 w}{\partial \theta^2} \right]$$

$$S_1 = \frac{Eh}{2(1 + \mu)} \left[\frac{1}{R} \frac{\partial u}{\partial \theta} + \frac{\partial v}{\partial x} \left(1 + \frac{h^2}{12R^2} \right) - \frac{h^2}{12R^2} \frac{\partial^2 w}{\partial x \, \partial \theta} \right] \qquad \text{(A2)}$$

$$S_2 = \frac{Eh}{2(1 + \mu)} \left[\frac{1}{R} \frac{\partial u}{\partial \theta} \left(1 + \frac{h^2}{12R^2} \right) + \frac{\partial v}{\partial x} + \frac{h^2}{12R^2} \frac{\partial^2 w}{\partial x \, \partial \theta} \right]$$

and

$$M_1 = D \left(\frac{\partial^2 w}{\partial x^2} + \frac{\mu}{R^2} \frac{\partial^2 w}{\partial \theta^2} - \frac{1}{R} \frac{\partial u}{\partial x} - \frac{\mu}{R^2} \frac{\partial v}{\partial \theta} \right)$$

$$M_2 = D \left(\mu \frac{\partial^2 w}{\partial x^2} + \frac{1}{R^2} \frac{\partial^2 w}{\partial \theta^2} + \frac{1}{R^2} w \right)$$

$$M_{12} = -\frac{D(1 - \mu)}{R} \left(\frac{\partial v}{\partial x} - \frac{\partial^2 w}{\partial x \, \partial \theta} \right)$$

$$M_{21} = -\frac{D(1 - \mu)}{R} \left(-\frac{1}{2R} \frac{\partial u}{\partial \theta} + \frac{1}{2} \frac{\partial v}{\partial x} - \frac{\partial^2 w}{\partial x \, \partial \theta} \right)$$

$$S_{\text{eff}} = S_1 + \frac{1}{R} M_{12} = \frac{Eh}{2(1 + \mu)} \left[\frac{1}{R} \frac{\partial u}{\partial \theta} + \frac{\partial v}{\partial x} \left(1 + \frac{h^2}{4R^2} \right) - \frac{h^2}{4R^2} \frac{\partial^2 w}{\partial x \, \partial \theta} \right]$$

$$Q_{\text{eff}} = \frac{1}{R} \frac{\partial}{\partial \theta} (M_{21} + M_{12}) + \frac{\partial M_{12}}{\partial x}$$

$$= -D \left(\frac{1}{R} \frac{\partial^2 u}{\partial x^2} - \frac{1 - \mu}{2R^3} \frac{\partial^2 u}{\partial \theta^2} + \frac{3 - \mu}{2R^2} \frac{\partial^2 v}{\partial x \, \partial \theta} \right. \qquad \text{(A3)}$$

$$\left. - \frac{\partial^3 w}{\partial x^3} - \frac{2 - \mu}{R^2} \frac{\partial^3 w}{\partial x \, \partial \theta^2} \right)$$

If the terms containing the pressure are omitted from Eqs (16) and (17) but of course not Eq. (18), then a great saving in computation time occurs with almost no loss of accuracy. The coefficients $\bar{U}_n$ and $\bar{V}_n$ can be expressed in terms of $\bar{W}_n$ for each n independently:

$$\bar{U}_n = u_n \bar{W}_n, \qquad \bar{V}_n = v_n \bar{W}_n \qquad \text{(A4)}$$

where

$$u_n = \frac{a12 * a23 - a13 * a22}{a11 * a22 - a12 * a21}$$

$$v_n = \frac{a13 * a21 - a11 * a23}{a11 * a22 - a12 * a21} \qquad \text{(A5)}$$

and

$$a11 = -\lambda^2 - n^2 \frac{1-\mu}{2}\left(1 + \frac{h^2}{12R^2}\right), \quad a12 = \frac{(1+\mu)}{2}n\lambda \qquad (A6)$$

$$a13 = \lambda\left(\mu - \frac{h^2}{12R^2}\frac{(1-\mu)}{2}n^2 + \frac{h^2}{12R^2}\lambda^2\right), \quad a21 = \frac{(1+\mu)}{2}\lambda n \quad (A7)$$

$$a22 = -n^2 - \frac{(1-\mu)}{2}\left(1 + \frac{h^2}{12R^2}\right)\lambda^2,$$

$$a23 = -n\left(1 + \frac{3-\mu}{2}\frac{h^2}{12R^2}\lambda^2\right)$$

$$a31 = \lambda\left(-\lambda^2 - \mu\frac{12R^2}{h^2} + \frac{(1-\mu)}{2}n^2\right) \qquad (A8)$$

$$a32 = n\frac{(3-\mu)}{2}\lambda^2 + \frac{12R^2}{h^2} \qquad (A9)$$

$$a33 = (\lambda^2 + n^2) - 2n^2 + 1 + \frac{12R^2}{h^2} \qquad (A10)$$

Thus, the size of the stability determinant is decreased. Substituting Eqs (A4)–(A10) into Eq. (18):

$$\sum_{n=0}^{n=\infty} \bar{W}_n \left\{ u_n(a31)J_1 - u_n\lambda\frac{R^3}{D}\sum_{r=0}^{\infty}a_r J_2 + v_n(a32)J_1 \right.$$

$$\left. -nv_n\frac{R^3}{D}\sum_{r=0}^{r=\infty}a_r J_2 + a33 J_1 - \frac{n^2 R^3}{D}\sum_{r=0}^{r=\infty}a_r J_2 \right\} = 0 \qquad (A11)$$

for $q = 0, 1, 2, \ldots$

The above infinite algebraic system is of the type:

$$\sum_{n=0}^{\infty} \bar{W}_n A_{\hat{q}n} = 0, \quad \hat{q} = 0, 1, 2, \ldots, \infty$$

$$[A]\{\bar{W}\} = 0 \quad \{\bar{W}\} = \{\bar{W}_1 \bar{W}_2 \ldots \bar{W}_N\}^t \qquad (A12)$$

The stability determinant that gives the buckling stagnation pressure is $\det[A] = 0$. The number N of terms kept in the infinite series (Eq. A12) is chosen to ensure convergence of the result. The order of the stability determinant is N by N.

Notation

a, R	Radius
D	Flexural rigidity, $Eh^3/12(1-\mu^2)$
E	Young's modulus
H, t	Wall thickness
L	Cylinder length
L_{ij}	Partial differential operator
M_1, M_2	Bending moments
M_{12}	Torsional moment
n	Number of harmonics included in the analysis
N_1, N_2	Direct membrane stress resultants
Q_1, Q_2	Transverse shear force
S_1, S_2	Shear membrane stress resultant
$p(\theta)$	External pressure, function of circumferential angle
p_0	External pressure at mid-height of cylinder, uniform pressure component
p_1	Linearly varying pressure
p_{cr}	Critical uniform pressure
p_{st}	Maximum (stagnation) pressure at buckling at $\theta = 0°$
R, a	Radius
t, h	Wall thickness
u	Axial displacement
v	Tangential displacement
w	Radial displacement
x	Axial coordinate
α	Circumferential extent of pressure, or Weingarten (1962) pressure factor
∂, ∇	Partial differential operator
μ	Poisson's ratio
ρ	Pressure factor $= p_1/(p_0 + p_1)$
θ	Angular position

References

Almroth, B.O. (1962). Buckling of a cylindrical shell subjected to non-uniform pressure. *Journal of Applied Mechanics, Transactions of ASME* **84**, 675–682.

Ansourian, P., Showkati, H., Sengupta, M. and Vodenitcharova, T. (1995). The behaviour of cylindrical shells under uniform and non-uniform external pressure. *Fourteenth Australasian Conference on the Mechanics of Structures and Materials*, Hobart, December, pp. 630–636.

Batdorf, S.B. (1947). A simplified method of elastic-stability analysis for thin cylindrical shells. NACA Report No. 874, March, pp. 1–26.

Blackler, M.J. and Ansourian, P. (1986a). Buckling behaviour of a full-scale tank under internal and external pressure. *Civil Engineering Transactions, Institution of Engineers, Australia.* **CE28**(3), 216–221.

Blackler, M.J. and Ansourian, P. (1986b). The influence of elastic end restraints on cylinder stability under wind loading. *Tenth Australasian Conference on Mechanics of Structures and Materials*, August, pp. 241–246.

Flügge, W. (1973). *Stresses in Shells*, 2nd edn. Springer Verlag, Berlin.

Gerard, G. and Baker, H. (1957). Handbook of structural stability, Part III – Buckling of curved plates and shells, NACA TN-3783.

Sengupta, M. (1997). The behaviour of cylindrical shells under external pressure. PhD Thesis, University of Sydney.

Schweizerhof, K. and Ramm, E. (1985). Stability of cylindrical shells under wind loading with particular reference to follower load effects. In *Design of steel bins for the storage of bulk solids* J.M. Rotter (ed.). Univ. Sydney, March, pp. 144–157.

Sturm, R.G. (1941). A study of the collapsing pressure of thin-walled cylinders. Bulletin 329, Eng. Exp. Stat., University of Illinois.

Thielemann, W. and Esslinger, M. (1967). Beul- und Nachbeulverhalten isotroper Zylinder unter Aussendruck. *Stahlbau* **36**(6), 161–175.

Vodenitcharova, T. and Ansourian, P. (1994). Influence of boundary conditions on the bifurcation static instability of circular cylindrical shells subject to uniform lateral pressure. Research Report R701, October, The University of Sydney, Department of Civil Engineering, Center for Advanced Structural Engineering, Sydney, Australia, pp. 1–136.

Vodenitcharova, T. and Ansourian, P. (1996). Buckling of circular cylindrical shells subject to uniform lateral pressure. *Engineering Structures* **18**(8), 604–614.

Vodenitcharova, T. and Ansourian, P. (1998). Hydrostatic, wind and non-uniform lateral pressure solutions for containment vessels. *Thin-Walled Structures* **31**(1–3), 221–236.

Weingarten, V.I. (1962). The buckling of cylindrical shells under longitudinally varying loads. *Journal of Applied Mechanics, ASME* **29**, 81–85.

Windenburg, D.F. and Trilling, C. (1934). Collapse by instability of thin cylindrical shells under external pressure. *Transactions of ASME* **56**, 819–825.

Chapter 7

Tall cylindrical shells under wind pressure

R. Greiner and W. Guggenberger

Introduction

Non-uniformly distributed external surface loadings – as wind pressure in particular – may bring about two additional effects compared to the standard case of buckling under uniform external pressure. The first one of these effects is related to the fact, that only a localized part of the circumference – the so-called stagnation zone – is under circumferential compression while the rest is under suction, so that buckles due to circumferential stresses occur only within a localized width in circumferential direction. The second effect is related to the overall load-carrying behaviour of the structure, which – particularly for cylinders of large length – creates additional membrane forces like axial compression and shear and by that may impair the buckling behaviour of the shell significantly, in a quantitative as well as in a qualitative sense.

The first effect of this non-uniformity was dealt with in Chapter 6. It is related to segments of cylindrical shells or to short or 'stocky' cylindrical shells, which do not produce considerable membrane forces due to the meridional load transfer to the ends. The second effect, in particular, is investigated in this chapter and it is related to the case of vertical cylinders supported at the lower end only. The practical applications are silos, tanks, cylinders or similar structures, which extend vertically like 'cantilevers' and are uniformly supported around the circumference at the lower edge and which are restrained by roofs or end-rings at the upper edge.

Investigated structures

The investigations, both analytically and experimentally, are much more complex than those of uniformly pressurised shells. While the buckling loads of the latter ones may be derived analytically or also numerically by linear buckling analyses, the former ones need nonlinear calculations on a high standard. Consequently, only limited numerical results on wind-loaded cylinders are available up to now. The same holds true for the experimental part, which requires tests in the wind-tunnel. All detailed investigations concern cylinders with constant wall thickness. Stepped walls may then approximately be treated in analogy to the procedure which is common for cylinders under uniform external pressure; see DIN 18800 (1990).

An application range that has already been comprehensively investigated is the range of thin-walled cylinders with moderate length, that is, cylinders of 'stocky' shape, like storage tanks or short silos, etc. For these cases also experimental evidence is available.

The range of long, slender cylinders, like tall silos, has been basically investigated for a specific parameter range only. However, a number of new effects have been revealed, which particularly concern the cases of highly slender cylinders, like chimneys, etc. For these cases, this chapter will give an overview on the state-of-the-art rather than agreed design recommendations.

A comprehensive chapter on the subject of wind-loaded cylinders is given by Greiner in the book *Silos*, edited by Brown and Nielsen (1998). There, further information on wind pressure distributions – supplementary to that given in the chapter above – are provided as well as information on the distribution of the membrane forces in unstiffened or stiffened cylinders. The following contribution deals with the buckling phenomenon alone.

Short and stocky cylinders

The buckling behaviour of thin-walled cylinders of the type of storage tanks was experimentally investigated by Resinger and Greiner (1981, 1982) in wind-tunnel tests. Following the idea of relating the wind buckling pressure, that is, the maximum stagnation pressure q_w, which causes buckling failure, to the buckling load q_u under uniform external pressure, leads to an increase of buckling resistance due to the non-uniformity of the pressure distribution. It may be described by the formula:

$$\frac{q_w}{q_u} = \frac{1}{(0.46 + 0.017 m_{cr})} = \frac{1}{0.46\left(1 + 0.1\sqrt{C_{BC} \cdot (r/l)\sqrt{(r/t)}}\right)} \tag{1}$$

This formula is valid in the range

$$1 \le \frac{q_w}{q_u} \le 1.6 \tag{2}$$

which covers cylinders with circumferential buckling wave numbers (under uniform external pressure) of

$$m_{cr} \ge 10 \tag{3}$$

The above formula (Eq. 1) was derived for an elastic imperfection reduction factor of $\alpha = 0.7$, which was confirmed by an additional series of test results for uniform external pressure.

These results are in good accordance with the analytically derived values for the wind buckling pressure of Chapter 6 by P. Ansourian – as long as the range of validity is considered (Eqs 2 and 3).

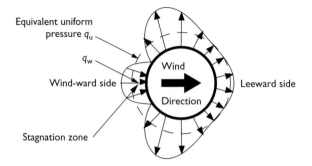

Figure 7.1 Wind pressure distribution.

This behaviour is illustrated in Fig. 7.5, where the ratio q_w/q_u is plotted over the parameter $\rho = m_{cr}$. It shows that in the range of longer or more slender cylinders, that is, $m_{cr} < 10$, wind buckling is influenced by effects that are not covered by the investigations discussed earlier.

In code regulations, as given in DIN 18800–4 (1990) or Eurocode 3-1.6 (1999), the actual wind pressure amplitide q_w is formally replaced by an 'equivalent uniform external pressure' q_u, which is calculated by using the reciprocal value of Eq. (1) as load reduction factor (Fig. 7.1). Therefore, wind pressure may be directly combined with additional uniform negative pressure created by internal suction effects, frequently to be taken into account in tank design. Further on, this concept was also adopted for cylinders with stepped wall thickness by replacing m_{cr} in Eq. (1) by the appropriate value of the shell with stepped wall thickness as specified in code regulations.

Long and slender cylinders

As shown in Figs 2–5, the buckling behaviour of wind-loaded cylinders changes significantly in that range of geometrical parameters, which is related to circumferential buckling wave numbers less than about 10. In this range of geometry the circumferential compression is not the only dominating physical effect, but there are also additional increasing membrane forces due to the overall load-carrying behaviour of cylinders with greater length, that is, greater slenderness. These cause a steep drop of the curve of the buckling resistance with increasing length.

A report on numerical studies of this problem is given by Greiner and Derler (1995). It shows that buckling occurs in the mid-height region of the shell in the stagnation zone, initiated by the axial compression stresses acting there. Realistic results require the use of geometrically non-linear analyses of the imperfect

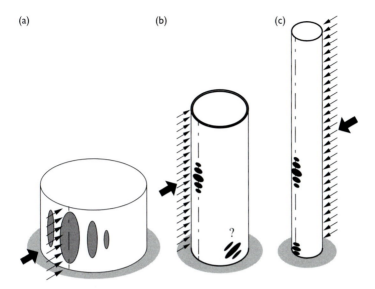

Figure 7.2 (a) Short, (b) long and (c) very long (=slender) wind-loaded cylindrical shell structures.

structure (GNIA). These results indicate that this type of buckling is highly influenced by the flattening (or ovalisation) of the cylindrical cross-section in the middle height of the shell, mainly produced by the effect of the suction forces perpendicular to the wind direction. Figure 7.3 gives a view of different buckling deformations as obtained by geometrically nonlinear analyses (GNA) of the perfect elastic structure. The lowest classical buckling eigenmode is also shown for comparison. This mode does not reflect the features of the nonlinear behaviour due to the lack of consideration of prebuckling deformations. In Fig. 7.4 the related nonlinear load displacement diagrams are plotted. Figure 7.5 shows the comparison between numerical results and the wind-tunnel tests.

Numerical investigations of cylinders with higher slenderness were performed by Schneider and Thiele (1998), who revealed that additional buckling modes may occur, both at the leeward side of the cylinder. One mode is located at the base area and it fails in the shape of the so-called 'elephant-foot' pattern. The other one occurs in the lower half of the shell, mainly initiated by the ovalisation of the shell combined with axial compression due to overall bending.

The windward buckling failure may be accounted for in design by combination of the non-uniform wind pressure with the axial membrane stress including ovalization. Also the buckling failure in the base area may be checked due to given design rules. The leeward buckling in the lower half of the shell, however, requires further studies for deriving design specifications.

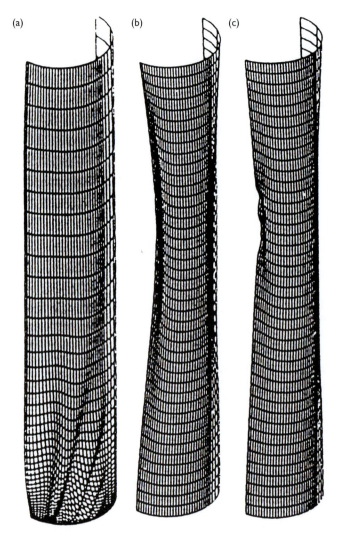

Figure 7.3 Finite element buckling modes of a long cylindrical shell with $R/t = 500$ and $L/R = 10$. (a) First classical buckling eigenmode. (b) GNA load maximum state. (c) GNA postbuckling state.

Ring stiffeners

Existing design formulae

Open top cylinders, in practice, need a restraint of the free edge of the shell by a ring stiffener, called upper end-ring or primary wind girder. Many present codes

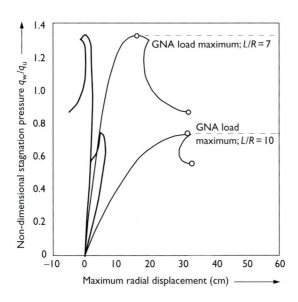

Figure 7.4 Pressure – displacement diagram (R/t=500; L/R=7 and 10; R=600 cm; GNA).

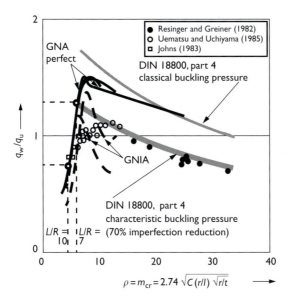

Figure 7.5 Buckling pressures of wind-loaded cylinders; numerical results and test data. (GNA = geometrically nonlinear elastic analysis; GNIA = with imperfections).

204 R. Greiner and W. Guggenberger

provide a design formula, given originally by the API-Standard, which specifies the required section modulus of the ring for in-plane bending by

$$Z = 0.058 \times 10^{-6} D^2 L \tag{4}$$

Another formula was developed by Blackler (1986) and Ansourian (1992), which results from the requirement of minimum stiffness under uniform external pressure on the basis of classical buckling eigenvalue analyses, but was recommended also for wind buckling by these authors:

$$I_R^* = 0.048 t^3 L \tag{5}$$

This formula is based on the assumption of clamped boundary of the shell at the bottom edge, which is – in a rigorous sense – usually not the case with practical tanks. The resulting ring section of this formula is very much smaller than that due to Eq. (4). However, this comparison does not allow clear conclusions, since for the API formula the technical background cannot reliably be identified since the magnitude and distribution of the wind pressure and the boundary conditions at the bottom end are not explicitly defined.

More advanced recent studies

Further studies – both experimentally and numerically – were carried out by Schmidt *et al.* (1998a,b), which made clear that the design of the primary wind girder is closely connected with the underlying concept of the shell design. The authors distinguish between 'design strategy 1', where the shell is designed in the usual way according to Eurocode 3, part 1.6 (1999) or DIN 18800, part 4 (1990), and 'design strategy 2', which makes use of the higher postcritical buckling resistance of very thin-walled shells in the elastic range. In the second case, the upper end-rings are of essential importance providing the statically necessary upper boundary of the locally buckled shell. In the first case, the upper end-ring may be regarded as stiffening ring just raising the buckling resistance onto the required level of safety.

If very thin walled cylindrical shells are designed to carry in the postcritical range (strategy 2), sufficiently strong upper end-rings are necessary to take over the wind load of approximately the upper half portion of the cylindrical shell wall in the deformed buckled configuration. To this end, edge-rings according to Eq. (4) were found sufficient in case of simply supported lower edges (with no axial restraint). For axially fixed lower edges, such rings are of course somewhat on the conservative side, because the boundary condition now provides a higher buckling resistance of the shell.

If the cylindrical shell with axially fixed lower edges is designed in accordance with strategy 1 – thus leading to higher wall thickness than in the previous case – upper end-rings according to the 'bifurcation-optimized' minimum stiffness

I_R^* (Eq. 5) were able to raise the buckling resistance in the tests to the level of cylindrical shells with radially fixed boundaries at both ends, which means that the aim of stiffening has been achieved in this sense. However, test results showed also that this lightly stiffened shell has a tendency to secondary snap-through into a global postbuckling mode similar to that of the unstiffened shell. Since good engineering practice may want to avoid such global buckling modes even in the postbuckling range, it is reccommended to increase the practical ring stiffness to a value of I_R^{**} of about 5–10 times of I_R^* (Eq. 6). This raises no economical problem at all, because even these increased stiffness values in general lead to extremely small ring dimensions.

$$I_R^{**} = 10I_R^* = 0.48t^3L \tag{6}$$

If strategy 1 is applied to cylindrical shells with simply supported lower boundary conditions (i.e. without any axial restraints which constitutes a statically determinate membrane support), this requires upper end-rings which are able to carry the whole external pressure on the upper half of the cylinder, because the isolated unsupported shell, without edge-rings, forms a highly unstable structure. In this latter case, the design according to Eq. (4) may be recommended, since it proved well applicable in the statically analogous case of cylinders designed for strategy 2.

References

Ansourian, P. (1992). On the buckling analysis and design of silos and tanks. *Journal of Constructional Steel Research* **23**, 273–284.

Blackler, M.J. (1986). Stability of silos and tanks under internal and external pressure. PhD Thesis, University of Sydney, October.

Brown, C.J. and Nielsen, J. (1998). *Silos – Fundamentals of Theory, Behaviour and Design.* Chapter 17, E & FN Spon, London, pp. 378–399.

DIN 18800 (1990). Part 4: Stability of Steel Shells.

Eurocode 3 (1999). Part 1.6, ENV 1993-1-6, Supplementary rules for the strength and stability of shell structures, September.

Greiner, R. and Derler, P. (1995). Effect of imperfections on wind-loaded cylindrical shells. *Thin-Walled Structures* **23**, 271–281.

Johns, D.J. (1983). Wind-induced static instability of cylindrical shells. *Journals of Wind Engineering and Industrial Aerodynamics* **13**, 261–270.

Resinger, F. and Greiner, R. (1981). Kreiszylinderschalen unter Winddruck – Anwendung auf die Beulberechnung oberirdischer Tankbauwerke. *Stahlbau* **50**, 65–72.

Resinger, F. and Greiner, R. (1982). Buckling of wind loaded cylindrical shells – application to unstiffened and ring-stiffened tanks. *Proceedings of the State of the Art Colloquium,* University of Stuttgart, Germany, 6–7 May.

Schmidt, H., Binder, B. and Lange, H. (1998a). Postbuckling strength design of open thin walled cylindrical tanks under wind load. *Thin-Walled Structures* **31**, 203–220.

Schmidt, H., Binder, B. and Lange, H. (1998b). Design of thin-walled open top cylindrical tanks under wind load considering the postbuckling load-carrying reserves. *Bauingenieur* **73**(5), 241–246 (in German).

Schneider, W. and Thiele, R. (1998). An unexpected failure mode of slender wind-loaded cylindrical shells. *Stahlbau* **67**, 870–875 (in German).

Uematsu, Y. and Uchiyama, K. (1985). Deflection and buckling behaviour of thin circular cylindrical shells under wind loads. *Journal of Wind Engineering and Industrial Aerodynamics* **18**, 245–261.

Chapter 8

Cylindrical shells under torsional and transverse shear

H. Schmidt and Th.A. Winterstetter

Introduction

Thin-walled cylindrical shells in civil engineering applications are often subjected to membrane shear stresses, mostly in connection with a transverse loading and overall bending. The most widely adopted approach for the design of such structures against loss of structural stability (shell buckling) is to check the maximum values of shear stresses against design values which are deduced from investigations on cylinders loaded by pure torsion. This is one of the three thinkable 'fundamental shell buckling cases' including axial compression, external pressure and torsion, which cover all buckling-relevant membrane stresses in a cylindrical shell. Buckling under torsion therefore was subject to numerous experimental and theoretical investigations.

Cylindrical shells under pure torsion

Theoretical investigations

In a tube with uniform wall thickness, torsion only induces uniform shear stresses, which can be calculated according to Eq. (1) ('Bredt's formula'). These shear stresses are equivalent to inclined principal tensile and compressive stresses (Fig. 8.1). Because of the latter stresses, thin-walled cylinders can suddenly lose structural stability when loaded in torsion.

$$\tau = \frac{M_T}{2\pi r^2 \cdot t} \tag{1}$$

Theoretical torsional buckling loads are much more difficult to calculate analytically than in the case of axial compression or external pressure, because no simple trigonometric shape functions can be used. First approximate solutions were obtained by Schwerin (1925) and Donnell (1933); more accurate solutions can be found, for example, in Timoshenko (1936), Kromm (1942), Batdorf (1947) and Flügge (1973). A detailed overview is given by Yamaki (1984); Fig. 8.2 is taken from his book (for the definition of boundary conditions see 'Notation' at the end of this chapter). According to Fig. 8.2, axially restrained cylinders

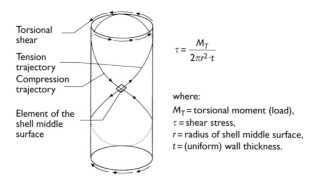

where:

M_T = torsional moment (load),
τ = shear stress,
r = radius of shell middle surface,
t = (uniform) wall thickness.

$$\tau = \frac{M_T}{2\pi r^2 \cdot t}$$

Figure 8.1 Stresses in a circular cylindrical shell subjected to torsion.

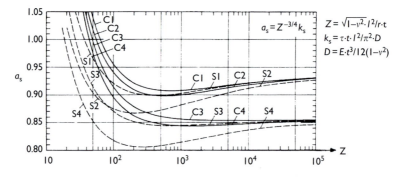

Figure 8.2 Critical torsional buckling stresses τ_{cr} (called τ in the diagram), calculated using different boundary conditions (reproduced from Elastic Stability of Cylindrical Shells, Yamaki, 1984).

(BCs C1, C2, S1, S2) have a slightly higher torsional buckling stress than axially free ones (about 9%).

Cylinders with different length parameters develop different buckling characteristics. Fig. 8.3 shows a re-calculation of the critical buckling stresses τ_{cr}. Three regions can be distinguished: very short cylindrical shells (left-hand side of Figure 8.3, low values of $(\ell/r)(\sqrt{r/t})$) behave like 'plate strips' under shear loading, very long cylinders buckle in only two circumferential helical waves, that is, virtually by helical ovalisation, without influence of boundary conditions.

For design purposes, these linear critical buckling stresses can be expressed approximately by the following formulas (see Fig. 8.3):

• Buckling of a plate strip:

$$\tau_{Rc} = 4.82E\left(\frac{t}{l}\right)^2 \qquad (2)$$

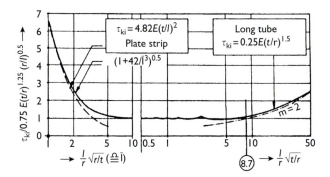

Figure 8.3 Critical torsional buckling stresses τ_{Rc} (called τ_{ki} in the diagram) for 'classical' boundary conditions S3 (Lindner *et al.* 1998; reproduced by kind permission of Beuth Verlag).

- Buckling of a 'medium-length' cylinder:

$$\tau_{Rc} = \frac{0.85\pi^2}{12(1-\nu^2)^{5/8}} E \left(\frac{t}{r}\right)^{5/4} \left(\frac{r}{l}\right)^{1/2} = 0.75E\left(\frac{t}{r}\right)^{5/4}\left(\frac{r}{l}\right)^{1/2} \quad \text{for } \nu = 0.3$$

(3)

The transition to the buckling of a plate strip can be expressed by adding a factor C_τ to Eq. (3), where:

$$C_\tau = \sqrt{1 + 42\left(\frac{r}{l}\right)^3\left(\frac{t}{r}\right)^{1.5}}$$

(4)

- Buckling of a long tubular cylinder with $(l/r) \geq 8.7(r/t)^{0.5}$:

$$\tau_{Rc} = \frac{E}{3 \cdot \sqrt{2}(1-\nu^2)^{3/4}} \left(\frac{t}{r}\right)^{3/2} = 0.25E\left(\frac{t}{r}\right)^{3/2} \quad \text{for } \nu = 0.3.$$

(5)

An even greater challenge than the calculation of the linear elastic buckling eigenvalues was the determination of the complete load-deflection path and the behaviour in the postbuckling range (Hutchinson and Koiter 1970; Budiansky 1974). Approximate solutions for cylinders under torsion were obtained, for example, by Batdorf (1947) and Nash (1957). A detailed overview can be found once again in Yamaki (1984).

Experimental investigations

First research was done mainly on behalf of the aerospace industry to determine experimental elastic buckling loads of thin-walled cylindrical shells and to check

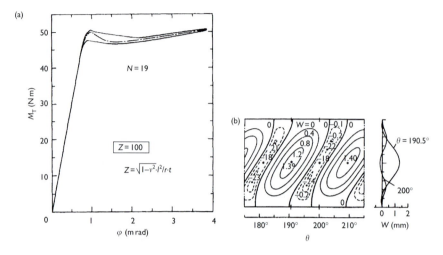

Figure 8.4 Elastic circular cylinder under torsion – experimental results (reproduced from Elastic Stability of Cylindrical Shells, Yamaki, 1984): (a) Load–deflection curve $M_T = f(\varphi)$; (b) buckled shape.

the theoretical results (Donnell 1933). Later, experiments with a special focus on the whole load-deflection path were conducted (Lundquist 1932; Nash 1957; Yamaki 1976). From these and other investigations, the experimental behaviour of thin-walled, elastic cylindrical shells under torsion can be described as follows (Fig. 8.4):

- practically linear prebuckling path;
- sharp, bifurcation-type buckling with sudden, large deflections;
- experimental buckling load 65–95% of the theoretically predicted linear bifurcation load;
- stable, smooth postbuckling path with a minimum of about 60% of the theoretical buckling load;
- geometry-dependent load increase after the postbuckling minimum;
- snap-back into original configuration during unloading.

In contrast to these findings, stocky shells made of materials with nonlinear stress–strain behaviour (steel, aluminium) buckle without sudden deflections, but with permanent plastic deformations of only a part of the shell wall. First experimental research on elastic–plastic torsional buckling was done by Stang *et al.* (1937) and Lee and Ades (1957). Table 8.1 and Fig. 8.5 show some results of a recently conducted test series with specimens having an r/t from 100 to 150 (Winterstetter and Schmidt 1999; Winterstetter 2000).

Table 8.1 Data of torsional buckling tests by Winterstetter and Schmidt (1999)

Spec. no.	r (mm)	l (mm)	t (mm)	r/t	l/r	E (kN/mm^2)	f_y (N/mm^2)	τ_{Ru}^{exp} (N/mm^2)
I/VI	100.1	199.8	1.061	94.3	2.00	201	165	101.0
II/I	100.2	199.5	0.676	148.3	1.99	203	168	115.5
III/I	100.2	399.5	1.068	93.8	3.99	209	161	97.7
IV/I	100.0	400.0	0.678	147.5	4.00	167	167	104.0

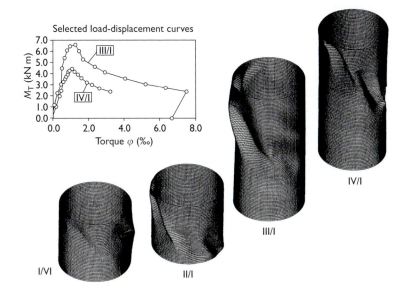

Figure 8.5 Results of elastic–plastic torsional buckling tests (Winterstetter and Schmidt 1999).

Design guidelines

From the very early stage on, there have been design proposals to use the linear critical buckling stresses expressed by Eqs (2)–(5), multiplied by a proper imperfection factor ('knock-down factor'), for very thin walled cylinders, to use the material shear yield strength for very thick walled cylinders and to use an empirical formula for the transition region (e.g. Stang *et al.* 1937; Lee and Ades 1957; Clark and Rolf 1964).

Today, most practical buckling design rules are semi-empirical ones, that is they work with some theoretical input (critical elastic buckling loads) and with slenderness-dependent reduction factors for the effects of imperfections and nonelastic material behaviour. As an example, the formulas of the Eurocode ENV

Table 8.2 Factors α_τ for shear buckling

Quality class	Description	α_τ
Class A	Excellent	0.75
Class B	High	0.65
Class C	Normal	0.50

1993-1-6 (CEN 1998) on stability of steel shell structures for shear buckling are presented hereafter.

- Non-dimensional ('relative') shell slenderness for shear buckling:

$$\lambda_\tau = \sqrt{\frac{f_{y,k}/\sqrt{3}}{\tau_{Rc}}} \quad \text{with } \tau_{Rc} \text{ according to Eqs (2)-(5).} \tag{6}$$

- Elastic imperfection factor for shear buckling according to Table 8.2.
- Squash limit slenderness and plastic limit slenderness for shear buckling:

$$\lambda_{\tau0} = 0.4 \tag{7}$$

$$\lambda_{\tau p} = \sqrt{2.5\alpha_\tau} \tag{8}$$

- Stability reduction factor:

$$\chi_\tau = 1.0 \quad \text{when } \lambda_\tau \leq \lambda_{\tau0} \tag{9a}$$

$$\chi_\tau = 1.0 - 0.6\frac{\lambda_\tau - \lambda_{\tau0}}{\lambda_{\tau p} - \lambda_{\tau0}} \quad \text{when } \lambda_{\tau0} \leq \lambda_\tau \leq \lambda_{\tau p} \tag{9b}$$

$$\chi_\tau = \frac{\alpha_\tau}{\lambda_\tau^2} \quad \text{when } \lambda_{\tau p} \leq \lambda_\tau \tag{9c}$$

- Characteristic shear buckling stress:

$$\tau_{Rk} = \chi_\tau f_{y,k}/\sqrt{3} \tag{10}$$

Figure 8.6 shows the results of these design formulas in comparison with other design standards and with selected experimental results. The scatter typical for shell buckling experiments is clearly visible, but the overall agreement of the test data with the design guidelines as lower-bound curves is quite good.

Cylindrical shells under non-uniformly distributed shear stresses

General

A lot of actions induce membrane shear stresses, which are not uniformly distributed like in the case of pure torsion, but vary in circumferential and

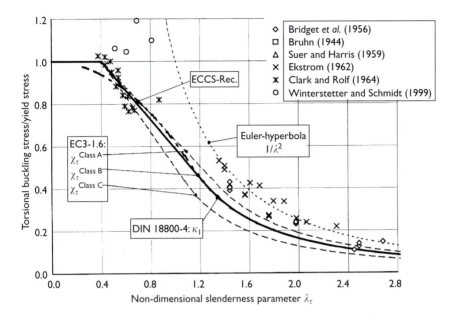

Figure 8.6 Buckling under pure torsion: comparison of design guidelines and experimental results.

axial direction and are connected with overall bending tensile and compressive membrane stresses. As far as long wind loaded shells for tower-type applications are considered, the lever arms and overall bending stresses are too high, and the r/t ratios are too small to make the shear buckling play the predominant role. But there are a lot of civil engineering structures where shear buckling design may be crucial, for example, silos, tanks and containments under seismic actions or even under wind loading, or horizontal pipelines with closely spaced supports.

The distribution of stresses in a tank wall during earthquake loading is very difficult to measure or to calculate. Experimental studies (e.g. Clough *et al.* 1979) and theoretical investigations (e.g. Rotter and Hull 1989; Rammerstorfer *et al.* 1990; Saal and Andelfinger 1995) agree that the lateral accelerations of the bulk solids or the liquids cause excessive shear stresses in the side walls and axial stresses in the upstream and downstream walls. Examples for similar stress distributions due to wind load can be found in Peil and Nölle (1988) and Derler (1993).

These highly nonlinear stress distributions can be covered by means of an idealised configuration. In this context, 'transverse loading and bending' means – according to the majority of theoretical and experimental investigations – a simple, cantilevered cylindrical shell under a transverse load T at the free end inducing a linearly increasing overall bending moment towards the bottom (Fig. 8.7).

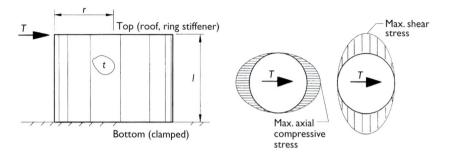

Figure 8.7 Cantilevered cylindrical shell loaded by a transverse edge load at the top, and treated by elementary beam theory.

Furthermore, the transverse load shear stresses are assumed to be sinusoidally distributed over the circumference, for example, introduced by proper ring stiffeners. For such cantilevered cylindrical shells, the stresses calculated by means of beam theory (Eqs 11 and 12) apply if the ratio l/r is greater than about 1. In the case of very short cylinders, there is not enough shell length left for the elementary beam bending stresses to develop; to the contrary, the transverse load induces a tension field-like shear stress concentration on the side walls.

$$\max \tau = \frac{T}{\pi r t} \tag{11}$$

$$\max \sigma_x = \frac{Tl}{\pi r^2 t}. \tag{12}$$

Cylinders with constant wall thickness subjected to a transverse load and bending

Theoretical investigations

Early elastic eigenvalue calculations using perfect shell geometry were presented by Lu (1965) and Schröder (1972); Fig. 8.8 is taken from the latter. The buckling of a cantilevered cylindrical shell under a transverse load with a given r/t ratio is height dependent, with two different and clearly distinguishable modes: Short cylinders develop a shear buckling mode at the sides, while long shells show the typical diamond-shaped, axial compressive bending buckling pattern. In between, there is nearly no transition area.

An interesting question is, how much the circumferentially sinusoidal shear distribution from transverse loading raises the shear buckling stress compared with pure torsion. Figure 8.9 shows the results of calculations from Yamaki (1984). Introducing the critical transverse loads T_{cr} into Eq. (11) and comparing the results

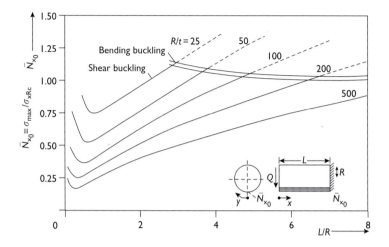

Figure 8.8 Buckling of cantilevered cylindrical shells with a transverse load T (called Q in the diagram) at the free end, according to linear elastic eigenvalue analysis (Schröder 1972; reproduced by kind permission of Wiley-VCH).

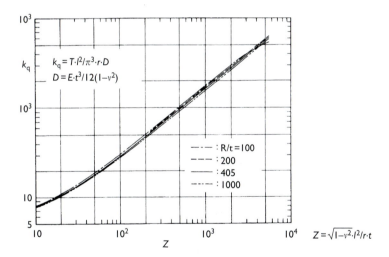

Figure 8.9 Critical shear buckling transverse loads T_{cr} (reproduced from Elastic Stability of Cylindrical Shells, Yamaki, 1984).

with Eq. (3) shows that the transverse load shear buckling stress is nearly the same as the shear buckling stress under pure torsion.

Much research work has been done in Japan and France related to the design of nuclear containments against horizontal seismic actions. Results of elastic–plastic

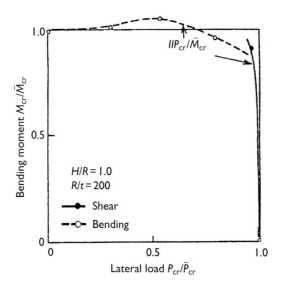

Figure 8.10 Buckling under combined lateral load and bending moment in a cantilevered cylindrical shell (Kokubo *et al.* 1993; reproduced by kind
permission of the Japanese Society of Mechanical Engineering).

buckling analyses using perfect shell geometry (GMNA) are reported by Dostal
et al. (1987); Obaia *et al.* (1992b); Kokubo *et al.* (1993) and Tsukimori (1996).
Figure 8.10, taken from one of these earthquake research papers, shows a characteristic plot of results on bending axial compression and shear buckling using a
parametric nonlinear finite element model with variable height and re-calculating
the membrane compressive and shear stresses according to beam theory from the
numerical transverse buckling loads. As a major result, there is nearly no interaction visible between the axial compressive buckling parts and the shear buckling
parts of the diagram, which agrees very well with the earlier results in Fig. 8.8.

   An important fact is the existence of a significant theoretical shear postbuckling
carrying capacity similar to the results of the tension field theory for the design of
girders with extremely thin-walled webs. Investigations of Canadian researchers
(Roman and Elwi 1988) show that very short, very thin-walled cylindrical shells
can develop such a helix-like tension field, provided that axial compressive buckling on the opposite sides of the shell does not occur firstly. Unfortunately, there
are large cross bending moment effects in the shell wall due to the typical ridge-
and-valley shear postbuckling pattern, which cause a very complicated stress state
in the shell walls and thus much earlier yielding compared to the membrane stress
state. Hence, further research effort is desirable to clarify the potential of postbuckling carrying capacity and proper range of geometries for which this postbuckling
strength reserve can be relied on.

Experimental investigations

Early experiments on the transverse load buckling of thin-walled elastic cylindrical shells were conducted by Lundquist (1935). A comprehensive later experimental study was undertaken by Yamaki *et al.* (1976, 1984). The observed behaviour can be summarised as follows (see also Fig. 8.11):

- buckling occurs suddenly, with deflections of the whole shell wall;
- the experimental buckling loads are ≥65% of the theoretical ones;

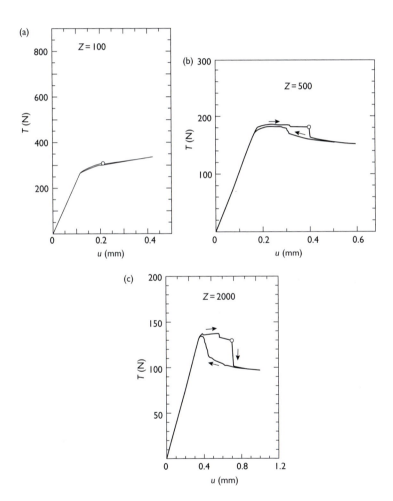

Figure 8.11 Shear buckling tests on thin-walled elastic cantilevered cylindrical shells: transverse load *T* versus transverse edge displacement *u* for (a) a short cylinder, (b) a medium-length cylinder and (c) a long cylinder (reproduced from Elastic Stability of Cylindrical Shells, Yamaki, 1984).

- short shells (low values of Z) develop a characteristic pure shear buckling pattern at the flank sides, while long shells buckle also with an oblique shear pattern at the flank sides, but combined with a diamond-shaped axial compression pattern at the back side;
- short shells show a distinct postbuckling carrying capacity, while long shells have a significant drop of load while buckling, followed by a descending postbuckling load-deflection path.

For steel shells with moderate r/t ratios, it is important to know the effects of nonlinear material behaviour. Tests on metal shells are reported by a wide variety of authors, mostly because of the importance of the design of containments for dangerous goods against seismic buckling (see, e.g. Galletly and Blachut 1985; Dostal *et al.* 1987; Obaia *et al.* 1992a; Kokubo *et al.* 1993). Contrary to the results in the case of elastic buckling, shells in the elastic–plastic region typically develop only partial wall deflections during buckling – either a few shear buckles at the flank sides or an elephant's foot bulge at the back side – and normally no postbuckling strength at all.

Figure 8.12 shows an evaluation of selected buckling experiments on steel shell models under a transverse edge load. The values for the maximum axial compressive and shear stresses at the clamped edge calculated from the measured buckling load are related to the characteristic buckling stresses of ENV 1993-1-6 (Quality Class B assumed). These related stresses are plotted as they would appear in

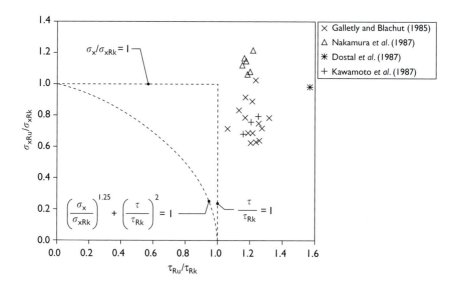

Figure 8.12 Selected transverse shear buckling experiments on metal shells compared to the design predictions of ENV 1993-1-6 (Class B) for general interactive axial compressive and shear buckling.

an interaction diagram, thus following the design rules of ENV 1993-1-6 where peak stress values at different locations over the shell surface are recommendedly treated as if acting at the same location together.

It is clearly visible, that the specific axial compressive and shear stress distributions in the present case – peak values offset by 90° – causes the absence of any interaction phenomenon. The design equation (13) proves too conservative. Instead, two single stress design checks following Eq. (14) could be used. However, these findings apply only in the discussed loading case and can not be extrapolated without further investigations to other cases of combined axial compressive and shear stresses.

- Buckling design equation for interaction of axial compressive and shear stresses according to ENV 1993-1-6 (without partial safety factors):

$$\left(\frac{\sigma_x}{\sigma_{xRk}}\right)^{1.25} + \left(\frac{\tau}{\tau_{Rk}}\right)^2 \leq 1 \tag{13}$$

- Proposed buckling design check for cylindrical shells with properly introduced transverse loads, e.g. via ring stiffeners (without partial safety factors):

$$\frac{\sigma_x}{\sigma_{xRk}} \leq 1, \qquad \frac{\tau}{\tau_{Rk}} \leq 1. \tag{14}$$

Cylinders with stepped wall thickness subjected to a transverse load and bending

Tanks, silos and containments are very often constructed with stepped wall thickness following the increasing hoop stress towards the bottom due to the pressure of the stored content. The buckling of such cylinders due to external pressure – local wind load or internal underpressure – has been investigated by Greiner (1981), and the results of his studies have been used as basis for several design codes. Similarly, cylinders with stepped wall thickness under uniform axial compression have recently been studied (Greiner and Yang 1996), without showing fundamental differences compared to the constant wall thickness case.

For bending and transverse loading, Schneider and Bohm (1998) investigated rather thick-walled tubular cylindrical shells as used for chimneys or wind turbine towers with varying wall thickness under wind load, but only with a special focus on plastic bending buckling modes.

Very recently, a buckling test series on thin-walled steel cylinders with stepped wall thickness as used for containments under the dominant influence of transverse shear membrane stresses has been finished at the University of Essen. These investigations were part of a long-term shell buckling research programme. A detailed description of the pilot test of this series is given by Külkens (2000). In the following, this pilot test is presented in brief.

Pilot test specimen

The specimen was a three-strake cylinder, and it was made of three sheets of mild steel plates having thicknesses of 0.2, 0.3 and 0.4 mm, respectively (Fig. 8.13(a)). The radius was 250 mm, thus leading to r/t ratios of 1250, 833 and 625, respectively. These dimensions are typical for tanks, silos or containments where shear buckling in the elastic region could be expected under transverse loading.

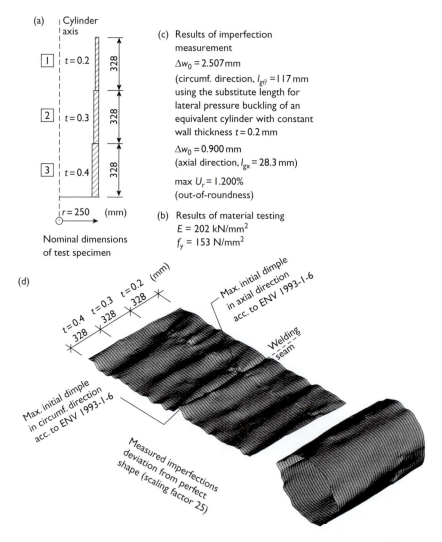

(a) Cylinder axis

1 $t = 0.2$ 328

2 $t = 0.3$ 328

3 $t = 0.4$ 328

$r = 250$ (mm)

Nominal dimensions of test specimen

(c) Results of imperfection measurement

$\Delta w_0 = 2.507$ mm
(circumf. direction, $l_{g\theta} = 117$ mm using the substitute length for lateral pressure buckling of an equivalent cylinder with constant wall thickness $t = 0.2$ mm

$\Delta w_0 = 0.900$ mm
(axial direction, $l_{gx} = 28.3$ mm)

max $U_r = 1.200\%$
(out-of-roundness)

(b) Results of material testing
$E = 202$ kN/mm^2
$f_y = 153$ N/mm^2

(d)

$t = 0.4$ 328 $t = 0.3$ 328 $t = 0.2$ 328   (mm)

Max. initial dimple in axial direction acc. to ENV 1993-1-6

Welding seam

Max. initial dimple in circumf. direction acc. to ENV 1993-1-6

Measured imperfections deviation from perfect shape (scaling factor 25)

Figure 8.13 Test specimen: (a) nominal dimensions, (b) material properties, (c) measured maximum values of shape deviations according to ENV 1993-1-6, (d) deviation from best-fit circles.

Flat sheets were cut to the desired dimensions, then joined to form a plain outside and laser-welded along the later circumferential junctions using a special welding-heat-distributing copper device. Afterwards the circumferential shape was produced by cold-rolling and laser-welding of the longitudinal seam. The shell edges were fixed into annular end ring plates using synthetic resin.

The material properties were evaluated from tension coupons by a specific testing procedure allowing for the determination of the long term 'quasi-static' yield strength (Fig. 8.13(b)). All geometrical dimensions and shape imperfections were measured, and the latter were analysed by Fourier decomposition. The measured maximum deviations according to the specific definitions in ENV 1993-1-6 are given in Fig. 8.13(c), and a plot of the characteristic imperfections is shown in Fig. 8.13(d).

Pilot test setup and procedure

The test setup is shown in Fig. 8.14. It consists of the test specimen including its annular end ring plates, the clamping device, the L-shaped loading frame and the servohydraulic compression jack with a lever arm. The horizontal transverse load was applied in a displacement-controlled manner at the top of the cantilevering test specimen using a tension rod with attached strain gauges for force measurement. The development of the wall deformations and the buckling pattern at different stages of the test was monitored by means of a non-contact measuring technique using an opto-electronic infrared scanner (see, e.g. Schmidt *et al.* 1998). Additionally, the imperfections of the unloaded shell prior to testing and the plastic deformations after testing were measured using a conventional contact measurement system.

The loading was applied stepwise with a very low strain rate, with quasi-static equilibrium states obtained by keeping the displacements constant for at least 10 min on every load step. The cylindrical shell was first loaded until elastic

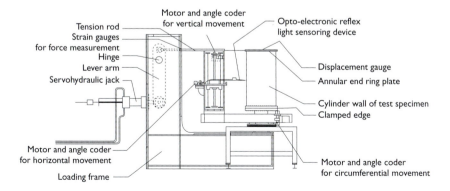

Figure 8.14 Test setup.

buckling occurred, then unloaded to observe the snap-back, and then loaded again far beyond the peak load until the plastic postbuckling pattern could clearly be identified.

Pilot test results

The plot of measured transverse force T versus horizontal overall top deflection u is shown in Fig. 8.15. In the first loading phase, the load could be increased beyond the level of first partial shear buckling at the flank sides within the upper, thinnest strake. It was stopped at the appearance of first axial compression buckles at the back side of the thinnest strake. During unloading, all buckles snapped back, indicating that the deformations had been (nearly) purely elastic. During the second loading phase, the first shear buckles developed earlier, but the total maximum load of the first phase could be reached again, before axial compression buckles caused a sudden load decrease. A stable postbuckling state could be reached at about 70% of the maximum shear force.

Figures 8.16 and 8.17 illustrate the permanent buckling deformations by means of a photograph of the specimen after testing and a plot of the measured deformations. The shear buckles at the flank sides of the cylinder and the axial compression buckles at the back side can clearly be identified, although they have joined when the axial compression buckles snapped in. Only the thinnest strake with $r/t = 1250$

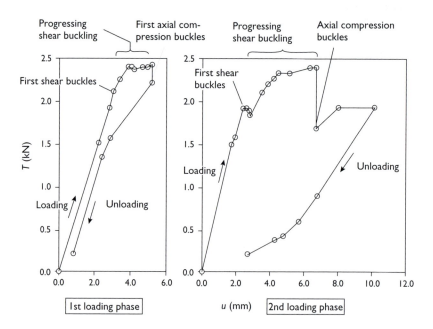

Figure 8.15 Test results: transverse force versus horizontal overall top deflection.

Subscripts

x axial direction
θ circumferential direction
u ultimate limit load value
R resistance value
k characteristic value
c critical value

References

Standards

CEN (1998). ENV 1993-1-6 (EC3-1.6): *Design of Steel Structures, Part 1-6: General Rules: Supplementary Rules for the Strength and Stability of Shell Structures.* CEN, Brussels, Belgium.
DIN (1990). DIN 18800 Teil 4. *Stahlbauten – Stabilitätsfälle, Schalenbeulen.* Deutsches Institut für Normung (DIN), Normenausschuß Bau (Ed.), Berlin, Germany.
ECCS (1988). European Convention for Constructional Steelwork (ECCS): *Buckling of Steel Shells – European Recommendations*, 4th edn. General Secretariat ECCS, Brussels, Belgium.

Papers and monographs

Batdorf, S.B. (1947). A simplified method of elastic stability analysis for thin cylindrical shells. NACA Rep. 874.
Bridget, F.J., Jerome, C.C. and Vosseller, A.B. (1956). Some new experiments on buckling of thin-wall construction. *Transactions of the ASME, Applied Mechanics* 56(6), 569–578.
Bruhn, E.F. (1944). Tests on thin-walled celluloid cylinders to determine the interaction curves under combined bending, torsion and compression or tension loads. NACA TN 951, Washington, DC.
Budiansky, B. (1974). Theory of buckling and post-buckling behaviour of elastic structures. *Advances in Applied Mechanics* 14, 1–65.
Clark, M. and Rolf, R.L. (1964). Design of aluminium tubular members. *Journal of Structural Division, ASCE* 6, 259–289.
Clough, R.W., Niwa, A. and Clough, D. (1979). Experimental seismic study of cylindrical tanks. *Journal of Structural Division, ASCE* 12, 2565–2590.
Derler, P. (1993). Zum Tragverhalten zylindrischer Behälter unter Windlast. PhD Thesis, TU Graz, Austria.
Donnell, L.H. (1933). Stability of thin-walled tubes under torsion. NACA Rep. 479.
Dostal, M., Austin, N., Combescure, A., Peano, A. and Angeloni, P. (1987). Shear buckling of cylindrical vessels benchmark exercise. *Transactions of the 9th International Conference on SMiRT*, Vol. E. Lausanne, France, pp. 199–208.
Ekstrom, R.E. (1962). Buckling of cylindrical shells under combined torsion and hydrostatic pressure. *Experimental Mechanics* 6, 192–197.
Flügge, W. (1973). *Stresses in Shells*, 2nd edn. Springer Verlag, Berlin.

Galletly, G.D. and Blachut, J. (1985). Plastic buckling of short vertical cylindrical shells subjected to horizontal edge shear loads. *Journal of Pressure Vessel Technology* **107**(5), 101–106.

Greiner, R. (1981). Zum Beulnachweis von Zylinderschalen unter Winddruck bei abgestuftem Wanddickenverlauf. *Der Stahlbau* **50**(6), 176–179.

Greiner, R. and Yang, Y. (1996). Effect of imperfections on the buckling strength of cylinders with stepped wall thickness under axial loads. In *Imperfections in Metal Silos Workshop*, Lyon, France, 19 April.

Hutchinson, J.W. and Koiter, W.T. (1970). Postbuckling theory. *Applied Mechanics Reviews* **23**, 1353–1366.

Kawamoto, Y., Yuhara, T., Tashimo, M., Sakurai, A. and Nakamura, H. (1987). Plastic buckling of short cylinders under transverse shearing loads. *Transaction of the 9th International Conference on SMiRT* (Amsterdam), Vol. E, pp. 225–230.

Kokubo, K., Nagashima, H., Takayanagi, M. and Mochizuki A. (1993). Analysis of shear buckling of cylindrical shells. *JSME International Journal of Mechanics and Materials Engineering* **36**(3), 259–266.

Kromm, A. (1942). Die Stabilitätsgrenze der Kreiszylinderschale bei Beanspruchung durch Schub-und Längskräfte. *Jahrbuch 1942 der deutschen Luftfahrtforschung*, pp. 602–616.

Külkens, S. (2000). Vergleichende Untersuchungen zum Beulsicherheitsnachweis kreiszylindrischer Schalentragwerke mit abgestufter Wanddicke unter Querschubbeanspruchung. Diploma Thesis, University of Essen.

Lee, L.H.N., Ades, C.S. (1957). Plastic torsional buckling strength of cylinders including the effects of imperfections. *Journal of Aeronautical Sciences* **24**(4), 241–248, 264.

Lindner, J., Scheer, J. and Schmidt H. (1998). *Beuth-Kommentar Stahlbauten – Erläuterungen zu DIN 18800 Teil 1 bis Teil 4*. Beuth-Verlag, Berlin.

Lu, S.Y. (1965). Buckling of cantilever cylindrical shell with a transverse end load. *AIAA Journal* **3**(12), 2350–2351.

Lundquist, E.E. (1932). Strength test of thin-walled Duralumin cylinders in torsion. NACA TN 427.

Lundquist, E.E. (1935). Strength tests of thin-walled Duralumin cylinders in combined transverse shear and bending. NACA TN 523, Washington, DC.

Nakamura, H., Matsuura, S. and Sakurai, A. (1987). Plastic buckling of short cylinders with axial temperature distribution under transverses shearing loads. *Transactions of the 9th International Conference on SMiRT* (Amsterdam), Vol. E, pp. 219–224.

Nash, W.A. (1957). Buckling of initially imperfect shells subject to torsion. *Journal of Applied Mechanics* **24**, 125–130.

Obaia, K.H., Elwi, A.E. and Kulak, G.L. (1992a). Tests of fabricated steel cylinders subjected to transverse loads. *Journal of Constructional Steel Research* **22**, 21–37.

Obaia, K.H., Elwi, A.E. and Kulak, G.L. (1992b). Ultimate shear strength of large diameter fabricated steel tubes. *Journal of Constructional Steel Research* **22**, 115–132.

Peil, U. and Nölle, H. (1988). Zur Frage der Schalenwirkung bei dünnwandigen, zylindrischen Stahlschornsteinen. *Bauingenieur* **63**, 51–56.

Rammerstorfer, F.G., Scharf, K. and Fisher, F.D. (1990). Storage tanks under earthquake loading. *Applied Mechanics Reviews* **43**, 261–282.

Roman, V.G. and Elwi, A.E. (1988). Postbuckling shear capacity of thin shell tubes. *Journal of Structural Engineering* **114**(11), 2511–2523.

Rotter, J.M. and Hull, T.S. (1989). Wall loads in squat steel silos during earthquakes. *Engineering Structure* **11**(7), 139–147.

Saal, H. and Andelfinger, E. (1995). Seismische Auslegung von verankerten, zylindrischen, oberirdischen Flachbodentankbauwerken aus Stahl. *Stahlbau* **64**(4), 97–103.

Schmidt, H., Binder, B. and Lange, H. (1998). Postbuckling strength of open thin-walled cylindrical tanks under wind load. *Thin-Walled Structures* **31**, 203–220.

Schneider, W. and Bohm, S. (1998). Tragverhalten schlanker, windbelasteter Kreiszylinderschalen mit abgestufter Wanddicke. *Leipzig Annual Civil Engineering Reports* **3**, 375–390.

Schröder, P. (1972). Über die Stabilität der querkraftbelasteten dünnwandigen Kreiszylinderschale. *ZAMM* **52**, 145–148.

Suer, H.S. and Harris L.A. (1959). The stability of thin-walled cylinders under combined torsion and external lateral or hydrostatic pressure. *Transactions of the ASME, Journal of Applied Mechanics* **3**, 138–140.

Schwerin, E. (1925). Die Torsionsstabilität des dünnwandigen Rohres. *ZAMM* **5**, 235–243.

Stang, A.H., Ramberg, W. and Back, G. (1937). Torsion tests of tubes. NACA Rep. 601, Washington DC.

Timoshenko, S.P. (1936). *Theory of Elastic Stability*. McGraw-Hill, New York.

Tsukimori, K. (1996). Analysis of the effect of interaction between shear and bending loads on the buckling strength of cylindrical shells. *Nuclear Engineering and Design* **167**, 23–53.

Winterstetter, Th.A. and Schmidt, H. (1999). Beulversuche an längsnahtgeschweißten stählernen KZS im elastisch-plastischen Bereich unter Axialdruck, Innendruck und Torsionsschub. Research Rep. 82, Dept. of Civil Eng., University of Essen.

Winterstetter, Th.A. (2000). Stabilität von Kreiszylinderschalen aus Stahl unter kombinierter Beanspruchung. Dr.-Ing. Thesis, University of Essen.

Yamaki, N. (1976). Experiments on the postbuckling behaviour of circular cylindrical shells under torsion. In *Buckling of Structures* (ed. B. Budiansky). Springer-Verlag, Berlin, pp. 312–330.

Yamaki, N. (1984). *Elastic Stability of Cylindrical Shells*. North-Holland, Amsterdam.

Chapter 9

Cylindrical shells under global shear loading

G. Michel, J.F. Jullien and J.M. Rotter

Introduction

Tanks must often be built in areas that are susceptible to earthquakes. The most dangerous accident under seismic action is probably the leakage of contained products, rather than buckling. For this reason, the progressive damage caused by repeated buckling under cyclically varying horizontal loading may be more important than a static buckling failure, but the latter is clearly an initial lower bound. For the safe and reliable design of cylindrical storage tanks against earthquakes, it is important to understand and predict the stability of vertical axis cylindrical shells under both static and cyclic horizontal loading.

A horizontal load applied to a cylinder with a vertical axis is conventionally termed a transverse shear load (not to be confused with shear transverse to the thickness of the shell). This transverse shear load or global shear causes a buckling mode characterised by diagonal shear wrinkles (Michel *et al.* 2000b) (Fig. 9.1). However, because it also causes global bending of the cylinder, diamond-shaped

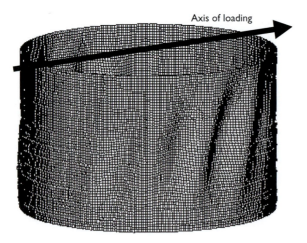

Figure 9.1 Buckling mode under transverse global shear load.

buckles due to axial compression near the lower end of the shell can also occur. At high internal pressures, this diamond shape could be replaced by a local plastic collapse near the base, or 'elephant's foot' buckle (Rotter 1990).

Unlike the simpler load cases for cylinder buckling (axial compression, external pressure and torsion), buckling under global shear cannot be evaluated using a simple treatment involving only one harmonic circumferential mode, because the prebuckling deformations are not axisymmetric. Consequently, practical evaluations of the buckling strength in shear always link shear buckling to the simpler case of buckling under torsion, and the critical shear stress is obtained from torsional buckling formulae. This provides a reasonably accurate prediction only because the individual buckles occupy a relatively small part of the circumference, permitting several buckle wavelengths to develop in a zone where the shear stress is relatively constant.

In the practical design of vessels under global shear, designers must consider possible interactions between the shell and contained liquid, together with the dynamic effects of earthquake loading (frequency, sloshing of fluid, etc.).

In the following, studies are outlined of the effects of geometric imperfections, changes of geometry and post-buckling responses, as well as the effects of coupling simple load cases with other conditions such as internal pressure, axial tension or thermal loading. Many questions still remain concerning the effects of dynamic loading on tanks filled with liquid and on possible detrimental couplings between vibration modes and buckling modes during dynamic oscillatory excitation.

Strength of perfect shells under global shear loading

Static loads

The first buckling tests of thin cylindrical shells under global shear were carried out by Lundquist (1935) on Duralumin cylinders. This study established the close relationship between the shear buckling stress and the torsional buckling stress. An approximate analysis was produced by Lu (1965) and a more precise formulation by Schröder (1972) assuming a membrane prebuckling stress state with simply supported boundaries. A similar analysis to that of Schröder was developed by Yamaki et al. (1980; 1983). In his classic text on the elastic stability of cylindrical shells, Yamaki (1984) described both experimental and theoretical studies of elastic buckling under transverse or global shear loading, together with the cases of coupling shear loadings with internal or external pressure. The comparison between the elastic shear buckling stress in pure torsion and the maximum shear stress at buckling under global shear is shown in Fig. 9.2.

Following the terminology used by the shell buckling Eurocode (ENV 1993-1-6 1999), the dimensionless length of the shell is here characterised by

$$\omega = \frac{L}{R}\sqrt{\frac{R}{t}} = \frac{L}{\sqrt{Rt}} \tag{1}$$

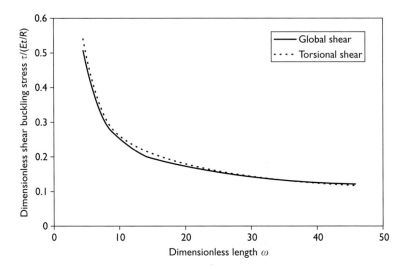

Figure 9.2 Elastic shear buckling stresses for torsional and global shear (reproduced from Elastic Stability of Cylindrical Shells, Yamaki, 1984).

in which L is the length of cylinder between boundaries or stiffening rings, R the radius, t the wall thickness and E is Young's modulus.

For longer cylinders ($10 < \omega < 8.7R/t$), the elastic critical shear stress may be expressed as

$$\tau_{\mathrm{cr,e}} = 0.75\sqrt{\frac{1}{\omega}}\,E\,\frac{t}{R} \tag{2}$$

and the global transverse shear force at elastic buckling is given by

$$Q_{\mathrm{e}} = \pi\,Rt\,\tau_{\mathrm{cr,e}} \tag{3}$$

This equation is identical to that for classical buckling under uniform torsion. A similar but more precise formula, giving a close representation over the range of lengths ($3 < L/\sqrt{Rt} < 70$) was proposed by Timoshenko and Gere (1961)

$$\tau_{\mathrm{cr,e}} = 4.82\frac{\sqrt{1+0.0239\omega^3}}{\omega^2}\frac{Et}{R} \tag{4}$$

The first elastic–plastic buckling study was undertaken by Galletly and Blachut (1985) who proposed an empirical design formula for plastic buckling based on experiments. They used steel shells with a radius/thickness ratio R/t between 125 and 190 and a height/radius ratio L/R between 0.73 and 1.2. The experimental shells were fabricated from steel sheet, and had clamped ends at the base and at the stiff ring where the transverse load was applied.

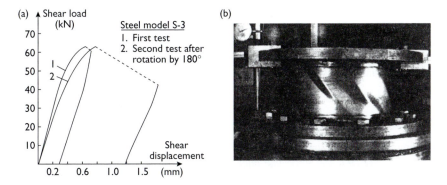

Figure 9.3 Plastic buckling under transverse shear: (a) Load–displacement curves; (b) Shear buckling mode (after Galletly and Blachut 1985; reproduced by kind permission of the ASME).

The load–displacement curves (Fig. 9.3(a)) and buckling modes (Fig. 9.3(b)) are both typical of those obtained in elastic shear buckling tests. Just after the critical load was reached, buckling waves appeared and a small asymmetry can be seen in the postcritical buckling waves where the welding line was orthogonal to the plane of the applied load. After each test a geometrical imperfection with an amplitude equal a least to five times the shell thickness remained. The shells were then turned through an angle of 180° and a second test was performed: the second buckling load was typically around 90% of the first value. Galletly and Blachut deduced from these results that geometrical imperfections play only a small role in plastic buckling under shear loading.

Their design rule for plastic shear buckling is an empirical extension of Yamaki's (1984) formula (Eq. 1), using the quadratic version of the Rankine formula to find the global shear force Q_p at plastic buckling:

$$\frac{1}{Q_p^2} = \frac{1}{Q_e^2} + \frac{1}{Q_y^2}$$

(5a)

or

$$\tau_p = \frac{\tau_e \tau_y}{\sqrt{\tau_e^2 + \tau_y^2}}$$

(5b)

in which the global shear force at yield under shear is given by:

$$Q_y = \pi R t \frac{\sigma_y}{\sqrt{3}}$$

(6)

A comparison between the experimental critical stress ($\tau_{\text{exp}} = Q_{\text{exp}}/\pi\,Rt$) and the plastic buckling value τ_{p} shows that Eq. (5) is always conservative for these tests.

It should be noted that Eqs (3) and (6) do not represent the relationship between global shear force and local stress precisely. These equations are used to transform the force into an apparent stress, and it is the force that is represented accurately in these equations.

Seismic loading

During recent earthquakes, several instances of storage tanks buckling have occurred (Figs 9.4 and 9.5). The buckling modes have two different forms, both occurring in the lower part of the tank: diamond-shaped buckles (Clough and Niwa 1982), and the elephant's foot buckle (Rotter 1990).

The controlling buckling mode is determined by the internal pressure, the tank aspect ratio (height/diameter), the shell slenderness ($R/t = 1650$ or 775) and the mechanical properties of the steel. The controlling mode can be quickly appreciated by comparing the elastic and local plastic buckling strengths for any given configuration (Rotter 1990).

The two buckling modes were also discussed by Fujita *et al.* (1990). Static and dynamic tests of cylindrical tanks loaded by a global transverse load when either filled with water or empty demonstrate the effect of the stored liquid.

Figure 9.4 Buckled storage tank after earthquake (diamond-shaped mode).

Figure 9.5 Buckled storage tank after earthquake (elephant foot mode).

Static loading

Overall shell dimensions: the interaction between shear and bending

A shell that is subjected to a global or transverse shear load also experiences global bending. Where the shell is long, it is clear that bending will dominate, but in very short shells, shear will certainly dominate. The boundary between these two effects, and any interaction between them, must be explored before good design rules can be developed. It is also important to determine whether the buckling will be elastic or plastic.

On the basis of a set of static tests, Matsuura *et al.* (1995) proposed that the form of the buckles could be found from the cylinder geometry alone. The test results are shown in Fig. 9.6, where all shells with an aspect ratio L/R less than 1.5 were found to buckle in shear. In thinner shells, shear buckling could also occur at larger aspect ratios.

Matsuura *et al.* (1995) also compared their shear buckling test results with the formula (Eq. 4) of Timoshenko and Gere (1961) and so validated the elastic shear critical load evaluation (Fig. 9.7).

Geometric imperfections

Under most loading patterns, shell buckling strengths are sensitive to the form and amplitude of geometric imperfections in the shell surface. Buckling under torsion (uniform shear) is often less imperfection-sensitive than axial compression

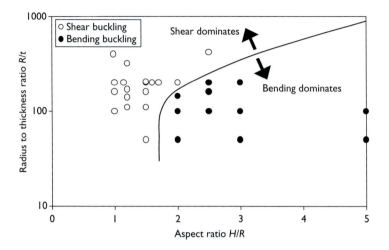

Figure 9.6 Diagram defining dominant buckling mode.

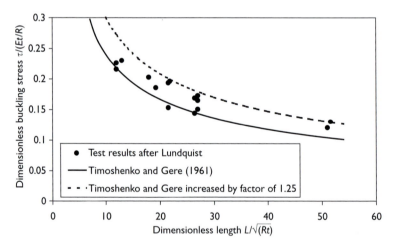

Figure 9.7 Comparison between Lundquist's test results and the Timoshenko design formula.

or external pressure, and buckling under global shear appears to be even less sensitive. Several forms of geometric imperfection have been studied, but all show this reduced sensitivity.

In their experimental study of plastic buckling, Galletly and Blachut (1985) noted that when a buckled shell was turned through an angle of 90° and tested again, the buckling load was only about 10% lower than the first load, even though

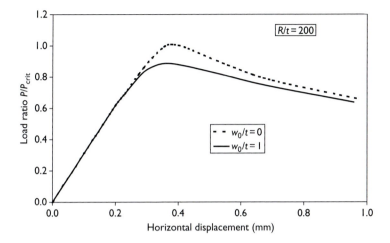

Figure 9.8 Postcritical behaviour, geometric imperfection influence (finite element simulations).

the amplitude of the geometric imperfections was approximately equal to the shell thickness.

A numerical study of global shear buckling using ABAQUS was performed by Kokubo *et al.* (1995) examining several imperfection forms. They concluded that the only important geometric imperfection form was affine to the elastic shear buckling mode. Even in this most pessimistic case, the buckling strength was found to be reduced by only about 20%. For thicker shells ($R/t = 200$), their nonlinear elastic–plastic simulations included geometric imperfections in the form of the perfect elastic shell eigenmode with an amplitude equal to shell thickness. They found that these imperfections did not affect the peak load dramatically, and that the postbuckling load path was not greatly affected by the imperfections (Fig. 9.8).

Using test results from cylindrical shells with R/t of either 100 or 210 (Fig. 9.9), Murakami *et al.* (1989) proposed a strength reduction factor α to account for the equivalent amplitude of the geometric imperfection under shear loading: if this amplitude is less than the thickness, the reduction factor is taken as $\alpha = 1.0$.

$$\alpha = \frac{1.18}{1 + 0.18(w_0/t)^{0.65}} \quad \text{for } w_0/t \geq 1 \tag{7}$$

in which w_0 is the amplitude of the geometric imperfection.

For plastic design, they recommended the same empirical interaction formula that had been proposed by Galletly and Blachut (1985) coupled with the elastic buckling equation (Eq. 4) of Timoshenko and Gere (1961) to calculate the elastic or plastic critical shear load.

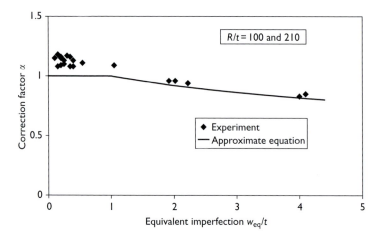

Figure 9.9 Load reduction associated with geometric imperfection.

Postbuckling behaviour

As identified by Koiter (1966), a falling postbuckling path indicates a sensitivity to geometric imperfections, whose severity depends on the dimensions, imperfection amplitude and loading. If the shell is not sensitive to geometric imperfections, the postbuckling response should be stable with either strengthening or only small reductions in load. These observations are also confirmed by the evidence from global shear buckling.

Yamaki (1984) studied the elastic postbuckling response under transverse shear loads coupled with hydrostatic pressure (Fig. 9.10). The results show that short shells generally have a stiffening postbuckling response, whilst longer shells display a significant and rapid strength loss following bifurcation, irrespective of the internal or external pressurisation level.

Effect of an end closure on shear buckling

All the above cylindrical shells were deemed to be terminated in simple boundary conditions, often defined by the test arrangement. Real shells must have a closure of some form. Where the closure includes a stiff ring, the simple conditions may be met, but where other closures are used, it is important to know whether the closure form has an impact on the buckling strength. To simulate the real boundary condition of a suspended tank, the effect on the shear buckling strength of a torispherical closure of the same thickness as the cylinder was investigated by Murakami *et al.* (1989). Two shell thicknesses were examined, producing radius to thickness ratios of 167 or 250 in a cylinder with length to radius ratio $L/R = 1$. The test set-up is illustrated on Fig. 9.11. The critical load was found to be lower

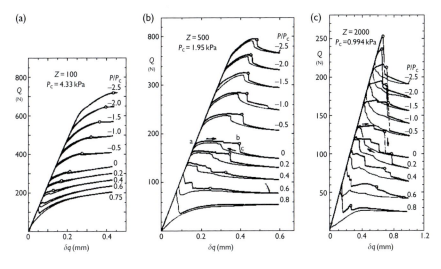

Figure 9.10 Effect of hydrostatic pressure on the postbuckling relation between the transverse load Q and transverse edge displacement: (a) $Z = 100$; (b) $Z = 500$; (c) $Z = 2000$ (reproduced from Elastic Stability of Cylindrical Shells, Yamaki, 1984).

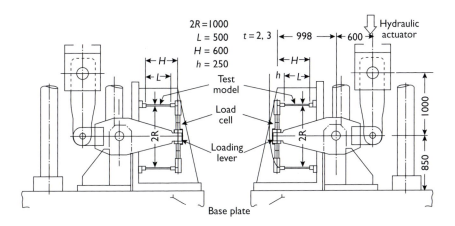

Figure 9.11 Test arrangement for Murakami's end closure tests.

for the shell with a torispherical closure than for the comparable shell with a rigid far-end (Fig. 9.12). It is clear from this study that the stiffness of the closure can adversely affect the shear buckling strength, but much more work is required to identify the range of geometries within which this effect is critically important.

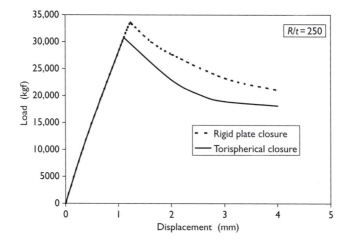

Figure 9.12 Influence of torispherical closure on buckling load.

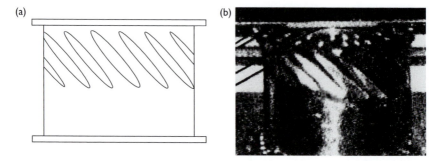

Figure 9.13 Buckling of filled cylindrical shells under static global shear at the top of the shell: (a) buckling mode; (b) shell after test.

Coupling between different loading types

Internal fluid pressure

Two cases of shear buckling in the presence of internal pressure due to fluids have been studied. In both cases, the cylinder was filled with fluid whose density was such that a significantly varying internal pressure existed within the cylinder. The two cases differ in the manner in which the global shear load is applied to the shell. In the first, Yamaki (1984) investigated the effect of an axially varying internal pressure on the shear buckling strength, where the global shear load was externally applied to the cylinder. Here, the high tensile circumferential membrane stresses towards the cylinder base resist the formation of shear buckle and cause a shorter buckle height, leading to an increase in buckling strength. Inclined buckles

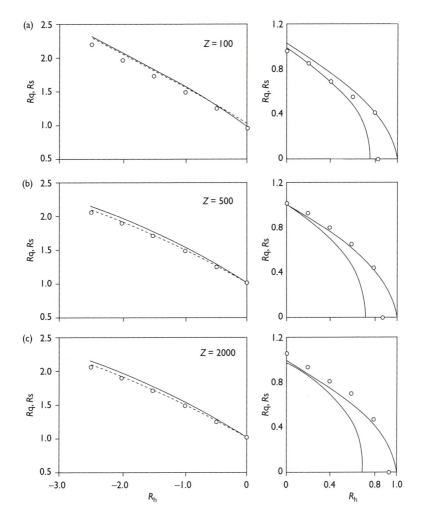

Figure 9.14 Effect of hydrostatic pressure on the critical shear load: (a) $Z = 100$;
(b) $Z = 500$; (c) $Z = 2000$.

appear in the upper part of the shell where the internal pressure is lower (Fig. 9.13).
Yamaki (1984) plotted interaction curves (Fig. 9.14) for the case of elastic buckling from which it can be deduced that a linear interaction appears to be always
conservative.

The second case relates to the buckling of filled tanks under earthquake loading.
Although the phenomenon may be thought of as dynamic, the period is so low
that the buckles are effectively quasi-static. The key difference here is that the
earthquake accelerations induce body force horizontal forces on the tank walls,

(a)   (b)

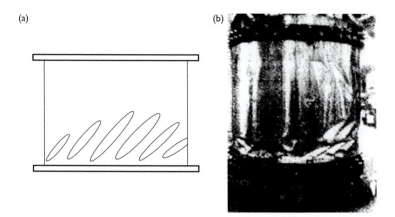

Figure 9.15 Buckling of a fluid-filled cylindrical shell under global shear at the bottom of the shell: (a) buckling mode; (b) shell after test.

inducing a shear that varies down the shell height. As a result, buckles appear near the bottom of the shell (Fig. 9.15) because this is the location of the highest shear stress. Using the relation between shear buckling and torsional buckling (see section 'Overall cell dimensions: the interaction between shear and bending'), the study of Jumikis and Rotter (1986) on buckling under non-uniform torsion can be used to obtain a close estimate of the shear buckling strength, which has been adopted into the European standard (ENV1993-4-1 1999), where the peak shear stress at buckling is given by

$$\tau_{max} = 1.4E \left(\frac{R}{\lambda_0} \right)^{0.5} \left(\frac{t}{R} \right)^{1.25} \tag{8}$$

in which the length λ_0 is found as

$$\lambda_0 = \frac{\tau_{max}}{(d\tau/dx)} \tag{9}$$

where $d\tau/dx$ is the rate of change of shear up the shell. It should be noted that the circumferential membrane tension caused by fluid pressure again increases the buckling load above this value.

Thermal loading coupled with shear

The coupling between thermal and shear loading was investigated by Nakamura *et al.* (1987) using steel cylindrical shells. The buckling load was found to fall as the material properties declined at high temperature (the proof stress $\sigma_{0.2}$ decreased faster than Young's modulus). For the geometries and loading conditions explored, the critical load was found at about 0.8–$0.9\sigma_{0.2}$.

Tension and shear load coupling

The condition of global shear loading coupled with axial tension in the shell was investigated by Nakamura *et al.* (1987). The axial tension was found to make only a small difference to the buckling load, but the postcritical behaviour was found to be more stable (Michel *et al.* 2000). This conclusion may, however, only be valid for relatively small axial tensile stresses.

Cyclic static shear loading

Many practical cases of global shear loading on cylinders involve cyclic loads, and the possible progressive deterioration of the shell due to previous cycles of loading must be taken into account. Two separate features of progressive damage are important: the plastic degradation of the material of the shell, and a steady increase in the amplitude of geometric imperfections.

When buckling occurs under elastic conditions (an elastic prebuckling stress state), the rapid growth of postbuckling deformations can easily lead to local plasticity in bending, causing both of the above phenomena. The extent and consequences of such local yielding depends very much on the maximum postbuckling displacement that is reached. In their study of elastically buckling cylinders ($R/t = 500$), Michel *et al.* (2000) permitted the shell to buckle during each cycle, but did not force the displacements very far into the postbuckling regime. They found that material degradation had a negligible influence on the buckling strength and response, even after 1000 cycles (Fig. 9.16).

By contrast, the strength of thicker shells (Kokubo *et al.* 1987) has been found to decrease very significantly with even a small number of cycles of loading (Fig. 9.17). A local failure can occur at the crest of each buckle due to exhaustion of the plastic deformation capacity.

Conclusions

Many aspects of cylindrical shell buckling under global shear loads have been explored. This type of loading generally leads to a relatively stable postcritical behaviour and a weak sensitivity to geometric imperfections, regardless of whether axial or shear stresses are the dominant consequence of the shear load.

Dynamic shear loading

Vibration response

Introduction

Vibration modes and buckling modes often have some common features, and it is natural to suppose that dynamic excitation may cause premature buckling through interaction between these phenomena. Before such an investigation can begin, it

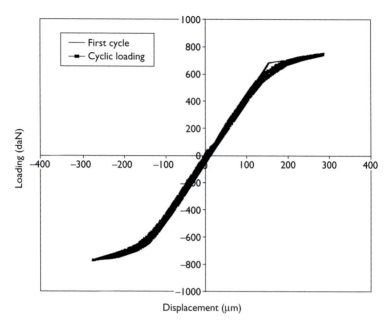

Figure 9.16 Load–deformation response under cyclic loading including buckling.

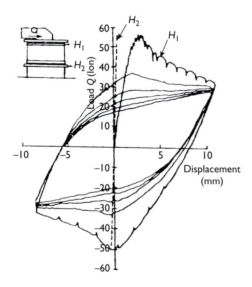

Figure 9.17 Cyclic shear loading causing plastic buckling ($R/t = 133$).

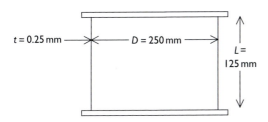

Figure 9.18 Dimensions of Michel's test specimens.

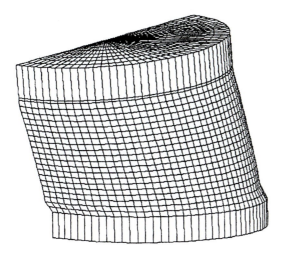

Figure 9.19 Lowest vibration eigenmode (80 Hz): beam mode 1.

is necessary to define the vibration modes and their corresponding frequencies for cylindrical shells under transverse excitation.

Michel *et al.* (2000a) studied the shell shown in Fig. 9.18, whose dimensions represent a scaled model of a fast breeder reactor vessel. A mass of 38 kg was added at the top to provide a realistic representation (self-weight of internal components) of the vibration characteristics of the industrial problem. Elastic analyses were performed using ABAQUS with a Young's modulus of 180 GPa and boundary conditions of a clamped base and a free top edge (where a transverse shear load was applied).

Perfect shell analyses

The lowest vibration eigenmode, with an associated frequency of 80 Hz, corresponds to beam shear/bending of the complete shell (Fig. 9.19). The lowest

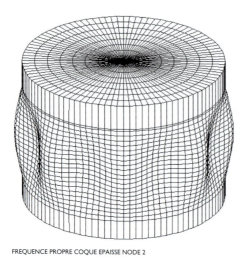

FREQUENCE PROPRE COQUE EPAISSE NODE 2

Figure 9.20 Lowest vibration eigenmode involving shell deformations (942 Hz).

vibration eigenmode that involved shell wall displacements (Fig. 9.20) is in the circumferential mode 14 and has an associated frequency of 943 Hz. In the following discussion, the former is referred to as a 'beam mode' and the latter as a 'shell mode'. The mass at the top influenced the frequency of the beam mode, but had no effect on the associated frequencies of the shell vibration modes.

Geometric imperfections

Most shell structures contain geometric imperfections of significant amplitude and they can modify both the frequency and the mode of vibration. A geometric imperfection similar to the elastic shear buckling mode was studied first. Where the amplitude of the geometric imperfection is less than the wall thickness, no significant change in the vibration mode was found (Michel *et al.* 2000a), but higher imperfection amplitudes caused a significant increase in the natural frequency of vibration in the buckling mode (Fig. 9.21). The translational beam eigenmode frequencies were found to be unaffected even by high imperfection amplitudes. An axisymmetric geometric imperfection was also explored, but no significant change in eigenmodes or eigenfrequencies was found.

Shear preload effect on eigenmodes and eigenfrequencies

The effect of a statically imposed axial compression on the vibration response of a shell was thoroughly explored by Singer *et al.* (1991) using stiffened shells.

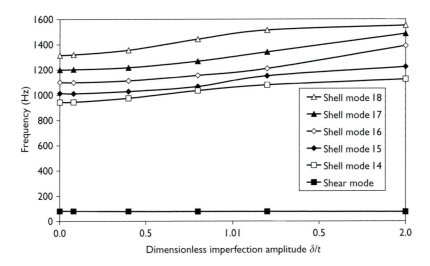

Figure 9.21 Effect of imperfection amplitude on eigenfrequency.

Both experiments and numerical simulations showed a progressive decrease in the natural frequency as the axial compression increased. A linear relationship between the square of the natural frequency ω^2 and the axial load was found, with its intersection at $\omega^2 = 0$ giving the static axial buckling load. This phenomenon was developed into a very effective method for assessing the static buckling load of complex shell structures without damaging them.

The possible extension of these ideas into the shear behaviour of cylinders was investigated by Michel *et al.* (2000a), who performed a numerical investigation into the effect of a transverse shear preload on the vibration eigenmodes and eigenfrequencies. The calculations began with a nonlinear incremental calculation to introduce the preload as a top horizontal displacement, including the associated changes of shell geometry. This was followed by a vibration eigenvalue analysis. The natural frequencies corresponding to shell eigenmodes decreased as the preload top displacement was increased, but the beam eigenfrequency was not affected by this shear preload (Fig. 9.22). The decline in ω^2 was not linear with shear displacement because of coupling between different eigenmodes.

When the imposed top translation reached 90% of the value for static buckling, the shell eigenmodes were radically modified (Fig. 9.23). The shell vibration eigenmodes then took on the form of a combination of the buckling mode and the vibration mode without preload. This behaviour is thus not similar to the condition of vibration under axial compression preload, due to a multimodal response under transverse shear load.

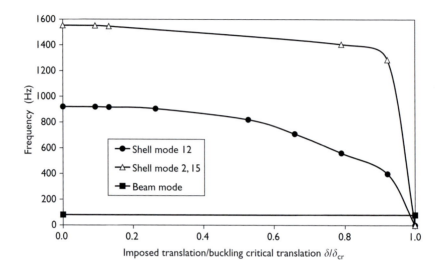

Figure 9.22 Effect of imposed shear displacement on natural frequency.

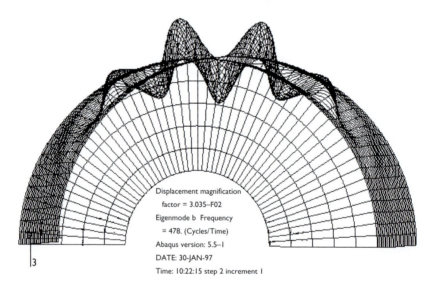

Figure 9.23 Lowest shell eigenmode in the presence of an imposed top displacement of 90% of the critical shear top displacement.

Dynamic buckling studies

Introduction: form of dynamic buckling analysis

Under earthquake conditions, many different vibration modes can be excited, so the frequency spectrum of shell natural frequencies is important. Combescure

(1989) performed numerical simulations with the INCA program of a ring under external pressure. He introduced an initial geometric imperfection δ_0 in the form of the buckling mode n and explored the elastic vibration behaviour. An oscillating load P was applied to the ring at the circular frequency ω_0

$$P(t) = P_0 \sin(\omega_0 t) \tag{10}$$

The results were characterised in terms of the ratio ω_0/ω_{cr} of the excitation frequency ω_0 to the buckling mode associated frequency ω_{cr}. This ratio, termed REY, permitted a classification of the different excitations as:

- REY $= 0.1$ slow excitation
- REY $= 1$ intermediate excitation
- REY $= 10$ rapid excitation

At REY $= 0.1$, the critical load is unaffected by the dynamic character of the excitation, but as REY increased, higher transitory dynamic pressures could be carried without buckling, finally rising as high as five times the static critical pressure. When the excitation frequency was equal to the associated buckling mode frequency, the critical load was found to depend on the level of damping. For a typical damping of 1%, the critical dynamic load was found as 50% larger than the static buckling load. Combescure (1989) devised a diagram (Fig. 9.24) to permit the dynamic load effect on buckling to be determined in terms of the excitation frequency.

A different approach was proposed by Gibert (1988), based on the concept of dynamic instability. This approach is linked to the Mathieu equation (Cheikh 1993). The dynamic response to the excitation of Eq. (10) is described by the equation

$$\ddot{u} + 2\varepsilon\omega_{cr}\dot{u} + \omega_{cr}^2(1 + 2\mu \cos \omega_0 t)u = 0 \tag{11}$$

in which μ is the load ratio ($= P_0/2P_E$), P_0 the excitation load amplitude, ω_0 the excitation frequency, P_E the static elastic critical load, ω_{cr} the frequency associated with the buckling mode and ε the structural damping. Gibert (1988) established a stability diagram (Fig. 9.25) in which the load ratio μ could be related to the frequency ratio Ω ($= \omega_0/\omega_{cr}$) and regimes of stability and instability could be identified (this chart for $\varepsilon = 0$).

Dynamic loading under transverse shear: concepts

For cylindrical shells under transverse shear, relatively few studies have explored dynamic buckling. In one of the first studies addressing the effects of an earthquake, Akiyami et al. (1985) used an energy approach and represented the shell as a 1 degree-of-freedom (d.o.f.) system. Their concept was to compare the energy E transferred to the structure by the earthquake with the elastic deformation

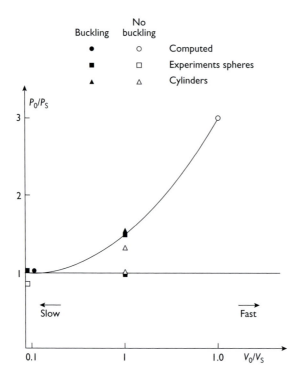

Figure 9.24 Transient buckling load capacity for slow and rapid oscillating loads on a ring.

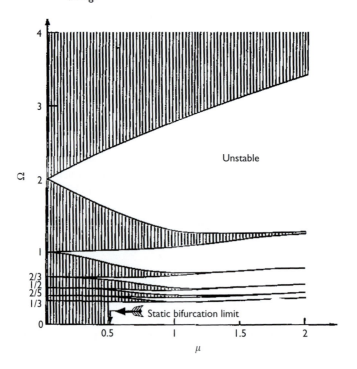

Figure 9.25 Stability diagram based on Mathieu equation.

energy W_e, the inelastic deformation energy W_p and the absorbed energy W_h.

$$E = W_e + W_p + W_h \tag{12}$$

The total energy absorption capacity of the structure E_D is calculated as the sum of the components before buckling W_e and in the postbuckling domain W_d

$$E_D = W_e + W_d \tag{13}$$

From this energy, the equivalent speed V_D is found as

$$V_D = \sqrt{\frac{2E_D}{M}} \tag{14}$$

where M is the structural mass.

The elastic energy W_e is determined as

$$W_e = \frac{Q_{cr}\delta_{cr}}{2} \tag{15}$$

in which Q_{cr} is the static elastic buckling load and δ_{cr} is the corresponding elastic deformation. The upper limit speed of elastic deformation V_E is found from W_e and the ratio V_D/V_E determined.

Load–displacement curves for static tests (with hysteresis) are used for calculating the absorbed energy V_h. Simulations with a 1 d.o.f. model were made for four different earthquakes. The results show that the postbuckling energy absorption capacity (plastification energy) of the shell is important in surviving an earthquake. A high energy absorption capacity in the postbuckling range greatly increases the security margin.

Dynamic loading under transverse shear: experiments

The above conceptual aids are of uncertain value without experimental verification. Several experiments have been conducted in attempts to simulate earthquake loading on cylindrical shells. The chief difficulty with these tests is the scaling of the structure and earthquake motions: it is difficult and expensive to conduct such tests at full scale. Two different types of experiments have been undertaken:

- cylindrical shells fixed to a shaking table and subjected to harmonic or seismic excitation;
- cylindrical shells clamped at the bottom and subjected to a transverse shear load at the top applied by an hydraulic actuator.

Dynamic loading under transverse shear: shaking table tests

The study of Kokubo *et al.* (1987) is briefly described here. In using a shaking table, it is helpful to fix a mass at the top of the shell (Fig. 9.26) to reduce the

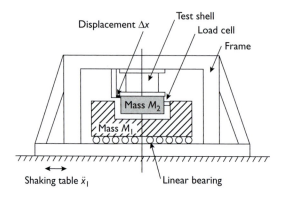

Figure 9.26 Experiments with a shaking table (Kokubo et al. 1987; reproduced by kind permission of A. A. Balkema).

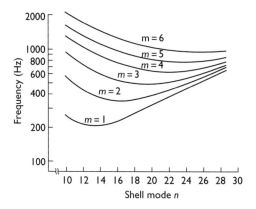

Figure 9.27 Shell natural frequencies (after Kokubo et al. 1987; reproduced by kind permission of A. A. Balkema).

beam eigenfrequency (the value $f_0 = 18.5 \, \text{Hz}$ was achieved here) and to separate it from the shell eigenfrequencies, which are much higher (Fig. 9.27).

Before buckling, the shell deformations have the same frequency as the excitation of the shaking table. At the instant of buckling a big radial deformation appears, and subsequently changes the frequency response of the shell. The eigenfrequency after buckling then becomes the eigenfrequency of the most prominent buckling mode.

The static critical load P_{cr} may be usefully compared with the instantaneous dynamic force from the top mass $P_{dyn} = M\ddot{x}_2$, where M is the mass and $\ddot{x}_2$ the peak acceleration of the top. For the study of Kokubo *et al.* (1987), this comparison

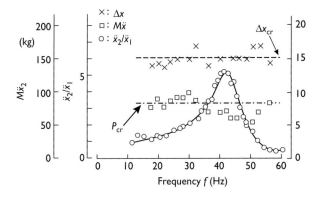

Figure 9.28 Acceleration and displacement during shaking table excitation (after Kokubo *et al.* 1987; reproduced by kind permission of A. A. Balkema).

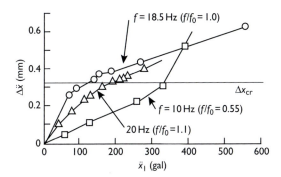

Figure 9.29 Postcritical displacement amplitudes at different excitation frequencies (after Kokubo *et al.* 1987; reproduced by kind permission of A. A. Balkema).

(Fig. 9.28) shows that the dynamic force was lower for frequencies between 30 and 50 Hz. The top shell acceleration was also amplified in this frequency range even though it does not correspond to eigenfrequencies.

The shell stiffness decreases after buckling, leading to smaller amplitude response when excited at the critical frequency, but increased response when the excitation is slower (Fig. 9.29).

A later experimental study by Hagiwara *et al.* (1989) used the same approach and confirmed the above description. The energy approach proposed by Akiyama was also validated, indicating that postbuckling shell energy absorption is very important if collapse is to be avoided.

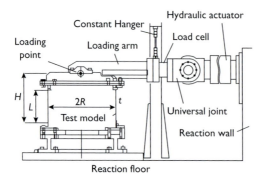

Figure 9.30 Experimental set-up (Hagiwara *et al.* 1993; reproduced by kind permission of the Japanese Society of Mechanical Engineering).

Dynamic loading under transverse shear: pseudo-dynamic tests

Hagiwara *et al.* (1993) used the alternative approach of applying a varying transverse shear load at the top applied by an hydraulic actuator (Fig. 9.30), sometimes called a pseudo-dynamic test.

Their rather thick test cylinders (radius to thickness ratios R/t of 100 and 200) produced load–displacement curves with significant nonlinearity both before and after buckling due to plasticity response, and buckling in the plastic regime. The postbuckling response was ductile, with increasing displacements not causing collapse (Fig. 9.31).

The study of Michel *et al.* (2000b) adopted the same experimental scheme and tested cylinders with a radius to thickness ratio of 450 and a length/radius ratio of 1. The material properties of each specimen were measured giving mean values of $E = 180\,\text{GPa}$, $v = 0.3$ and $\sigma_{0.2} = 450\,\text{MPa}$. The two ends of the shell were clamped to rigid plates, with the top one connected to a horizontal hydraulic actuator applying controlled displacements. To simulate a nuclear vessel supported from its top edge, a constant tensile load could be applied by another actuator. The shear force, tensile force, displacements and accelerations were measured at the top of the shell. A special internally gauged load sensor, linked to the actuator and the rigid top plate permitted the shear force on the shell to be measured under dynamic conditions without the inertia mass effect. A high-speed CCD camera was used to obtain images of the buckling and vibration modes during high frequency tests.

In a dynamic test, the displacement amplitude was fixed and a sweep through a range of frequencies (1–100 Hz) was performed.

$$u = u_0 \sin(\omega t) \tag{16}$$

Initially, the displacement amplitude u_0 was set at 10% of the static buckling value. Several cycles were applied at each excitation frequency. When the

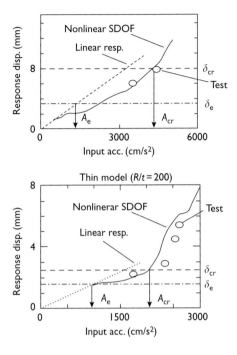

Figure 9.31 Nonlinear response to different accelerations (Hagiwara *et al.* 1993; reproduced by kind permission of the Japanese Society of Mechanical Engineering).

excitation frequency reached 73 Hz, the shear force declined and the radial displacement increased. At frequencies between 75 and 80 Hz, the shear force temporarily rose again, but at higher frequencies it dropped away. The fall in shear force was not an inertial effect because it was not fully correlated with the ram load. The transfer function of top displacement to shear force (x_0/F) (Fig. 9.32) confirms this behaviour.

A second test on the same specimen with the displacement amplitude at 70% of the static critical value showed that the shell behaviour was unaffected by cyclic loading. A larger force drop occurred at the lowest beam eigenfrequency (73 Hz) but the radial displacement amplification is larger (Fig. 9.33).

The photographic images assisted the interpretation of the tests, leading to the following conclusions:

- At low load levels, a beam mode is always encountered during a sweep over the frequency spectrum. There is consequently always a classical vibration response with a minor drop in load when the beam eigenfrequency is passed.
- At higher load levels (here 70% of static critical), the mode of vibration appears to be a coupled mode between the buckling mode and the unstressed

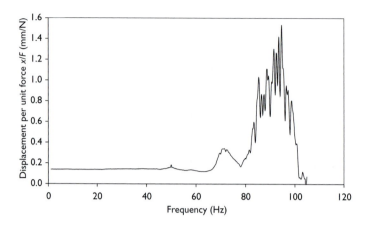

Figure 9.32 Transfer function relating displacement and force.

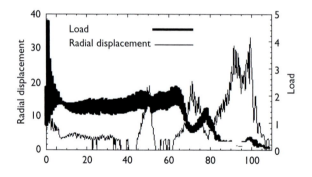

Figure 9.33 Shear load and radial displacement evolution at $(u_0 = 0.7\delta_{cr})$ FFT signals.

shell vibration eigenmode (Fig. 9.34). This interaction only occurs when the excitation frequency is close to the beam eigenfrequency and consequently the shell stiffness becomes very low.

Measurements of residual deformation after testing show that a modal wave (mode 14: the lowest shell eigenfrequency) remains imprinted on the shell in addition to the buckling mode. This modal wave confirms the coupling between instability and vibration, causing local plastic deformations.

Finite element simulations were performed under load control, where a cyclic load of fixed amplitude is applied to the shell top and the time evolution of displacements and the reaction load were determined. In some calculations, the top displacement progressively increases over several cycles, reaching the critical

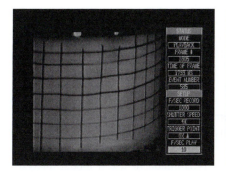
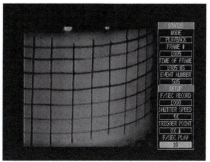

Figure 9.34 Shell deformation during a test at 73 Hz with a displacement amplitude equal to 80% of the static critical one.

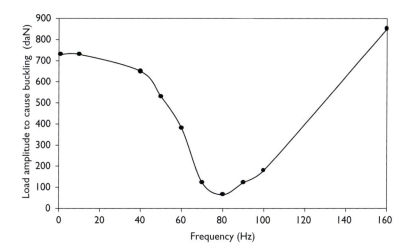

Figure 9.35 Minimum imposed load in order to obtain instability (ABAQUS simulations).

displacement after five cycles. At lower loads, the top displacement amplitude does not grow significantly, so no buckling occurs. The minimum load amplitude to cause buckling at each frequency was determined (Fig. 9.35), and has the same form as the experimental shear load curve (Fig. 9.33). This curve, which can be similarly determined for any geometry and loading condition, permits the relationship between critical load and excitation frequency to be found, including the coupling between buckling and vibration modes.

Conclusions

The problem of buckling under transverse shear loading occurs in many different applications in shell structures, but has been less studied than the simpler loading cases of axial compression and external pressure.

Under static loading conditions, all studies of buckling under a shear load showed two phenomena:

- there is a weak sensitivity to geometric imperfections;
- the postcritical behaviour is relatively stable.

Under cyclic and dynamic loading, it is common to use a modal analysis and ignore the coupling that occurs between vibration and buckling behaviour. Whilst this is satisfactory for low amplitude loads and frequencies far from the eigenfrequencies, this coupling can have a substantial effect and should be included.

Studies of the seismic response of thin cylindrical shells undertaken by Japanese researchers have shown that the structural behaviour can be quite different from that under static loading. Nonlinear vibration studies of the shell are necessary to understand its dynamic response and to prevent coupled instability between vibration and buckling modes.

References

Akiyama, H., Shimizu, S., Yuhara, T. and Morishita, M. (1985). Ultimate limit state of metal containment shell buckling under seismic loading. SMIRT8, Brussels, p. 7.

Cheikh, L. (1993). Flambage dynamique d'un anneau parfait soumis à une pression radiale sinusoïdale. CEA, Rapport DMT 93/313, Saclay, France, p. 63.

Clough, R.W. and Niwa, A. (1982). Buckling of cylindrical liquid-storage tanks under earthquake loading. *Earthquake Engineering and Structural Dynamics* **10**, 107–122.

Combescure, A. (1989). Buckling under time-dependent loads. CEA, Rapport DEMT/90-043, Saclay, France, p. 4.

ENV 1993-1-6 (1999). Eurocode 3: Design of steel structures, Part 1.6: General rules – Supplementary rules for the strength and stability of shell structures, Eurocode 3 Part 1.6. CEN, Brussels.

ENV 1993-4-1 (1999). Eurocode 3: Design of steel structures, Part 4.1: Silos, Eurocode 3 Part 4.1. CEN, Brussels.

Fujita, K., Ito, T. and Wada, H. (1990). Experimental study on the dynamic buckling of cylindrical shell due to seismic excitation. In *Flow-structure Vibration and Sloshing*, Vol. 191. ASME Pressure Vessels and Piping Division Publication, New York, pp. 31–36.

Galletly, G.D. and Blachut, J. (1985). Plastic buckling of short vertical cylindrical shells subjected to horizontal edge shear loads. *Journal of Pressure Vessel Technology* **107**, 101–106.

Gibert, R.J. (1988). Vibrations des structures – Interactions avec les fluides, sources d'excitation aléatoires. Editions EYROLLES, Paris, p. 251.

Hagiwara, Y., Akiyama, H., Kokubo, K. and Sawada, Y. (1989). Post buckling behavior during earthquakes and seismic margin of FBR main vessels. *ASME PVP*, Vol. **175**, pp. 101–106.

Hagiwara, Y., Akiyama, H., Kawamoto, Y. and Nakagawa, M. (1993). Dynamic buckling and nonlinear response of fast breeder reactor main vessels under earthquake loading. *JSME International Journal, Series B* **36**, 476–484.

Jumikis, P.T. and Rotter, J.M. (1986). Buckling of cylindrical shells under non-uniform torsion. *Proceedings of the Tenth Australasian Conference on the Mechanics of Structures and Materials*, Adelaide, August, pp. 211–216.

Koiter, W.T. (1966). General equations of elastic stability for thin shells. *Proceedings of the Symposium on the Theory of Shells*, Donnell Anniversary volume, University of Houston, pp. 187–227.

Kokubo, K., Nakamura, N., Madokoro, M. and Sakurai, A. (1987). Buckling behaviors of short cylindrical shells under dynamic loads. SMIRT 9, Lausanne, Vol. E, pp. 167–172.

Kokubo, K., Nagashima, H., Takayanagi, M. and Mochizuki, A. (1995). Analysis of shear buckling of cylindrical shells. *JSME International Journal, Series A* **153**, 305–317.

Lu, S.Y. (1965). Buckling of a cantilever cylindrical shell with a transverse end load. *AIAA Journal* **3**, 2350–2351.

Lundquist, E.E. (1935). Strength tests of thin-walled Duralumin cylinders in combined transverse shear and bending. NACA Technical note No. 523. NASA, Washington.

Matsuura, S., Nakamura, H., Kawamoto, Y., Murakami, T., Ogiso, S. and Akiyama, H. (1995). Shear-bending buckling strength of FBR main vessels. SMIRT13, Porto Alegre, Vol. E, pp. 457–462.

Michel, G., Combescure, A. and Jullien, J.F. (2000a). Finite element simulation of dynamic buckling of cylinders subjected to periodic shear. *Thin-Walled Structures* **36**(2), 111–135.

Michel, G., Limam, A. and Jullien, J.F. (2000b). Buckling of cylindrical shells under static and dynamic shear loading. *Engineering Structures* **22**, 535–543.

Murakami, T., Yoguchi, H., Hirayama, H., Sawada, Y. and Nakamura, H. (1989). Buckling of short cylinders with elliptical head and core support structure under transverse shearing loads. SMIRT10, Anaheim, pp. 223–228.

Nakamura, H., Matsuura, S. and Sakurai, A. (1987). Plastic buckling of short cylinders with axial temperature distribution under transverse shearing loads. SMIRT9, Lausanne, pp. 219–224.

Rotter, J.M. (1990). Local inelastic collapse of pressurised thin cylindrical steel shells under axial compression. *Journal of Structural Engineering, ASCE* **116**, 1955–1970.

Schröder, P. (1972). Über die Stabilität der querkraftbelasteten dünnwandigen Kreiszylinderschale. *ZAMM* **56**, T145–T148.

Singer, J., Weller, T. and Abramovich, H. (1991). The influence of initial imperfections on the buckling of stiffened cylindrical shells under combined loading. *Buckling of Shell Structures, on Land, in the Sea and in the Air* (ed. J.F. Jullien). Elsevier Applied Science, London, pp. 1–10.

Timoshenko, S.P. and Gere, J.M. (1961). *Theory of Elastic Stability*, 2nd edn. McGraw-Hill, New York.

Yamaki, N. (1984). *Elastic Stability of Circular Cylindrical Shells*. Elsevier Applied Science Publishers, Amsterdam.

Yamaki, N., Naito, K. and Sato, E. (1980). Buckling of circular cylindrical shells under combined action of a transverse edge load and hydrostatic pressure. In *Thin-Walled Structures* (eds J. Rhodes and A.C. Walker). Granada Publishing Ltd, pp. 286–298.

Yamaki, N., Naito, K. and Sato, E. (1983). Buckling of circular cylindrical shells subjected to both transverse edge load and hydrostatic pressure. *Memoirs of the Institute for High Speed Mechanics*, Tohoku University, Vol. **49**, pp. 41–59 (in Japanese).

Chapter 10

Cylindrical shells under combined loading

Axial compression, external pressure and torsional shear

H. Schmidt and Th.A. Winterstetter

Introduction

Circular cylindrical shells made of steel are used in a variety of engineering structures, for example, tanks, chimneys, offshore platforms, pipelines, silos or wind turbine towers. The design against tensile hoop stresses (containments) or against tubular bending stresses (tower-type structures) often results in very thin shell walls which implies the danger of failure by instability due to compressive and/or shear forces (shell buckling).

The behaviour of cylindrical shells under each of the fundamental loads of axial compression, external pressure and pure torsion (Fig. 10.1) is well understood. However, in almost all cases the actions on the structure do not induce such a simple fundamental membrane stress state, but a combination of them, including membrane stress variations in axial and circumferential direction and stress peaks at different locations. The actual stress distribution is often very difficult to calculate and differs considerably from beam theory (e.g. Peil and Nölle 1988 for chimneys or Chapter 7 of this book). Therefore, in general, the whole load–deflection and stability characteristics of a complicated load combination would have to be investigated. Only if the specific failure modes due to different stress peaks over the shell wall can be separated clearly – for example, plastic buckling due to either

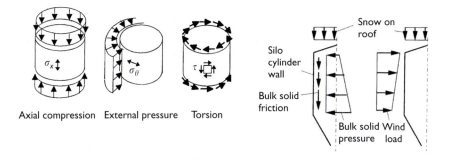

Figure 10.1 Fundamental loads and example of combined loading.

axial compressive or shear stresses in the case of cylindrical shells under bending and transverse shear, see Chapters 8 and 9 of this book – then it is dispensable to perform a buckling interaction check.

But in general, the locations of peak stresses are, to a certain extent, variable due to simplifying assumptions for the load induction and to approximative calculation methods. Furthermore, it is a characteristic of shell buckling phenomena that after buckling a large part of the shell wall is deformed, including areas of different stress peaks even if they are located at a large distance from each other. Thus, it is generally not overconservative to perform a buckling design check with the maximum values of the individual membrane stress components present under an acting load combination – even if they do not occur at the same location.

The buckling design formulas for interactive loading given in design standards are often deduced from a few sets of test results and are therefore of a comparatively simple nature, in many cases not covering effects due to inelastic material behaviour in a biaxial stress state. Furthermore, while the code predictions for fundamental loads are quite similar, they differ considerably for the interactive buckling.

This chapter provides a comprehensive picture of the behaviour of cylindrical steel shells under combined loading. Theoretical as well as experimental results have been collected to provide insight into interactive buckling and to put design recommendations forward.

Cylindrical shells under fundamental loads

Axial compression

Theoretical linear analyses of the elastic buckling of axially compressed perfect cylindrical shells show three different regions (e.g. Flügge 1973):

- Very short shells buckle, theoretically, like a plate strip of infinite length.
- Cylinders of medium length buckle at a buckling stress that depends mainly on the ratio r/t with a multi-modal buckling pattern, that is, there are many eigenmodes with eigenvalues near the lowest critical load. Different boundary conditions have only a small influence on the buckling load except when the circumferential restraint is absent. Most engineering applications fall into this range.
- Very long cylinders fail by column buckling.

Boundary conditions (BCs) different to pure membrane BCs cause geometrically nonlinear prebuckling bending deformations. Thus, the calculated buckling loads of perfect elastic shells reduce to about 90% (clamped) and 85% (simply supported) of the classical critical buckling load (e.g. Almroth 1966; Yamaki 1984).

When the r/t ratio becomes sufficiently small, the theoretical buckling stress of the elastic perfect cylinder is higher than its yield strength when made of steel.

The buckling is no more elastic but called 'elastic–plastic', or 'plastic' in the case of very thick cylinders. These effects due to inelastic material behaviour can be calculated using a nonlinear material law. The experimentally observed buckling pattern changes from the periodic 'diamond-shaped' elastic buckling pattern with certain buckling wave numbers in axial and circumferential direction to the so-called 'elephant's foot' pattern due to the moments near the edge restraints.

Experimentally observed axial buckling loads are much lower than the theoretical ones. The main causes were found to be unavoidable initial imperfections which disturb the stress and displacement state of the shell. The theoretical multi-modal bifurcation characteristics allow the initially imperfect shell to adjust to a proper lowest-possible buckling mode. Furthermore, in the case of elastic axial buckling there is no postcritical strength reserve.

Modern design codes cover these effects using a 'semi-empirical' design concept. Approximative formulas for the theoretical linear buckling stresses are used to calculate a non-dimensional slenderness parameter. From this, a slenderness-dependent, empirically based reduction factor is derived which takes account of nonelastic material behaviour and the influence of initial imperfections as observed in experiments. The shell buckling reduction curves of several design codes for axial compression are shown in Fig. 10.2. For direct comparison, the curves are, as good as possible, brought to a comparable safety level; that is, they include any specific built-in safety factors exceeding the standard value of $\gamma_M = 1.1$. The tolerance levels for initial imperfections – if specified – are similar for all curves, which is important in the case of ENV 1993-1-6 (CEN 1998), which gives three quality classes. The differences between the individual curves result mainly from different experimental data bases, different safety philosophies and different reduction factor concepts.

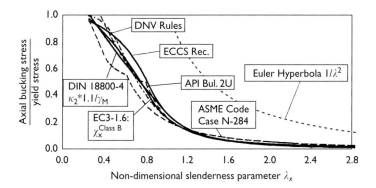

Figure 10.2 Axial compression – comparison of design code reduction curves.

External pressure

'External pressure' in the present chapter means 'external lateral pressure', that is, only circumferential compressive stresses are induced. Thus, a hydrostatic stress state would have to be treated as a load combination of axial compression and external pressure.

According to classical linear stability analysis, there are three regions for external pressure buckling similar to the case of axial compression. In the again practically most important range of medium-length cylinders, the buckling load depends on the geometric ratios l/r and r/t and strongly on the existence of an axial restraint (about 50% increase). The buckled shape has one half-wave in the longitudinal direction and $m \geq 2$ full waves in the circumferential direction. In contrast to axial compression, nonlinear prebuckling deflections have no significant effects on the buckling strength. Real thin-walled elastic imperfect shells buckle at stresses of at least about 65% of the critical buckling stress of the perfect shell. When buckling occurs in the plastic region, the buckling stress is more and more influenced by the yield limit, but the general buckling characteristics are not affected (Stracke 1987).

Design code guidelines for external pressure buckling are shown in Fig. 10.3. Again, the differences result from different safety philosophies and reduction factor concepts; especially important is the fact that some codes do not distinguish between axially restrained and axially free edges.

Finally, this chapter deals only with buckling, that is, it does not cover effects like the external pressure postbuckling carrying capacity of thin-walled cylindrical shells as described by Schmidt *et al.* (1998).

Torsion

Again, there are three regions for pure torsional buckling according to classical linear stability analysis. In the practically most important range of medium-length

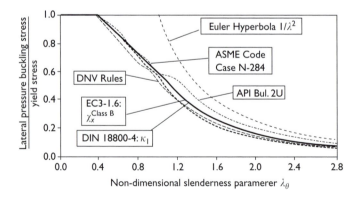

Figure 10.3 External lateral pressure – comparison of design code reduction curves.

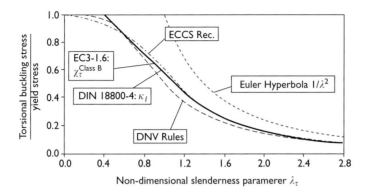

Figure 10.4 Torsion – comparison of design code reduction factor curves.

cylinders, the buckling load depends on the geometric ratios l/r and r/t, similar to external lateral pressure buckling. The cylinder buckles in a helix-like pattern with one half-wave in the longitudinal direction and $m \geq 2$ full waves in the circumferential direction. Again, nonlinear prebuckling deflections have no significant effects on the buckling strength, and the elastic imperfect buckling load is at least about 65% of the critical buckling stress of the perfect shell, which corresponds to the postbuckling minimum. Finally, there exists a slight torsional postbuckling carrying capacity which is not nearly as significant as in the external pressure case – see Chapter 5 of this book.

Figure 10.4 shows the torsion reduction curves given in several buckling design standards. Because pure torsional buckling is often of minor importance, only a few codes deal with it. The agreement is satisfactory.

Cylindrical steel shells under combined loading

In general, the design of a specific steel shell structure has to be performed with the information describing the perfect state by geometry, boundary conditions and loads. The ultimate limit state of the structure can usually be expressed by an acting load combination (or the resulting stress distribution, respectively), multiplied by a magnification factor. The ratio of stress over resistance (both including partial safety factors if necessary) must be smaller than one.

As described in the preceding section, the buckling of cylindrical steel shells under fundamental loads is well understood, and safe but not uneconomical design strength predictions are available. To formulate proper interaction design rules, these fundamental load predictions should be used. Because three basic membrane stress states exist, all possible combined loading path vectors form a space in three dimensions (3-D); if one stress component is omitted or held constant, we obtain a 'buckling curve' in 2-D representing a cross section through the 3-D 'buckling surface'. Our object is to determine the shape of the curves and of the whole buckling surface and to check, by comparison, the proper description in existing design rules. Eventually the need for more correct descriptions will be stated. Firstly, theoretical results are presented for several calculation levels with successively growing accuracy. Thereafter, interactive shell buckling experiments published over the past five decades are re-examined. Finally, existing code design rules are presented and discussed.

Theoretical results

Linear analysis (LA)

A comprehensive set of FE linear eigenvalue calculations has been performed to study the influence of the relevant parameters – shell dimension ratios $(r/t, l/r)$ and boundary conditions – on the shape of the interaction curves. Some results are shown in Fig. 10.5. The following effects can be noted:

- The interaction between axial compression and external pressure for the practically very important case S3 (simply supported) is nearly linear. Circumferential tensile stresses do not raise the axial compression buckling strength. These results are in good agreement with the analytical linear eigenvalue calculations based on Flügge's (1973) shell theory.
- The interaction between axial compression and torsion for the case S3 is nearly linear, too. This agrees well with the results presented by Kromm (1942) and Batdorf et al. (1947).
- The parabolic shape of the interaction curve for external pressure and torsion confirms the results given by Ho and Cheng (1963) or Simitses (1967).
- Different boundary conditions do affect the shape of the interaction curves. As stated above, the axial restraint leads to considerable buckling strength gains in the case of external pressure, but not in the case of axial compression. Therefore, the shape of the axial compression – external pressure buckling interaction curve for an axially restrained shell is more convex than the shape for an axially free cylinder. The shapes of the interaction curves for axial compression and torsion and for externral pressure and torsion differ much less.
- Internal pressure does not affect the axial compression buckling strength, but raises the torsional buckling strength considerably; axial tension raises the external pressure and torsional buckling loads, too.

Other calculations with varying shell geometric ratios r/t and l/r show that the shapes of the interaction curves depend hardly on the shell dimensions. The results stated above are valid at least for the whole range of 'medium-length' cylinders.

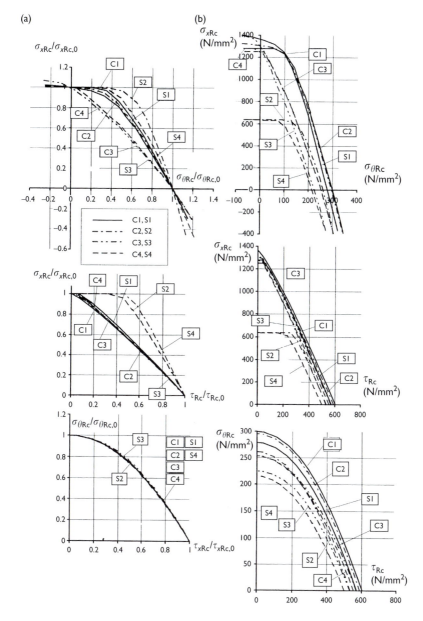

Figure 10.5 Linear buckling analysis (LA) – interactive buckling stress curves: (a) related to fundamental load buckling stresses; (b) in absolute values (N/mm²); $r/t = 100, l/r = 1$.

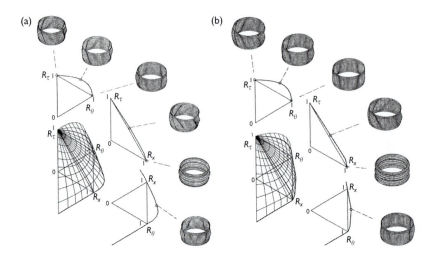

Figure 10.6 Linear buckling analysis (LA) – interactive 'buckling surfaces' for combined loading: R = buckling stresses $\sigma_{(\cdot)Rc}$ related to fundamental buckling stresses $\sigma_{(\cdot)Rc,0}$. Geometry: $r/t = 100, l/r = 1$. Material: $E = 210,000\,\text{N/mm}^2$, $\nu = 0.3$. BC's: (a) C1 and (b) S3.

Finally, complete buckling interaction surfaces have been calculated as shown in Fig. 10.6 for the cases C1 and S3 which are of practical importance. In the figure, $R_{(\cdot)}$ denotes the quotient of $\sigma_{(\cdot)Rc}$ (combined loading buckling stress component) over $\sigma_{(\cdot)Rc,0}$ (fundamental load buckling stress for the respective component). These shapes are continual, suggesting a corresponding continual description for shell buckling design. For illustration and better understanding, the three curves of each two combined stress components and some selected buckling modes are added in Fig. 10.6.

Geometrically nonlinear analysis (GNA)

A nonlinear calculation takes the elastic prebuckling bending deformations into account which occur due to rigid boundary conditions. The axial compression buckling load of the elastic perfect shell is reduced by up to 15% of the linear critical load, as mentioned earlier. The buckling under lateral pressure and under torsion is not affected. Thus, the shape of the interaction curves including axial compression components gets slightly more convex, as shown in Fig. 10.7. A detailed overview is given by Yamaki (1984) for the cases of axial compression and external pressure and of external pressure and torsion. The interaction between axial compression and torsion is discussed in detail by Tennyson *et al.* (1978). New GNA interaction curves for axial compression and torsion calculated by Winterstetter (2000) are shown in Fig. 10.8.

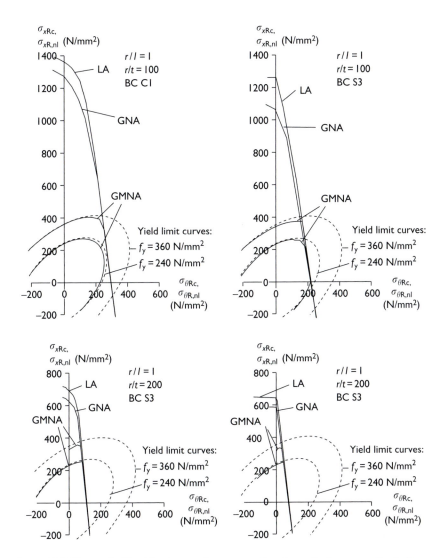

Figure 10.7 Theoretical buckling interaction curves of perfect cylinders according to linear and nonlinear buckling analyses (LA, GNA, GMNA): axial compression and external pressure, different geometries, different boundary conditions.

Geometrically and materially nonlinear analysis (GMNA)

Cylindrical steel shells with large wall thickness may buckle by 'snap-through', that is, by reaching an equilibrium divergence state in which the buckling deformation follows the initial deformation path, rather than by sudden bifurcation into a periodically buckled state whose displacement field vector is orthogonal to

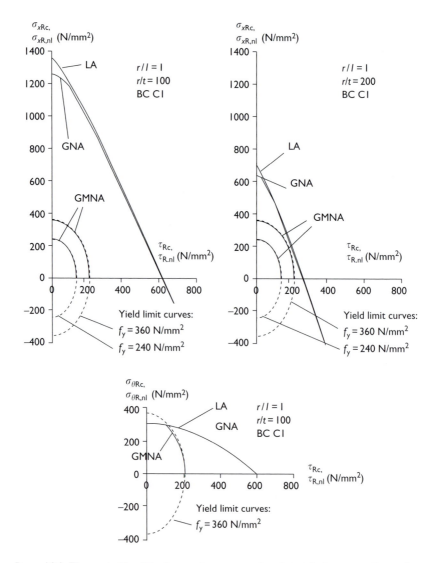

Figure 10.8 Theoretical buckling interaction curves of perfect cylinders according to linear and nonlinear buckling analysis (LA, GNA, GMNA): axial compression and torsion, external pressure and torsion.

the initial displacement vector. These differences are caused by inelastic material behaviour which governs the behaviour if the theoretical bifurcation load is much higher than the plastic material failure load. For that reason, possible interaction curves must follow asymptotically the material yield surface which may be described for mild steel by the well-known von Mises–Huber–Hencky yield rule (Eq. 1, for a biaxial stress state).

$$\sigma_x^2 - \sigma_x\sigma_\theta + \sigma_\theta^2 + 3\tau_{x\theta}^2 = f_y^2 \tag{1}$$

A main feature of this yield behaviour of metal components in a multi-axial stress state is that stress components with the same sign (e.g. biaxial compression) 'support' each other, while stresses with different signs or additional shear stresses lower the yield carrying capacity in each direction.

A set of calculations investigating the interaction of axial compression and external pressure is reported by Galletly and Pemsing (1985). The f_y-related curves calculated therein at different levels of yield strength show essential differences, even more with respect to elastic buckling. To clarify the GMNA behaviour, further calculations have been performed at the University of Essen using an ideal bilinear elastic–plastic material law (Winterstetter 2000). They are included in Figs 10.7 and 10.8.

In general, GMNA calculations yield buckling interaction curves which approach asymptotically either the elastic GNA interaction curve (where elastic buckling occurs below the yield limit) or the yield limit curve (where the elastic buckling stress state is beyond the yield limit). The GMNA curves of thick-walled shells follow completely the shape of the yield limit curve, thus including load-reducing effects of the elastic–plastic material subject to a biaxial stress state.

Geometrically and materially nonlinear analysis, imperfect geometry (GMNIA)

The main cause for the severe discrepancies between the theoretical axial compression buckling load of a perfect thin cylindrical shell and the experimetal results has long been shown by many authors to be the geometrical imperfections. Thus, if one wants to calculate the ultimate load carrying capacity of a cylinder directly and with the highest-possible accuracy, an imperfect model has to be employed.

Many works have been done on calculating the load–deflection path and buckling strength of imperfect cylindrical shells under the fundamental loads (e.g. Hutchinson 1965; Booton and Tennyson 1979; Galletly and Pemsing 1985; Rotter 1997). Though, only a few reports exist that deal with interactive buckling, and if so, only a small range of possible dimensions and load ratios is covered.

From a structural engineer's point of view, the buckling strength of a particular shell which is to be designed is of interest. The actual imperfections are of course unknown at the design stage, so they have to be replaced by so-called substitute 'equivalent geometric imperfections'. For a design by global numerical GMNIA, one has to assume a proper equivalent geometric imperfection pattern and a sufficient amplitude for the design to be safe.

Some recently issued shell buckling design codes (DASt 1992; CEN 1998) give recommendations on which patterns and amplitudes should be used. As a guide, one possible unfavourable pattern is the critical buckling mode (eigenvector of the LA). The amplitude should be determined by the larger value of the initial dimple tolerance or the wall thickness, scaled by a factor of 1.5 (DASt 1992) or 1.6 (CEN 1998). These imperfection amplitudes are relatively large in order to provide a conservative design with respect to the standard 'full strength' reduction curves.

Imperfection patterns adapted to constructional detailing should be examined, too. For example, local axisymmetric imperfections can cause a stronger reduction in axial compression buckling strength than eigenform-affine equivalent imperfections with the same amplitude (Hutchinson 1965; Teng and Rotter 1992). Though, only the latter imperfection type is used herein. Further calculations with more refined imperfection patterns are reported by Winterstetter (2000) and Winterstetter and Schmidt (2002).

Figure 10.9 shows several GMNIA axial compression – external pressure interaction buckling curves. A circular cylindrical steel shell with $E = 210,000\,N/mm^2$,

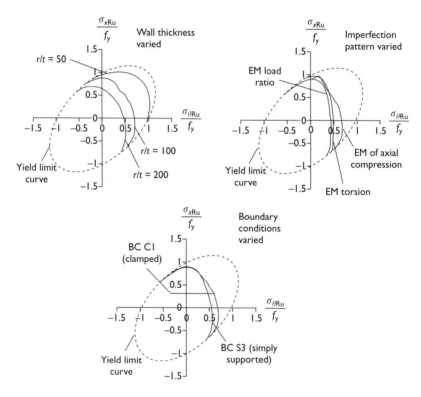

Figure 10.9 Theoretical buckling interaction curves of imperfect cylinders according to nonlinear analysis (GMNIA): axial compression and external pressure. Geometry: $l/r = 1.0$. Material: $E = 210,000\,N/mm^2$, $f_y = 240\,N/mm^2$, $\nu = 0.3$. Imperfection amplitude: $w_0 = 1.5\,t$. Top left: BC C1, imp. pattern eigenmode ax. comp. Top right: BC C1, $r/t = 100$. Bottom: $r/t = 100$, imp. pattern eigenmode ax. comp.

$f_y = 240 \, \text{N/mm}^2$ and $l/r = 1.0$ was chosen to investigate the effects that result from a variation of the wall thickness, the imperfection pattern and the boundary conditions.

On the top left of Fig. 10.9, the wall thickness is varied while the BCs are the same (C1). The linear elastic buckling modes for pure axial compression have, in the three investigated cases, $n \geq 3$ half-waves in the longitudinal direction and $m \geq 5$ full waves in the circumferential direction. Thus, the wave numbers were sufficient to induce the local bending effects that lower the carrying capacity of imperfect thick shells which buckle inelastically. These axial compression eigenmodes were used as the equivalent imperfection pattern. Analogously to practical shells which have one single determined imperfection pattern due to the production process without respect to any actual load combination, the same equivalent imperfection pattern was used for any loading path in this figure. The shape of the interaction curve is more convex the smaller the r/t ratio becomes; thus, the growing influence of the biaxial stress state is visible.

On the top right of Fig. 10.9, the imperfection pattern is varied. The boundary conditions are C1 and the r/t ratio is 100. Three patterns were investigated: the pure axial compression eigenmode (EM) as described for the top left figure, the EM for pure torsion and the EM of each load path (stress ratio), respectively. In the latter case the imperfection pattern varies within the set of calculations for one specific interaction curve. As visible in the range of dominating axial compressive stresses, the EM belonging to the combined loading is not necessarily the equivalent imperfection pattern which yields the lowest buckling load. The EM for pure axial compression yields higher pure external pressure buckling loads, because the high imperfection wave number in the axial direction is very different from the external pressure buckling mode of the perfect shell and therefore has a stiffening effect rather than a knock-down effect. Contrary to that, the torsion EM with only one half-wave in axial direction yields a rather good approximation for the pure external pressure buckling load.

On the bottom of Fig. 10.9, two different boundary conditions – C1 and additionally, S3 – are investigated. The imperfection pattern is held constant as the EM for pure axial compression. The effect on the axial compression buckling strength is very small, though the effect on the pure external pressure buckling load due to the axial restraint is significant.

Figure 10.10 shows a similar study for the interaction of axial compression and torsion. The larger the r/t-ratio, the more similar are the GMNIA interaction curves to the straight LA/GNA curves. The more the failure occurs in the range of plastic buckling, the more the interaction curves follow the yield limit curve asymptotically.

An investigation of interactive buckling under external pressure and torsion is omitted herein because it is of minor practical importance and because the shapes of the parabola-type LA curves and the bi-quadratical GMNA curves (see Fig. 10.8, bottom) are very similar. No significant influences of the imperfections on the real buckling interaction could be identified.

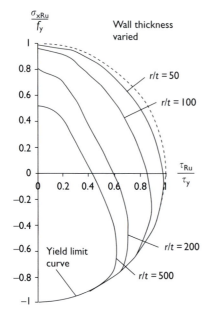

Figure 10.10 Theoretical buckling interaction curves of imperfect cylinders according to nonlinear analysis (GMNIA): axial compression and torsion. Geometry: $l/r = 1.0$. Material: $E = 210,000\,\text{N/mm}^2$, $f_y = 240\,\text{N/mm}^2$, $\nu = 0.3$. Imperfection pattern: Eigenmode for pure torsion. Imperfection amplitude: $w_0 = 1.5\,t$.

In summary, a GMNIA analysis yields interaction curves which, though including effects of biaxial stress states, are not as convex as when calculated using GMNA. This is a consequence of the assumed equivalent geometric imperfections.

Experimental results

Evaluation of shell buckling tests

Many shell buckling tests have been carried out to determine the behaviour of cylindrical shells under combined loading experimentally, see for discussion the overviews given by Babcock (1974) and Singer (1964). A literature study has been performed to evaluate all published load interaction buckling test data known to the authors. Over 2200 individual buckling experiment results have been collected. The data base is very heterogeneous, containing test results on small-scale models made of plastic as well as large welded steel cylinders. Unfortunately, often information is missing concerning material properties or boundary conditions, so some data could only be related approximately to yield limit (as is obvious in the case of mylar or plexiglass) or design buckling strength predictions. Thus, conservative estimates have been used for items that remain uncertain (e.g. BC S3 was assumed if there were doubts concerning the axial restraint of the test shell). The whole evaluation procedure including source and buckling relevant parameters of each test is reported by Winterstetter (2000).

Results of fundamental load reference tests

Every interactive buckling test series consists of a set of experiments, which includes tests under fundamental loads for reference. In Fig. 10.11, these results of tests with only one basic stress component are plotted against some of the European design code reduction curves as discussed in the section 'Cylindrical shells under fundamental loads'.

The figures show the wide scatter typical of shell buckling experiments. In general, the fundamental stress test data are in good agreement with the reduction curves. Thus, the values obtained from the whole of interaction test series can be expected to give reasonable results.

Results of combined loading tests

In Figs 10.12 and 10.13, all interaction test results are plotted against the reported/assumed yield limit. No validated test data were found concerning experiments on steel shells with combinations of all three fundamental load types. A lack of interaction tests on thicker cylinders under load combinations including torsion was also detected. Thus, a series of combined loading shell buckling tests under axial compression and torsion was performed at the University of Essen. The results are reported by Winterstetter and Schmidt (1999), a summary of which is given by Schmidt and Winterstetter (1999). The cases with tensile stress components (e.g. combined axial tension and external pressure) are included in Figs 10.12 and 10.13, though they remain unconsidered within this chapter. These effects are discussed by Winterstetter (2000) and by Winterstetter and Schmidt (2002).

In the case of axial compression and external pressure (Fig. 10.12), the test data cover the whole range of purely elastic up to plastic interactive buckling. In the other two cases of axial compression and torsion and external pressure and torsion (Fig. 10.13), no experiments were found in the region of plastic buckling, as mentioned above.

The well-known strength-raising effect of internal pressure for elastic axial compression buckling can be observed as well as the reducing effect of biaxial stresses with unequal signs and the strengthening effect of biaxial compression in the plastic region. In general, experimental elastic interactive buckling occurs in agreement with the theoretical LA/GNA buckling curve shapes, at least in the cases of near-perfect test shells often used. Steel cylinders with moderate to large wall thickness buckle more in agreement with the very convex plastic interaction curve.

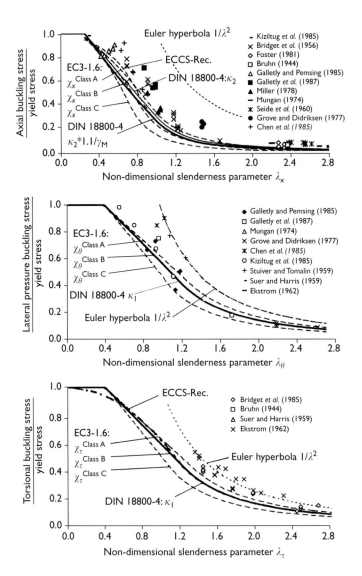

Figure 10.11 Buckling under fundamental loads: comparison of the fundamental load reference test data of the collected interactive buckling experiment series with European design code reduction curves.

Some selected experimental interactive buckling curves are shown in Fig. 10.14. They are constructed from the buckling stresses of several test series with nominally identical specimens or with one elastically buckling specimen subject to various load combinations.

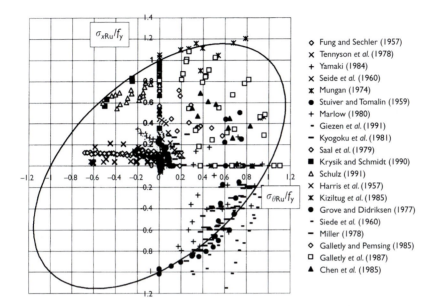

Figure 10.12 Shell buckling experiments under combined axial and pressure loading: evaluated ultimate load test data, related to reported/assumed yield limit.

In Fig. 10.14b, the interaction between axial compression and torsion is shown. In the case of elastic buckling, where LA interaction is nearly linear (see Fig. 10.8), the experimental curves of Foster (1981) or Bridget *et al.* (1956) are rather convex, probably because of the different imperfection sensitivities of the two basic fundamental buckling types. In Fig. 10.14a, the slenderness-related shapes of the interactive buckling curves dealing with axial compression and external pressure with their marked convexity are confirmed again.

In general, the test results show that the interactive buckling behaviour of a cylindrical steel shell with given dimensions depends strongly on the amount of plasticity involved in the buckling process.

Design rules for combined loading in codes and recommendations

As demonstrated in Figs 10.2–10.4, existing codified design rules differ moderately when predicting the buckling stresses under fundamental loads. The predictions for interactive buckling differ much more. Figures 10.15 and 10.16 show an evaluation of some design codes and standards. For example, the ECCS Recommendations (ECCS 1988) specify for the combination of axial compression and external pressure (Fig. 10.15) a linear interaction according to Eq. (2), the German DIN 18800–4

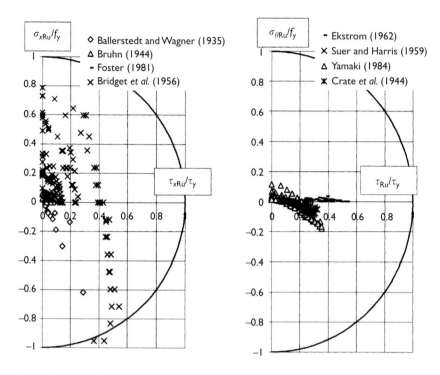

Figure 10.13 Shell buckling experiments under combined loading including torsion: evaluated ultimate load test data, related to reported/assumed yield limit.

(DIN 1990) and the European Prestandard ENV 1993-1-6 (CEN 1998) give a slightly curved interaction (Eq. 3), and the ASME Code (ASME 1980) distinguishes between plastic buckling (assumption: no interaction at all) and elastic buckling (parabolic interaction according to Eq. 4).

$$\frac{\sigma_{xd}}{\sigma_{xRd}} + \frac{\sigma_{\theta d}}{\sigma_{\theta Rd}} = 1 \qquad \text{(ECCS Recommendations)} \qquad (2)$$

$$\left(\frac{\sigma_{xd}}{\sigma_{xRd}}\right)^{1.25} + \left(\frac{\sigma_{\theta d}}{\sigma_{\theta Rd}}\right)^{1.25} = 1 \qquad \text{(DIN 18800-4, ENV 1993-1-6)} \qquad (3)$$

$$\frac{\sigma_{xd}}{\sigma_{xRd}} + \left(\frac{\sigma_{\theta d}}{\sigma_{\theta Rd}}\right)^{2} = 1 \qquad \text{(ASME Code Case N, elastic buckling)} \qquad (4)$$

The DnV Rules use a more sophisticated concept described by Odland (1991) which in essence yields a slenderness-related curvature of the interactive buckling

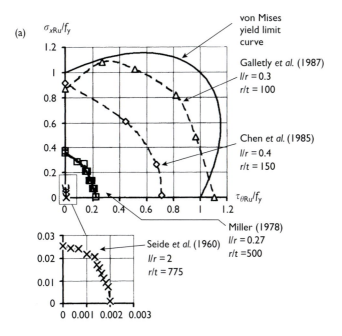

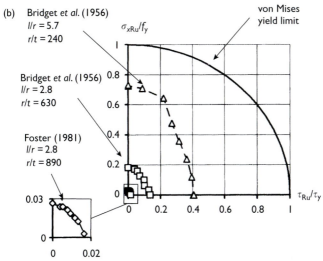

Figure 10.14 Selected experimental interactive buckling curves: (a) axial compression and external pressure; (b) axial compression and torsion.

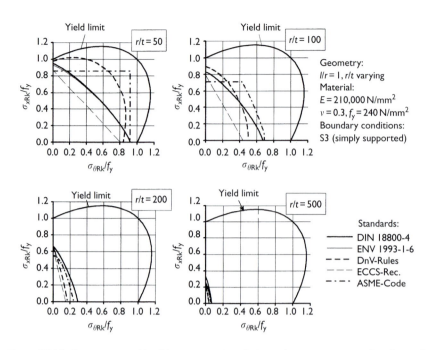

Figure 10.15 Comparison of buckling interaction design rules in international codes and standards: axial compression and external pressure.

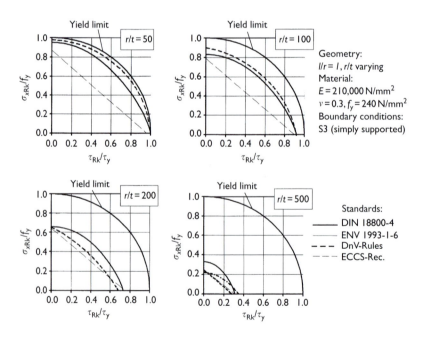

Figure 10.16 Comparison of buckling interaction design rules in international codes and standards: axial compression and torsion.

curves. Unfortunately, this concept does not allow for the strengthening effects of internal pressure on the axial compression buckling of thin cylinders.

Concluding remarks

This chapter has examined the interactive buckling of cylindrical shells under combinations of the fundamental loads both theoretically and experimentally. The interaction has been shown to be slenderness-dependent, that is, the slenderness and therefore the amount of yielding involved in the buckling process determine the convexity of the individual interaction curves. For several shell geometry parameters and load combinations, interaction curve shapes have been presented and compared with design rules given in international standards. It was found that the quite simple interaction formulations used in most standards are not sufficient to give a proper description of the theoretical and experimental results. A slenderness-related formulation is used by the DnV Rules, but some effects are not covered by this. A general, more accurate method for the description of interactive buckling of steel cylinders was part of a combined loading research programme at the University of Essen which has been completed very recently (Winterstetter 2000; Winterstetter and Schmidt 2002).

Notation

Buckling stress components

σ_{xRc}, $\sigma_{\theta Rc}$, τ_{Rc}	linear critical buckling stress components under combined loading (linear eigenvalues of load combination)
$\sigma_{xRc,0}$, $\sigma_{\theta Rc,0}$, $\tau_{Rc,0}$	linear critical buckling stresses under fundamental loads (linear eigenvalues)
σ_{xRnl}, $\sigma_{\theta Rnl}$, τ_{Rnl}	nonlinear buckling stress components, either from GNA or from GMNA
σ_{xRu}, $\sigma_{\theta Ru}$, τ_{Ru}	ultimate buckling stress components, either from GMNIA or from tests
σ_{xRk}, $\sigma_{\theta Rk}$, τ_{Rk}	characteristic buckling stress components (design resistance values, without safety factor)

Slenderness parameters

$$\lambda_{(\cdot)} = \sqrt{\frac{f_y}{\sigma_{(\cdot)Rc}}} \qquad (\cdot) = x \text{ or } \theta$$

$$\lambda_{\tau} = \sqrt{\frac{f_y}{\sqrt{3}\tau_{Rc}}}$$

Material parameters

E modulus of elasticity
f_y uniaxial yield stress
ν Poisson's ratio

Geometric parameters

r radius of the cylinder middle surface
t wall thickness
l length between supports

Boundary conditions (BCs)

C1 $w = w,_x = u = v = 0$
C2 $w = w,_x = u = N_{x\theta} = 0$
C3 $w = w,_x = N_x = v = 0$
C4 $w = w,_x = N_x = N_{x\theta} = 0$
S1 $w = w,_{xx} = u = v = 0$
S2 $w = w,_{xx} = u = N_{x\theta} = 0$
S3 $w = w,_{xx} = N_x = v = 0$
S4 $w = w,_{xx} = N_x = N_{x\theta} = 0$

Subscripts

x axial direction
θ circumferential direction
u   ultimate limit state load component
R   resistance value
k   characteristic value
c   critical value
0   value for fundamental buckling stress

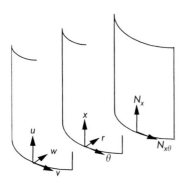

References

Standards

API (1987). *Bulletin on Stability Design of Cylindrical Shells*. Bulletin2U, 1st edn. American Petroleum Institute, Washington, DC.

ASME (1980). Section III Code Case N-284. *Metal Containment Shell Buckling Design Methods*. American Society of Mechanical Engineers, New York.

CEN (1998). ENV 1993-1-6 (EC3-1.6): *Design of Steel Structures, Part 1–6: General Rules: Supplementary Rules for the Strength and Stability of Shell Structures*. CEN, Brussels, Belgium.

DASt (1992). DASt-Richtlinie 017 (Entwurf). *Beulsicherheitsnachweise für Schalen – spezielle Fälle*. Deutscher Ausschuß für Stahlbau, Stahlbau-Verlagsgesellschaft, Düsseldorf, Germany.

DIN (1990). DIN 18800 Teil 4. *Stahlbauten – Stabilitätsfälle, Schalenbeulen*. Deutsches Institut für Normung (DIN), Normenausschuß Bau (Ed.), Berlin, Germany.

ECCS (1988). European Convention for Constructional Steelwork (ECCS): *Buckling of Steel Shells – European Recommendations*, 4th edn. General Secretariat ECCS, Brussels, Belgium.

Papers and monographs

Almroth, B.O. (1966). Influence of edge conditions on the stability of axially compressed cylindrical shells. *AIAA Journal* **4**, 134–140.

Babcock, C.D., Jr (1974). Experiments in shell buckling. In *Thin-Shell Structures – Theory, Experiment and Design* (eds Y.C. Fung and E.E. Sechler). Prentice-Hall, New Jersey, pp. 345–369.

Ballerstedt, W. and Wagner, H. (1935). Versuche über die Festigkeit dünner unversteifter Zylinder unter Schub-und Längskräften. *Luftfahrtforschung* **12**, 309–312.

Batdorf, S.B., Stein, M. and Schildcrout, M. (1947). Critical combinations of torsion and direct axial stress for thin-walled cylinders. NACA TN 1345, Washington, DC.

Booton, M. and Tennyson, R.C. (1979). Buckling of imperfect anisotropic circular cylinders under combined loading. *AIAA Journal* **17**, 278–287.

Bridget, F.J., Jerome, C.C. and Vosseller, A.B. (1956). Some new experiments on buckling of thin-wall construction. *Transactions of the ASME, Applied Mechanics* **56**(6), 569–578.

Bruhn, E.F. (1944). Tests on thin-walled celluloid cylinders to determine the interaction curves under combined bending, torsion and compression or tension loads. NACA TN 951, Washington, DC.

Chen, Y., Zimmer, R.A., de Oliveira, J.G. and Jan, H.Y. (1985). Buckling and ultimate strength of stiffened cylinders: model experiments and strength formulations. OTC Paper 4853, Houston, TX, pp. 113–124.

Crate, H., Batdorf, S.B. and Baab, G.W. (1944). The effect of internal pressure on the buckling stress of thin-walled circular cylinders under torsion. NACA ARR No. L4E27.

Ekstrom, R.E. (1962). Buckling of cylindrical shells under combined torsion and hydrostatic pressure. *Experimental Mechanics* **8**, 192–197.

Flügge, W. (1973). *Stresses in Shells*, 2nd edn. Springer, New York.

Foster, C.G. (1981). Interaction of buckling modes in thin-walled cylinders. *Experimental Mechanics* **27**, 124–128.

Fung, Y.C. and Sechler, E.E. (1957). Buckling of thin-walled circular cylinders under axial compression and internal pressure. *Journal of Aeronautical Sciences* **24**, 351–356.

Galletly, G.D. and Pemsing, K. (1985). Interactive buckling tests on cylindrical shells subjected to axial compression and external pressure – a comparison of experiment, theory and various codes. *Proceedings of IMechE* **199**(CA), 259–280.

Galletly, G.D., James, S., Kruzelecki, J. and Pemsing, K. (1987). Interactive buckling tests on cylinders subjected to external pressure and axial compression. *Transactions of the ASME, Journal of Pressure Vessel Technology* **109**, 10–18.

Giezen, J.J., Babcock, C.D. and Singer, J. (1991). Plastic buckling of cylindrical shells under biaxial loading. *Experimental Mechanics* **37**, 337–343.

Grove, T. and Didriksen, T. (1977). Buckling experiments on four large ring-stiffened cylindrical shells subjected to axial compression and lateral pressure. Det norske Veritas Report No. 77-431, Oslo, Norway.

Harris, L.A., Suer, H.S., Skene, W.T. and Benjamin, R.J. (1957). The stability of thin-walled unstiffened circular cylinders under axial compression including the effects of internal pressure. *Journal of Aeronautical Sciences* **8**, 587–596.

Ho, B.P.C. and Cheng, S. (1963). Some problems in stability of heterogenous aeolotropic cylindrical shells under combined loading. *AIAA Journal* **1**(7), 1603–1607.

Hutchinson, J. (1965). Axial buckling of pressurized imperfect cylindrical shells. *AIAA Journal* **3**, 1461–1466.

Kiziltug, A.Y., Grove, R.B., Peters, S.W. and Miller, C.D. (1985). Collapse tests of short tubular columns subjected to combined loads. Final Report, Pennzoil/Industry Project 1985, CBI Industries Inc., Plainfield Ill.

Kromm, A. (1942). Die Stabilitätsgrenze der Kreiszylinderschale bei Beanspruchung durch Schub-und Längskräfte. *Jahrbuch 1942 der deutschen Luftfahrtforschung*, pp. 602–616.

Krysik, R. and Schmidt, H. (1990). *Beulversuche an längsnahtgeschweißten stählernen KZS im elastisch-plastischen Bereich unter Meridiandruck-und innerer Manteldruckbelastung*. Res. Rept. 51, Dept. Civ. Eng., University of Essen.

Kyogoku, T., Tokimasa, K., Nakanishi, H. and Okazawa, T. (1981). Experimental study on the effect of axial tension load on the collapse strength of oil well casing. *Proceedings of the 13th Annual Offshore Technology Conference*, Houston, TX, vol. 3, pp. 4065–4115 Paper OTC 4108, pp. 387–395.

Marlow, R.S. (1980). Collapse performance of fabricated cylinders under combined axial tension and external pressure. Final Report, Southwest Research Institute Project No. 03-6139, San Antonio, TX.

Miller, C.D. (1978). Summary of buckling tests on fabricated steel cylindrical shells in USA. *Buckling of Shells in Offshore Structures* (ed. J.E. Harding, P.J. Dowling and N. Agelidis). Granada, London, pp. 429–473.

Mungan, I. (1974). Buckling stress states of cylindrical shells. *ASCE Journal of the Structural Division* **100**(11), 2289–2306.

Odland, J. (1991). Design codes for offshore structures, buckling of cylindrical shells. In *Buckling of Shell Structures, on Land, in the Sea and in the Air* (ed. J.F. Jullien). Elsevier Science, London, UK, pp. 277–285.

Peil, U. and Nölle, H. (1988). Zur Frage der Schalenwirkung bei dünnwandigen, zylindrischen Stahlschornsteinen. *Bauingenieur* **63**, 51–56.

Rotter, J.M. (1997). Design standards and calculations for imperfect pressurized axially compressed cylinders. *Proceedings of the International Conference on Carrying Capacity of Shell Structures* (eds. Krupka/Schneider). Brno, pp. 354–360.

Saal, H., Kahmer, H. and Reif, A. (1979). Beullasten axialgedrückter KZS mit Innendruck – neue Versuche und Vorschriften. *Stahlbau* **48**, 262–269.

Schmidt, H., Binder, B. and Lange, H. (1998). Postbuckling strength design of open thin-walled cylindrical tanks under wind load. *Thin-Walled Structures* **31**, 203–220.

Schmidt, H. and Winterstetter, Th.A. (1999). Buckling interaction strength of cylindrical steel shells under axial compression and torsion. *Proceedings of the ICASS '99* (eds Chan/Teng). Hong Kong, China. Vol. 2, pp. 597–604.

Schulz, U. (1991). Zylinderschalen mit und ohne Innendruck im elastisch-plastischen Bereich. *Stahlbau* **60**, 103–110.

Seide, P., Weingarten, V.I. and Morgan, E.J. (1960). Final Report on the Development of Design Criteria for Elastic Stability of Thin Shell Structures. STL/TR-60-0000-19425, Los Angeles.

Simitses, G.J. (1967). Instability of ortothropic cylindrical shells under combined torsion and hydrostatic pressure. *AIAA Journal* **5**(8), 1463–1469.

Singer, J. (1964). On experimental technique for interaction curves of buckling of shells. *Experimental Mechanics* **10**, 279–287.

Stracke, M. (1987). Stabilität kurzer stählerner Kreiszylinderschalen unter Außendruck. Dr.-Ing. Thesis, University of Essen, Deutscher Verlag für Schweißtechnik, Düsseldorf.

Stuiver, W. and Tomalin, P.E. (1959). The failure of tubes under combined external pressure and axial load. *Proceedings of the SESA XVI* vol. 2, pp. 39–48.

Suer, H.S. and Harris, L.A. (1959). The stability of thin-walled cylinders under combined torsion and external lateral or hydrostatic pressure. *Transactions of the ASME, Journal of Applied Mechanics* **3**, 138–140.

Teng, J.G. and Rotter, J.M. (1992). Buckling of pressurized axisymmetrically imperfect cylinders under axial load. *Journal of Engineering Mechanics* **118**, 229–247.

Tennyson, R.C., Booton, M. and Chan, K.H. (1978). Buckling of short cylinders under combined loading. *Transactions of the ASME, Journal of Applied Mechanics* **45**, 574–578.

Winterstetter, Th.A. (2000). Stabilität von Kreiszylinderschalen aus Stahl unter kombinierter Beanspruchung. Dr.-Ing. Thesis, University of Essen.

Winterstetter, Th.A. and Schmidt, H. (1999). Beulversuche an längsnahtgeschweißten stählernen KZS im elastisch-plastischen Bereich unter Axialdruck, Innendruck und Torsionsschub. Research Rep. 82, Dept. of Civil Eng., University of Essen.

Winterstetter, Th.A. and Schmidt, H. (2002). Stability of circular cylindrical steel shells under combined loading. *Thin-Walled Structures* **40**, 893–909.

Yamaki, N. (1984). *Elastic Stability of Cylindrical Shells*. North-Holland, Amsterdam.

Chapter 11

Stiffened cylindrical shells

J. Singer

Introduction

Stiffened shells, and in particular stiffened circular cylindrical shells, are usually very efficient structures that have many applications in civil, aerospace, marine, petrochemical and mechanical engineering.

A good example in aeronautical construction was the classical aircraft fuselage of the 1930s. This widely used 'semimonocoque' aeroplane shell structure was typified by a very thin skin, which often buckled in flight, supported by much stiffer longitudinal and transverse reinforcing elements (see, e.g. Hoff 1967). The structural element that buckled was therefore a plate, or an unstiffened shell segment, supported by relatively rigid stiffeners. Buckling was primarily 'local buckling' and the final strength of the structure was then estimated with the aid of the concepts of effective width and tension field (see, e.g. Wagner 1929; von Kármán *et al.* 1932; Cox 1933; Kuhn 1933; Singer *et al.* 1997, 2000, chapter 8). With increase in the speed of flight, skins became thicker and stiffeners closer, till skin and reinforcements were of similar rigidity and interacted during buckling. The buckling behaviour of the modern aerospace structure differs, therefore, from that of the earlier semimonocoque, in that it is primarily not a local phenomenon, but involves the entire structure, being predominantly 'global buckling' or 'general instability'.

The efficient aerospace shell structure today is a stiffened shell, usually closely stiffened, the stiffening being stringers and rings, sandwich construction, corrugated reinforcement, composite material stiffening, 45° waffle-stiffeners or integrally machined stringers and rings. These integrally stiffened shells have gained prominence in aircraft, missiles and space launchers in recent years and have, therefore, been the subject of extensive analytical and experimental studies, in particular the most commonly employed type of shells – stiffened cylindrical shells (see, e.g. the summaries in Milligan *et al.* 1966; Singer and Baruch 1966; Singer 1972; Esslinger and Geier 1975, Sections 3.5 and 3.6; Bushnell 1985, chapters 4–6).

In civil, marine or mechanical engineering applications, the stiffeners were generally more widely spaced, causing local buckling to precede any general instability. The purpose of ring stiffeners is then primarily to enhance the local buckling strength of the cylinder, whereas the longitudinal stiffeners, the stringers, serve mainly to increase its axial or bending strength. In recent years, however, optimization has led to wider use of closely stiffened cylindrical shells also in marine and civil engineering, accompanied by even more extensive studies of their properties (e.g. Singer 1976; Dowling and Harding 1982; Green and Nelson 1982; Walker *et al.* 1982a,b; Croll 1985; Kendrick 1985). Useful information can also be found in Ellinas *et al.* (1984, Part III) or Galambos (1998, sections 14.4 and 14.5).

Steel offshore structures represent a typical application of cylindrical shells in civil and marine engineering. With their increase in size, relatively thin walls appeared. As these shell elements are subjected to loads that produce considerable compressive stresses, buckling becomes their most probable mode of failure. The quest for structural efficiency soon eliminated unstiffened shells in steel offshore structures, as had happened earlier in aerospace structures, and made the designers turn to stiffened shells.

Since the main loads in offshore structures are axial compression, bending and external pressure, both stringers and rings are employed for the stiffening of their cylindrical shell elements. Though the welded construction employed in steel offshore structures and other marine applications implies different fabrication tolerances than those imposed in the aerospace industry, and therefore different imperfection reduction factors pertain, much of the accumulated aerospace experience can be utilized. Indeed, extensive technology transfer from aerospace to steel offshore structures took place in the 1970s (e.g. Singer 1976; Walker and Kemp 1976; Walker *et al.* 1982a).

There exist, however, some important differences between the stiffened shells used in aerospace applications and those employed in offshore structures. These differences are associated with the different materials of construction and their stress–strain characteristics, the geometries and number of stiffeners, the methods of manufacture (like flame-cutting and welding) and the consequent initial imperfections and residual stresses. Hence, in the early 1980s, the British Department of Energy and Science Research Council commissioned an extensive research and test programme on small scale and large stiffened cylindrical shells for offshore applications at four UK universities (see, e.g. Dowling and Harding 1982; Dowling *et al.* 1982; Green and Nelson 1982; Walker *et al.* 1982a,b). This four-pronged research provided a significant data base for analysis and design of steel offshore shells. Similar research programmes that were carried out simultaneously in other countries (see, e.g. Miller 1982; Valsgård and Foss 1982) verified and enhanced this data base. Although ring-stiffened cylindrical shells are employed more often in these structures, axial loads, bending moments, shear forces and concentrated loads that occur frequently have recently led to wider use of stringer stiffening, as well as combined ring- and stringer-stiffening.

One may note in passing that whereas the postbuckling behaviour of stiffened shells has usually been studied primarily to infer from it their imperfection sensitivity, for offshore structures it may have another significance. The postbuckling behaviour and the load carrying capacity after large buckling distortions may be important here as design criteria for the case of very rarely occurring excessive loads, such as in an exceptional storm or in a collision, where reparable damage would perhaps be acceptable.

Another important application of stiffened cylindrical shells in marine engineering is submarines. Here, the primary load is external pressure, and, therefore, ring-stiffening dominates. The buckling behaviour of ring-stiffened cylindrical shells under external pressure has, therefore, worried submarine designers for decades and has motivated extensive theoretical and experimental studies at several research centres like the US Navy David Taylor Model Basin (see, e.g. Galletly *et al.* 1958; Krenzke and Reynolds 1967), the UK Naval Construction Research Establishment (see, e.g. example Kendrick 1953, 1985) and many others. Indeed, many of the 'classical' external pressure experiments of the 1920s and 1930s were actually on ring-stiffened cylindrical shells (see, e.g. Tokugawa 1931; Saunders and Windenburg 1932; Wenk *et al.* 1954). The problems encountered later in submarine design motivated further extensive research in the following decades on the buckling behaviour of stiffened cylindrical shells subjected to external pressure (see, e.g. Galletly *et al.* 1958; Ross *et al.* 1971; Yamamoto *et al.* 1989; Ross 1990; Ross *et al.* 1999).

One should note that submarine hulls are usually relatively thick-walled and, therefore, they mostly buckle inelastically. The relevant stiffened cylindrical shells under external pressure can, therefore, fail in one of three modes: inelastic local shell instability, elastic or inelastic general instability or axisymmetric collapse. Because of the difficulty to accurately predict the inelastic buckling pressures of full-scale submarine hulls, the modern design approach is, therefore, to place the rings close enough to rule out (local) shell instability and then to make the rings strong enough to prevent general instability (see, e.g. Ross *et al.* 1999). This limits possible failure to axisymmetric collapse, which can be predicted more reliably.

From the examples briefly discussed, one notes that the similarities in the buckling behaviour of the various applications of stiffened cylindrical shells outweigh the differences. Hence, also their buckling and postbuckling analyses are very similar.

Analysis of stiffened cylindrical shells

Global and local buckling, and axisymmetric collapse, of stiffened shells

Summarising their buckling and collapse behaviour, stiffened cylindrical shells subjected to external pressure or axial compression can fail in one of three modes: *local shell instability* (elastic or inelastic), *global (general) instability* (elastic or inelastic) or *axisymmetric plastic collapse*.

For the basic case of *ring-stiffened* shells under external pressure, typical local shell instability (or inter-ring buckling) is shown in Fig. 11.1, typical global (general) instability (or overall collapse) is represented in Fig. 11.2, whereas axisymmetric local plastic collapse (yielding) is depicted in Fig. 11.3. When

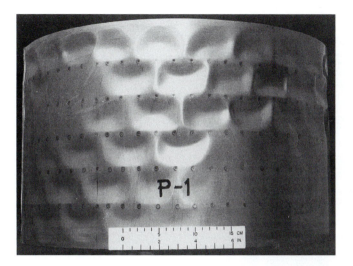

Figure 11.1 Local shell instability of a ring-stiffened cylindrical shell subjected to external pressure. The shell tested at Technion was made of stainless steel, had
a radius of 275 mm, $(R/h) \cong 344$ and had five spot-welded Z-shape rings
spaced at 50 mm distance (i.e. $a/h \cong 63$). Because of the wide spacing,
the shell buckled in the local shell mode. As the pressure increased, the
local buckles spread between the rings, eventually covering most of the
shell and transformed into inelastic deformations.

subjected to axial compression, ring-stiffened cylindrical shells behave in a very
similar manner. The postbuckling pattern of an axially compressed, widely spaced,
ring-stiffened cylinder is shown in Fig. 11.4, exhibiting typical local shell buckling
between the rings. Failure by general instability is typical of shells with closely
spaced ring-stiffening, as represented in Fig. 11.5. For thicker walled ring-stiffened
cylindrical shells under axial compression, the collapse mode may be axisymmetric
plastic collapse for both widely and closely spaced stiffening.

Stringers, the longitudinal stiffeners, are most effective in increasing the axial
or bending strength of cylindrical shells, and their optimization has resulted in
closely spaced stiffening, which generates general instability failure. Figure 11.6
depicts such a typical global instability in an axially compressed, stringer-stiffened
7075-T6 aluminium alloy cylindrical shell. When the stringers are more widely
spaced, local shell instability will precede global collapse in axially compressed
cylindrical shells. Figure 11.7 shows local buckling in such a Technion 7075-
T6 aluminium alloy stringer-stiffened shell, whose stringers were spaced slightly
wider than would have been necessary to avoid local panel buckling, that is, with a
Koiter measure of total curvature (Koiter 1956) of $\cong 0.8$. The behaviour of the shell
under successive axial compressive loads is presented in this figure, exhibiting the
formation of local buckling waves, till finally a sudden collapse, with a completely
different general instability pattern, occurs. More details are available in Chapter 13

Figure 11.2 General instability (overall collapse) of a ring-stiffened cylindrical shell subjected to external pressure (from Krenzke and Reynolds 1967).

Figure 11.3 Axisymmetric local plastic collapse (yielding) in a ring-stiffened cylindrical shell subjected to external pressure (from Krenzke and Reynolds 1967).

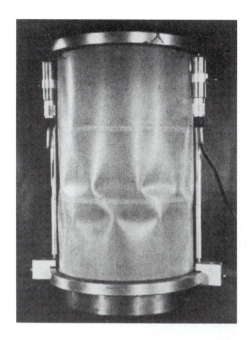

Figure 11.4 Local shell instability of an axially compressed Mylar cylindrical shell with widely spaced rings, tested at DLR Braunschweig (from Esslinger and Geier 1970; reproduced by kind permission of Springer-Verlag).

Figure 11.5 General instability of a Technion closely spaced, ring-stiffened 7075-T6 aluminium alloy cylindrical shell subjected to axial compression (from Rosen and Singer 1975).

Figure 11.6 General instability of a Technion closely spaced, integrally stringer-stiffened 7075-T6 aluminium alloy cylindrical shell subjected to axial compression (from Singer 1979; reproduced by kind permission of Springer-Verlag).

of Singer *et al.* (1997, 2001). One should note that such a transition at high loads from local buckling to a general instability collapse may occur also for different stiffener configurations and loadings. In thicker-walled shells with widely spaced stringers, however, no transition to general instability will occur but local elastic buckling will directly transform into local plastic mechanisms that lead to collapse in a local plastic mode (see Fig. 11.8).

Furthermore, the buckling and postbuckling behaviour of ring- or stringer-stiffened cylindrical shells, subjected to bending, torsion or combined loading, or of shells stiffened both circumferentially and axially, is more complex, but it can essentially also be divided into three similar types of failure modes.

Since most optimal configurations of stiffened cylindrical shells have closely spaced stiffeners, general instability is their dominant failure criterion. Hence, for many years buckling analyses of stiffened shells considered equivalent orthotropic shells (see, e.g. Bodner 1957; Thielemann 1960; Becker and Gerard 1962).

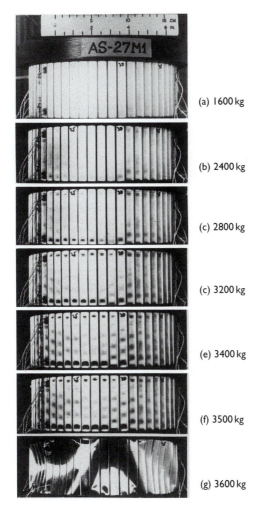

(a) 1600 kg

(b) 2400 kg

(c) 2800 kg

(c) 3200 kg

(e) 3400 kg

(f) 3500 kg

(g) 3600 kg

Figure 11.7 Buckling and postbuckling behaviour at successive loads of an axially compressed integrally stringer-stiffened 7075-T6 aluminium alloy cylindrical
shell, with more widely spaced stringers, tested at Technion. The shell
buckled locally between the stringers, with local buckling waves deep-
ening and extending as the load increased, till finally a sudden global
collapse occurred, with an entirely different pattern representing general
instability.

This approach, however, did not permit taking into account the eccentricity of
stiffeners (whether they are on the outside or inside of the shell), whose effect had
already been observed in the 1930s and 1940s (Flügge 1932; van der Neut 1947).
As stiffeners became heavier, in particular in the large space launch vehicles and

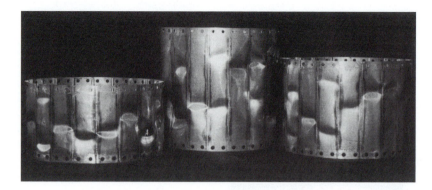

Figure 11.8 Collapse in a local plastic buckling mode of axially compressed welded structural steel cylindrical shells, with widely spaced stringers, tested at University College London. These shells were 1/20th scale models of typical designs for vertical legs of semi-submersible offshore structures. Buckling commenced with local elastic buckles that transformed into local plastic mechanisms (from Walker and Kemp 1976).

marine structures, designers became aware of the importance of these eccentricity effects and their influence.

The eccentricity effects for stringer- and ring-stiffened cylindrical shells have been extensively studied, been given a physical explanation and comprehensive parametric studies have been carried out (see, e.g. Thielemann and Esslinger 1965; Singer *et al.* 1966, 1967). Outside stiffeners have usually been found to stiffen the shell more than inside ones, but the opposite may sometimes be true, depending on the loading and shell geometry, and in a certain range of shell geometries an inversion of the eccentricity effect occurs, which is most pronounced in the case of ring-stiffened shells under external pressure or torsion (see Baruch *et al.* 1966; Singer and Baruch 1966; Singer *et al.* 1966).

Smeared-stiffener theory

'Smeared'-stiffener theory, in which the stiffeners are 'smeared', or 'distributed', over the entire shell, in a manner that takes into account the eccentricity of stiffeners, has been found to be a satisfactory approach for closely stiffened shells that fail by general instability, since the effect of the discreteness of stiffeners is usually negligible.

The smeared stiffener theory is characterised by the mathematical model representing the stiffeners. The model can be employed in a simplified linear theory, as used in most studies (e.g. Thielemann and Esslinger 1965; Baruch *et al.* 1966; Singer *et al.* 1966; Bushnell 1984), or in more sophisticated numerical solutions, which take into consideration the nonlinear effects caused by the edge restraints. In the early 1960s, a simple 'smeared'-stiffener theory was derived at Technion (see Baruch and Singer 1963), which has been widely used on account of its simplicity (see, e.g. Block *et al.* 1965; Burns 1966; Card and Jones 1966; Simitses 1968).

The 'smeared' rings and stringers are introduced to the equilibrium equations, which are represented in terms of displacements, through the force and moment expressions. The stiffeners are 'smeared' to form a *cut* layer. For example, external rings are replaced by a layer of many parallel rings that cover the whole exterior of the shell, touch each other but are not connected with each other. One may note that the model slightly underestimates the torsional stiffness, but this is preferable to its overestimation that occurs when the stiffeners are replaced by a concentric orthotropic continuous layer, as in Seggelke and Geier (1967).

Therefore, the main assumptions of the model are (as in Baruch and Singer 1963; Singer *et al.* 1966, 1967), using the notation of Fig. 11.9:

1   The stiffeners are 'distributed' over the whole surface of the shell.
2   The normal strains $\varepsilon_x(z)$ and $\varepsilon_\phi(z)$ vary linearly in the stiffener as well as in the sheet. The normal strains in the stiffener and in the sheet are equal at their point of contact.
3   The shear membrane force $N_{x\phi}$ is carried entirely by the sheet.
4   The torsional rigidity of the stiffener cross-section is added to that of the sheet (the actual increase in torsional rigidity is larger than that assumed).

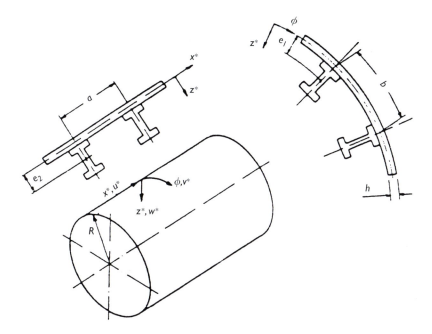

Figure 11.9 Notation.

The middle surface of the shell is chosen as reference line (Fig. 11.9), and the expressions for forces and moments in terms of displacements are (Singer *et al.* 1967, 1997, 2001; Singer 1972):

$$N_x = [Eh/(1-v^2)][u,_x\,(1+\mu_1) + v(v,_\phi\,-w) - \chi_1 w,_{xx})$$
$$N_\phi = [Eh/(1-v^2)][(v,_\phi\,-w)(1+\mu_2) + vu,_x\,-\chi_2 w,_{\phi\phi}\,]$$
$$N_{x\phi} = N_{\phi x} = [Eh/2(1+v)](u,_x\,+v,_\phi\,)$$
$$M_x = -(D/R)[w,_{xx}\,(1+\eta_{01}) + vw,_{\phi\phi}\,-\zeta_1 u,_x\,]$$
$$M_\phi = -(D/R)[w,_{\phi\phi}\,(1+\eta_{02}) + vw,_{xx}\,-\zeta_2(v,_x\,-w)]$$
$$M_{x\phi} = +(D/R)[(1-v) + \eta_{t1}]w,_{x\phi}$$
$$M_{\phi x} = -(D/R)[(1-v) + \eta_{t2}]w,_{x\phi} \tag{1}$$

where h is the thickness of the shell, μ_1, μ_2, η_{01}, η_{02}, η_{t1} and η_{t2} are the changes in stiffnesses due to stringers and rings and χ_1, χ_2, ζ_1 and ζ_2 are the changes in stiffnesses caused by the eccentricities of the stringers and rings (see Eqs 2–5). Note that u, v and w are the additional displacements during buckling, which are non-dimensional, the physical displacements having been divided by the radius of the shell. (Notice that thus x and ϕ are the non-dimensional axial and circumferential coordinates, respectively.) As usual, $D = Eh^3/12(1-v^2)$, P is the axial compressive load and p the external pressure.

The changes in stiffnesses are defined as follows: The increase in effective cross-sectional area of the shell due to stringers and frames, respectively,

$$\left.\begin{aligned}\mu_1 &= (1-v^2)(A_1 E_1/bhE)\\ \mu_2 &= (1-v^2)(A_2 E_2/ahE)\end{aligned}\right\} \tag{2}$$

The changes in extensional stiffness caused by the eccentricities of stringers and frames, respectively,

$$\left.\begin{aligned}\chi_1 &= (1-v^2)(E_1 A_1 e_1/EbhR)\\ \chi_2 &= (1-v^2)(E_2 A_2 e_2/EahR)\end{aligned}\right\} \tag{3}$$

The increases in bending and twisting stiffness of the shell due to stringers and frames, respectively,

$$\left.\begin{aligned}\eta_{01} &= E_1 I_1/bD\\ \eta_{02} &= E_2 I_2/aD\\ \eta_{t1} &= G_1 I_{t1}/bD\\ \eta_{t2} &= G_2 I_{t2}/aD\end{aligned}\right\} \tag{4}$$

The changes in bending stiffness caused by the eccentricities of stringers and frames, respectively,

$$\left.\begin{aligned}\zeta_1 &= E_1 A_1 e_1 R/bD\\ \zeta_2 &= E_2 A_2 e_2 R/aD\end{aligned}\right\} \tag{5}$$

In these changes of stiffnesses, in Eqs (2)–(5), it is assumed that the stiffeners are closely spaced and, therefore, the entire shell is active. One could, however, analyse in a similar manner a shell with wider stiffener spacing where only part of the shell can be considered active (the effective width), by multiplying the terms in the above expressions relating to the shell proper by an appropriate coefficient.

Linear smeared-stiffener theory

As a first approximation, one usually considers perfect shells and employs 'classical' linear theory, which uses a membrane prebuckling solution. Within certain bounds of validity, which are obtained from experimental results, this linear theory predicts the buckling loads adequately and yields the primary characteristics of the buckling behaviour of closely stiffened cylindrical shells as well as the influence of various geometric parameters.

The linear smeared-stiffener theory developed at Technion, sometimes called the Baruch–Singer equations (see, e.g. Burns 1966), employs the linear Donnell stability equations (discussed in detail, e.g., in Donnell 1933; Brush and Almroth 1975), into which the 'smeared' stiffeners are introduced via the force and moment expressions, Eq. (1). The substitution of these expressions into the Donnell stability equations transforms them into three equations in terms of displacements:

$$[Eh/(1-v^2)]\{(1+\mu_1)u_{,xx} + [(1-v)/2]u_{,\phi\phi} + [1+v)/2]v_{,x\phi}$$
$$- \chi_1 w_{,xxx} - v w_{,x}\} = 0$$
$$[Eh/(1-v^2)]\{[(1+v)/2]u_{,x\phi} + (1+\mu_2)v_{,\phi\phi} + [(1-v)/2]v_{,xx}$$
$$- (1+\mu_2)w_{,\phi} - \chi_2 w_{,\phi\phi\phi}\} = 0$$
$$(-D/R)\{\zeta_1(-u_{,xxx}) + \zeta_2(2w_{,\phi\phi} - v_{,\phi\phi\phi}) + (1+\eta_{01})w_{,xxxx}$$
$$+ (2+\eta_{t1}+\eta_{t2})w_{,xx\phi\phi} + (1+\eta_{02})w_{,\phi\phi\phi\phi}$$
$$+ 12(R/h)^2[(1+\mu_2)(w - v_{,\phi}) - v u_{,x}] + (PR/\pi D)(w_{,xx}/2)$$
$$+ (pR^3/D)[(w_{,xx}/2) + w_{,\phi\phi}]\} = 0 \qquad (6)$$

For classical simple supports (SS3: $w = M_x = N_x = v = 0$), Eqs (6) are solved by the same displacements as for isotropic cylinders,

$$u = A_n \cos m\beta x \sin n\phi$$
$$v = B_n \sin m\beta x \cos n\phi \qquad (7)$$
$$w = C_n \sin m\beta x \sin n\phi$$

(see Singer et al. 1967). Notice that m, n are integers denoting the number of half-waves in the axial and the number of full waves in the circumferential direction, respectively, and $\beta = (\pi R/L)$. For classical clamped ends

(C2: $w = w,_x = u = N_{x\phi} = 0$) a closed form solution is not possible, but the same displacements and procedure (applying the Galerkin method to the third equation) as for unstiffened shells could be employed (e.g. Batdorf 1947; Singer et al. 1967). Other linear analyses start with slightly different assumptions (e.g. Hedgepeth and Hall 1965; Seggelke and Geier 1967). Furthermore, in the mid-1960s and early 1970s, generalisations and improved solution procedures that consider different in-plane boundary conditions were developed as well (e.g. Block et al. 1965; Card and Jones 1966; Soong 1967; Weller 1971, 1978; Singer and Rosen 1974, 1976; Weller et al. 1979). Computer programs for shells of revolution with different types of stiffening, based on finite differences, were also generated at that time (e.g. Bushnell et al. 1968).

The main results of linear smeared-stiffener theory have been summarised as conclusions about the effectiveness of stiffeners and the relative importance of shell and stiffener geometric parameters (Baruch et al. 1966; Singer et al. 1966, 1967; Singer 1972; Singer et al. 1997, 2001, chapter 13). The shell geometry is usually represented by one parameter, the Batdorf parameter $Z = (1 - v^2)^{1/2}(L^2/Rh)$, and the stiffener parameters are spacing, shape, cross-sectional area and eccentricity. Spacing is determined by local buckling and is hence outside the realm of the smeared-stiffener theory, whereas the theory yields the influence of the other parameters that is discussed in detail below.

The shape of the stiffeners is very important, since it determines their bending and torsional stiffness. Whereas the influence of the bending stiffness is self-evident, the importance of the torsional stiffness has not always been realised (see, e.g. Hedgepeth and Hall 1965; Seggelke and Geier 1967). Calculations at the Technion, covering a wide range of geometries (e.g. Baruch et al. 1966; Singer et al. 1967; Weller and Singer 1977), have brought out the importance of the torsional rigidity in the case of axial compression loading or torsion. It should be pointed out that the magnitude of this shape effect is also strongly dependent on the shell geometry and has a pronounced maximum at a certain Z for each stiffener geometry.

Among the parameters determining the efficiency of stiffeners, there is strong interdependence between the effects of stiffener and shell geometry and a weaker one between the various stiffener geometric parameters. The cross-sectional area of the stiffeners, or the corresponding non-dimensional area ratio, is usually the prime geometric parameter determining their effectiveness. It is, however, also the one that affects the weight directly and hence structural efficiency considerations may dictate relatively weak stiffeners.

As an example, consider Fig. 11.10 (reproduced from Singer et al. 1966) in which the influence of the ring area on the structural efficiency in buckling of a ring-stiffened cylindrical shell under hydrostatic pressure was studied. The structural efficiency is expressed here as the ratio of the calculated buckling load of the ring-stiffened shell to that of an equivalently thickened shell (an unstiffened shell thickened to the same weight). Though ring-stiffening is very effective in this case, it is apparent that the heavy rings, say with a ring area ratio $(A_2/ah) = 0.8$, are

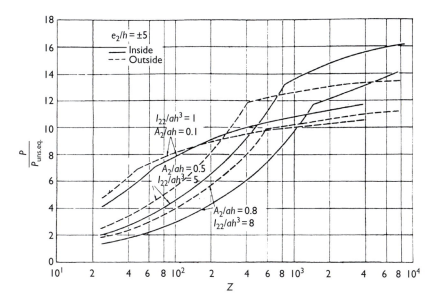

Figure 11.10 Influence of ring area on structural efficiency of closely ring-stiffened cylindrical shells under hydrostatic pressure (from Singer et al. 1966; this material
has been reproduced from the *Proceedings of the Institution of Mechanical
Engineers*, vol. 8, p. 372, Figures 7 and 9, 1966 'Inversion of the eccentricity effect in stiffened cylindrical shells buckling under external pressure' by
Singer, Baruch and Harari by permission of the Council of the Institution of
Mechanical Engineers).

never the most efficient ones. For small Z the weakest rings are best and for large
Z there appears to be an optimum area ratio.

It may be mentioned that the cross-section of the stiffeners is usually constant
for the whole shell, but noticeable weight savings of 10–20% were shown to be
possible by variation of the stiffener cross-section along the shell (e.g. Singer and
Baruch 1966; Harari *et al.* 1967). One should note that also experimentally the
stiffener cross-sectional area, or the corresponding non-dimensional area ratio, has
been shown to be the prime parameter in the determination of the applicability of
the linear theory (which assumes a membrane prebuckling state), as pointed out
below.

Eccentricity and discreteness effects

The eccentricity of the stiffeners has two effects on the behaviour of the shell. One
is an increase in the effective moment of inertia of the stiffener that is independent
of the position of the stiffener; and the other – the eccentricity effect proper – is
the result of the coupling of membrane forces with bending moments, which is

determined by the position of the stiffeners. This eccentricity effect varies with different types of loading and geometries. It is made up of two opposing contributions: the primary effect – the change in actual bending stiffness of the stiffener due to the membrane forces in the shell, and the secondary effect – the change in actual extensional stiffness of the shell due to bending moments in the stiffeners (for details see Singer *et al.* 1966, 1967). As a result of the interplay of these opposing contributions, the eccentricity effect depends very strongly on the geometry of the shell that determines which bending moments and membrane forces carry the brunt of the load.

Under axial compression, *stringers* are the most effective stiffeners. Figure 11.11 (reproduced from Singer *et al.* 1967) shows the structural efficiencies of stringer-stiffened cylindrical shells under axial compression. For a typical stringer geometry, the ratio of buckling load of stiffened shells to that of equivalent-weight unstiffened shells is plotted versus the Batdorf parameter $Z = (1 - v^2)^{1/2}(L^2/Rh)$ for outside, inside and centrally placed stringers and for classical simple supports and clamped ends. The effectiveness of the stringers is large in the low Z range but appears to diminish rapidly for longer and thinner shells. However, the curves in Fig. 11.11 are conservative, since the equivalent unstiffened shells were also

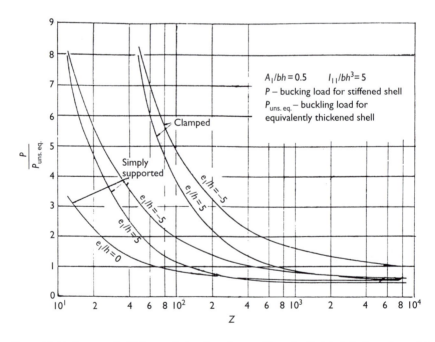

Figure 11.11 Structural efficiency of closely stringer-stiffened cylindrical shells under axial compression (from Singer *et al.* 1967; reprinted from the *International Journal of Solids and Structures*, vol. 3, Singer, Baruch and Harari, on the stability of eccentricity shells and arial compression, pp. 458–460, 1967 with permission from Elsevier Science).

computed with linear theory, which is unrealistic for them. Re-evaluation, after application of empirical corrections to the buckling loads of the unstiffened shells, yielded much higher efficiencies and showed that at least outside stringers are always more effective than equivalent thickening of the shell.

The *eccentricity effect* is very pronounced in axially compressed stringer-stiffened shells and depends strongly on the geometry of the shell, as can be seen in Fig. 11.12 (reproduced from Singer *et al.* 1967) in which (P^{out}/P^{in}) is plotted against Z. The aerospace designer may note that the eccentricity effect has a pronounced maximum that occurs for values of Z common in aerospace practice. This maximum, the shape of the curve and the inversion of the eccentricity effect, which occurs here for extremely short shells, are the results of the interplay between the opposing primary and secondary contributions that make up the total effect. For clamped ends the eccentricity effects are similar, except for a shift in the maximum to a higher Z.

One may note that by the mid-1960s the effects of stiffener eccentricity were not only found to be very significant in many theoretical studies, but were also observed experimentally! For example, in NASA Langley tests (Card and Jones 1966), externally stringer-stiffened cylinders were found to carry loads up to twice those sustained by their internally stiffened counterparts.

Rings are much less effective than stringers in stiffening of cylindrical shells, subjected to axial compression, when alone. However, they may increase the buckling load considerably when they act together with stringers (see, e.g. Burns 1966; Singer and Baruch 1966; Bushnell *et al.* 1968). Furthermore, the behaviour

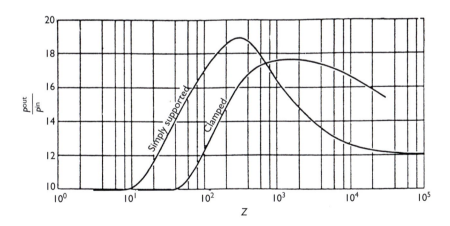

Figure 11.12 Variation of eccentricity effect with shell geometry (represented by the Batdorf parameter Z) in closely stringer-stiffened cylindrical shells under axial compression (from Singer *et al.* 1967; reprinted from the *International Journal of Solids and Structures*, vol. 3, Singer, Baruch and Harari, on the stability of eccentricity shells and arial compression, pp. 458–460, 1967 with permission from Elsevier Science).

of ring-stiffened shells under axial compression is also important, since shells primarily designed to withstand lateral pressure (which will be predominantly ring-stiffened) may be subjected to axial loads under certain conditions. Ring stiffening was, therefore, extensively studied and is discussed in detail in Singer (1972). For example, in the case of torsion loading, rings were found to be more efficient than stringers, except for short shells, and large eccentricity effects were found for rings.

In ring-stiffened cylinders under axial compression, the usual non-axisymmetric buckling occurs with inside rings and the positive eccentricity reduces the buckling load below that for centrally placed rings. With outside rings, however, the increase in buckling load that would result from the negative eccentricity in usual non-axisymmetric buckling is not realized, since the shell buckles now in an axisymmetric mode (ring shape pattern) that is unaffected by eccentricity and yields a lower buckling load. Hence, for outside rings, the buckling load can be computed from a simple formula (Eq. 8), if the axial wave number is treated as a continuous variable:

$$P_{cr} = [3(1 - v^2)]^{-1/2} 2\pi h^2 E[1 + (A_2/ah)]^{1/2} = P_{cl}[1 + (A_2/ah)]^{1/2} \qquad (8)$$

This simple formula indicates to the designer the amount of stiffening that rings may provide under axial compression.

For lateral or hydrostatic pressure loading, rings are usually the most effective stiffeners. As was seen in Fig. 11.10, rings stiffen the shell in this case considerably. It may be useful to recall here the difference between buckling under lateral and hydrostatic pressure and to point out that hydrostatic pressure is actually a particular case of combined loading with the corresponding possible buckling mode change. Whereas under lateral pressure ring-stiffened shells always buckle with one longitudinal wave and many circumferential waves, under hydrostatic pressure a different buckle pattern with many longitudinal waves, which often is also axisymmetric, can appear. The transition from the $n = 1$ mode to the $n \neq 1$ mode occurs when the axial stress component due to the hydrostatic pressure becomes dominant in short shells, or in shells with very stiff rings. This transition appears as a discontinuity in curves of buckling pressure versus Z (see, e.g. Fig. 11.10) and its position also depends on the eccentricity of the rings. The designer should note that when the $n \neq 1$ mode is critical, the buckling pressure may be considerably lower than it would have been with $n = 1$. Hence, for hydrostatic pressure, a combination of rings and stringers may sometimes be more effective, since the stringers may postpone the transition to the $n \neq 1$ mode.

Figure 11.10 shows clearly that outside rings are more effective for short shells. Then, with increase in Z an inversion of the eccentricity effect occurs and inside rings are stronger. This inversion of the eccentricity effect and the corresponding Z or the 'range of inversion' are discussed in detail in Singer et al. (1966) or Singer (1972). Here they are only summarized for typical dimensions in Fig. 11.13 (reproduced from Singer et al. 1966). For stringer-stiffened shells, no inversion

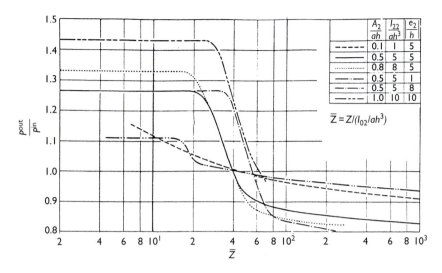

Figure 11.13 Variation of eccentricity effect with modified shell geometry parameter, $\bar{Z} = Z/(I_{02}/ah^3)$ in closely ring-stiffened cylindrical shells under hydrostatic pressure loading (from Singer et al. 1966; this material has been reproduced from the *Proceedings of the Institution of Mechanical Engineers*, vol. 8, p. 372, Figures 7 and 9, 1966 'Inversion of the eccentricity effect in stiffened cylindrical shells buckling under external pressure' by Singer, Baruch and Harari by permission of the Council of the Institution of Mechanical Engineers).

of eccentricity occurs under lateral pressure, and even under hydrostatic pressure such an inversion may appear only for extremely short shells. Hence, in practice, outside stringers are always better than inside ones for both lateral or hydrostatic pressure loading.

The *discreteness effect* of the eccentric stiffeners was investigated at Technion and at NASA Langley by a *discrete stiffener theory* (see Baruch 1965; Block 1968; Singer and Haftka 1968, 1975). Instead of being 'smeared', the stiffeners were now considered as linear discontinuities represented by the Dirac delta function. The force and moment expressions of Eq. (1) were modified accordingly and the remainder of the analysis was similar to the 'smeared'-stiffener theory. The representation is, however, satisfactory only as long as the width of stiffener is small compared to the spacing. One may note that, in addition to the general instability of the shell, such a 'discrete' analysis includes also consideration of the local instability between stiffeners.

For buckling under hydrostatic pressure, appreciable load reductions were found for discrete rings, even when the number of rings was not small (Singer and Haftka 1968). In the case of axial compression, the discreteness effect of rings could usually be neglected (Singer 1969; Singer et al. 1971). However, if the number of rings in the shell is very close to the number of axial half-waves in the predicted axisymmetric buckle pattern, a recheck should be made with the discrete stiffener theory. Such a recheck (in Singer *et al.* 1971) confirmed the negligibility of discreteness effects even when the number of rings was close to the number of axial half-waves. The only exception was the case in which the rings were strong enough to make local buckling dominant. The discrete theory then yields buckling loads very close to the local ones, and when these are considerably lower than the general instability loads, appreciable discreteness effects appear (see, e.g. Singer *et al.* 1971, table 3). Hence, in ring-stiffened shells designed for failure by general instability, discreteness effects are negligible.

In the case of stringer-stiffened shells the discreteness effect was again found to be negligible for practical stringer spacings used in aerospace applications, which are usually very close, while significant discreteness effects are only likely for stringers with very large torsional stiffness in shells with $Z < 400$ (Singer and Haftka 1975).

Boundary effects

The above analysis with 'smeared'-stiffener theory considered 'classical' simple supports or clamped ends. Since, however, very significant effects of the in-plane boundary conditions on the buckling loads had been found in unstiffened cylindrical shells, their influence in stiffened shells was also investigated. Some earlier studies of orthotropic and stiffened shells had already considered different in-plane boundary conditions (e.g. Thielemann and Esslinger 1964; Seggelke and Geier 1967; Singer *et al.* 1971). But in the 1970s complete parametric studies for the different boundary conditions (see Table 11.1) were carried out at the Technion (Weller 1971, 1978; Weller *et al.* 1979).

These parametric studies showed that for *ring-stiffened* shells under axial compression (Weller *et al.* 1979), the effects of the in-plane boundary conditions are similar to those observed in isotropic shells, which are well known. The 'weak in shear' boundary conditions SS1 and SS2 (with $N_{x\phi} = 0$ instead of $v = 0$), which rarely occur in practice, reduce the buckling load to about half the classical value,

Table 11.1 Notation for boundary conditions

Simple supports	SS1: $w = M_x = N_x = N_{x\phi} = 0$	Classical
	SS2: $w = M_x = u = N_{x\phi} = 0$	
	SS3: $w = M_x = N_x = v = 0$	
	SS4: $w = M_x = u = v = 0$	
Clamped	C1: $w = w,_x = N_x = N_{x\phi} = 0$	Fully clamped
	C2: $w = w,_x = u = N_{x\phi} = 0$	
	C3: $w = w,_x = N_x = v = 0$	
	C4: $w = w,_x = u = v = 0$	

whereas axial restraint ($u = 0$ instead of $N_x = 0$) has a negligible effect, yielding practically identical results for SS3 and SS4. As a matter of fact, if one compares typical curves of (P_{SS1}/P_{SS3}), etc., versus (L/R) for ring-stiffened shells with those for isotropic shells, the similarity is apparent. The only significant differences occur for heavy internal rings, where the weakening effect of the 'weak in shear' boundaries tends to disappear.

For *stringer-stiffened* cylindrical shells under axial compression, however, the effects of in-plane boundary conditions differ appreciably from those in isotropic shells. The effects are different for external and internal stringers, and depend also on the stringer geometry. In general, the axial restraint ($u = 0$ instead of $N_x = 0$) is the predominant factor, whereas the 'weak in shear' boundary conditions have only a minor influence. Figure 11.14 (from Weller 1971) shows the variation with the shell geometry parameter Z of the ratios of buckling loads for typical heavily stringer-stiffened cylindrical shells under axial compression. 'Heavy' stringers were chosen in this example, to emphasise the rather different effects of in-plane boundary conditions for stringer-stiffened shells.

With internal stringers, axial restraint ($u = 0$, SS4 or SS2) raises the buckling loads by about 45% and considerably above as the shells become longer and thinner, irrespective of the tangential boundary conditions. On the other hand, (SS1/SS3) is approximately unity, confirming the negligibility of the v condition. With external stringers, axial restraint is less effective and even ineffective in a certain important range of shell geometries, but the weakening $N_{x\phi} = 0$ is still practically absent, except for a relatively minor decrease in a limited range of Z.

For weaker stringers, the importance of the axial restraint diminishes and as the isotropic shell is approached, the tangential constraint obviously gains in

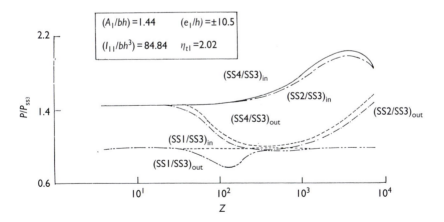

Figure 11.14 Variation with shell geometry (represented by the Batdorf parameter Z) of ratios of buckling loads of closely stringer-stiffened shells under axial compression, with different in-plane boundary conditions (from Weller 1971).

importance. However, even for relatively weak stringers, with say $(A_1/bh) \cong 0.1$ and $(e_1/h) \cong \pm 0.5$, the effects are far from the ones observed in isotropic shells.

These pronounced effects of axial constraints appear primarily for simple supports (with $M_x = 0$). Axial restraints also affect the buckling loads for clamped ends (with $w,_x = 0$), but the effects differ slightly. An example of the effect of axial or rotational restraints on buckling for typical stringer-stiffened cylindrical shells of two lengths and varying area ratios is presented in Fig. 11.15 (from Singer and Rosen 1974). Comparison between the fully clamped C4 and the SS4 curves in the figure shows that, whereas rotational restraint is effective only for short shells, axial restraint is even more effective for long ones.

The predominant influence of the boundary conditions, and in particular the axial restraint, on the buckling of stringer-stiffened cylindrical shells, lead to many further studies (see, e.g. Singer and Rosen 1974, 1976; Weller 1978). Similar effects were also observed on the lower natural frequencies of vibrations whose mode shapes resembled the buckling modes (e.g. Sewall and Naumann 1968; Rosen and Singer 1974). This, and the rather complicated effect of the boundary conditions, motivated extensive correlation studies between vibration and buckling at the Technion for more than two decades (see, e.g. Rosen and

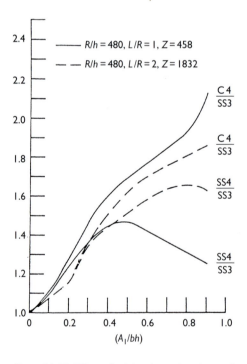

Figure 11.15 Effect of axial and rotational restraints on buckling load ratios of stringer-stiffened cylindrical shells with varying area ratio (from Singer and Rosen 1974).

Singer 1974, 1976; Singer and Rosen 1974, 1976; Singer 1983), aiming at a non-destructive experimental determination of the actual boundary conditions. The resulting Vibration Correlation Technique is discussed in detail in chapter 15 of Singer *et al.* (1997, 2001).

The influence of in-plane boundary conditions was also studied for ring-stiffened shells under hydrostatic pressure. For weak rings, axial restraint has the predominant effect as in unstiffened shells (Sobel 1964). As the rings become stronger, the effect of the 'weak in shear' boundary conditions becomes dominant, since then the shell has a relatively smaller resistance to the axial component of the hydrostatic pressure. For medium stiffening, say $(A_2/ah) \cong 0.5$ and $(e_2/h) \cong \pm 0.5$, the ring-stiffened shell under hydrostatic pressure is affected by the in-plane boundary condition as an axially compressed isotropic shell (yielding for SS1 and SS2 buckling loads about half the classical ones), except for long shells, $Z > 1000$.

Nonlinear prebuckling deformations may also be considered as boundary effects. They were shown to have only a small effect on the buckling load of clamped isotropic cylindrical shells (Almroth 1966a) and also for simply supported isotropic shells their effect was usually found to be small and at most 15% (Almroth 1966b). For *ring-stiffened cylinders* under axial compression the effect has also been found to be small (e.g. Block 1968; Peterson 1969; Singer *et al.* 1971). A similar small effect has been observed for ring-stiffened corrugated cylinders (Block 1968). Also for *stringer-stiffened* shells the effect is generally found to be small, except for short shells (see Almroth and Bushnell 1968; Singer *et al.* 1971). A similar conclusion was also reached for certain types of orthotropic shells by Kobayashi (1968).

Hence, nonlinear prebuckling deformations are apparently not a major factor in the determination of the buckling load of stiffened cylindrical shells, except for short stringer-stiffened cylindrical shells (e.g. Bushnell *et al.* 1968; Singer and Rosen 1974, 1976). Figure 11.16 (from Singer and Rosen 1974) shows an example of the variation of the influence of prebuckling deformations on the buckling load ratio ($\rho_{PRE} = P_{PRE}/P_{MEMB}$, where P_{PRE} is the buckling load calculated with consideration of nonlinear prebuckling deformations and P_{MEMB} is that computed by linear theory), for two types of stringer-stiffened shells, which illustrates this conclusion.

Load eccentricity, on the other hand, is a very important boundary effect for stringer-stiffened shells, and has been actively investigated at Technion, NASA Langley and Lockheed Palo Alto (e.g. Almroth and Bushnell 1968; Singer 1972; Weller *et al.* 1974). One should note that the load eccentricity considered in cylindrical shells is *not* the off-centre loading that sometimes occurs in columns, but a type of *concentric load eccentricity* that produces concentric end moments about the mid-surface of the shell. The load eccentricity is usually defined as the radial distance between the line of axial-load application and the shell mid-surface.

The load eccentricity effect can be summarised qualitatively by pointing out that moments that tend to bend the stringer-stiffened cylindrical shell into a barrel shape (and thus give rise to tensile prebuckling hoop stresses) increase the buckling load.

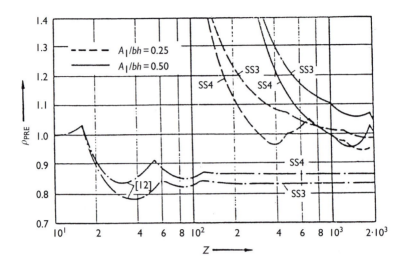

Figure 11.16 Effect of nonlinear prebuckling deformations on the buckling load ratio of stringer-stiffened cylindrical shells – variation with shell geometry (from Singer and Rosen 1976; reproduced by kind permission of Springer-Verlag).

Hence, for external stringers, for example, loading through the mid-surface of the skin will yield higher buckling loads than application of the load at the stringer centroids, whereas for internal stringers the opposite occurs. One may add that for medium-length shells of $(L/R) \cong 1$, the effect may amount to up to 20–40% in practical shells.

An experimental study was carried out at the Technion in the early 1970s (Weller *et al.* 1974) to verify the predicted influence of eccentricity of loading on the buckling of integrally stringer-stiffened cylindrical shells. The experiments were accompanied by parametric studies, comparing the results obtained by different nonlinear analyses (Block 1966, 1968; Stuhlman *et al.* 1966; Segelke and Geier 1967; Almroth and Bushnell 1968; BOSOR – Almroth *et al.* 1968; Stein 1968; Chang and Card 1971).

The load application for three types of edges of the integrally machined, stringer-stiffened cylindrical shells tested at Technion is shown in Fig. 11.17. These were the three main types of edges investigated, but additional different edges were also employed (Weller *et al.* 1974).

The experimental results of the Technion study confirmed, in general, the influence of the eccentricity of loading predicted by the nonlinear theories of Almroth *et al.* (1968) and Block (1968) and others. However, the actual manner of load application and the details of the end fixtures determined the severity of the load eccentricity effects. One should note that all the various analyses show that no load eccentricity effect exists for clamped stringer-stiffened shells. Also in the

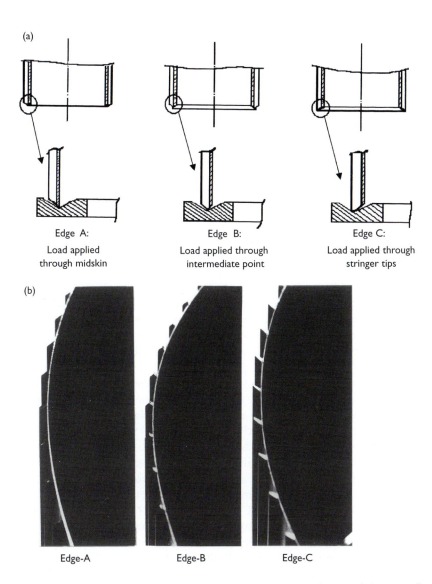

Figure 11.17 Details of load application for three different types of edges of the integrally stringer-stiffened cylindrical shells tested at Technion in axial compression to verify the load eccentricity effect (from Rosen and Singer 1976): (a) schematic: edge A – load applied through shell midsurface, edge B – load applied through intermediate point, edge C – load applied through tips of stringers; (b) the three types of machined edges (reproduced by kind permission of the Society for Experimental Mechanics, Inc.).

experiments, when the ends of the shells were closer to clamped end conditions, no load eccentricity effect could be discerned.

It should also be mentioned here that the nature of buckling of eccentrically loaded stringer-stiffened shells differed completely from that of the concentrically loaded ones (those loaded through their midskin). The eccentrically loaded shells buckled at a lower load, but buckled gradually and in a softer manner.

'Adequacy' and bounds of validity of linear smeared-stiffener theory

The conclusions about the influence of eccentricity, boundaries and stiffener and shell geometric parameters on the buckling loads of stiffened shells, arrived at in the previous sections, are meaningful only if the linear smeared-stiffener theory predicts the buckling load adequately and if the discreteness of stiffeners has no noticeable effect. The latter, the discreteness effects, were shown in the section on 'Eccentricity and discreteness effects' by analysis to be usually negligible, but the adequacy of linear theory can be conclusively settled only by tests. Extensive experimental studies were, therefore, carried out at the Technion Aerospace Structures Laboratory and other research centres on the buckling of closely spaced stiffened cylindrical and conical shells in the 1960s and 1970s (e.g. Singer 1969; Singer et al. 1971, 1997, 2001, Chapter 13; Singer and Rosen 1974, 1976; Weller and Singer 1974, 1977).

The applicability of linear theory is conveniently expressed by the ratio of the experimental buckling load P_{exp} to that predicted by linear theory P_{cr}, sometimes called 'linearity' $\rho = (P_{exp}/P_{cr})$ in stiffened shells, in preference to the more common term 'knock down factor' used in unstiffened shells (on account of the large reductions in predicted loads necessary there to correlate them with the observed ones), since in closely stiffened shells ρ is usually closer to unity.

For *ring-stiffened shells*, the ring area ratio (A_2/ah) was found to be the predominant ring and shell geometry parameter, and, therefore, their 'knock down factor' (or 'linearity') is plotted in Fig. 11.18 as a function of (A_2/ah). Shells of different materials, and tested by different investigators (Almroth 1966b; Singer 1969; Singer et al. 1971; Weller and Singer 1974; Rosen and Singer 1975), are plotted in the figure. Since for ring-stiffened shells the boundary effects are relatively small, the results from different test setups fit within a reasonable scatter band.

For $(A_2/ah) > 0.3$ there is ±10–15% scatter about a mean $\rho = 0.95$ (except one point that is 17% below 0.95), or one can state roughly that $\rho = 0.8$. As the ring area ratio decreases below 0.3, ρ decreases, at first slowly, but below 0.15 rather rapidly. Hence, the applicability of linear theory appears to be bounded here by $(A_2/ah) \geq 0.2$.

Also for *stringer-stiffened shells*, the stringer area ratio (A_1/bh) appeared to be the predominant stringer and shell geometry parameter, determining their 'knock down factor' (or 'linearity'). Hence, ρ is also plotted here as a function

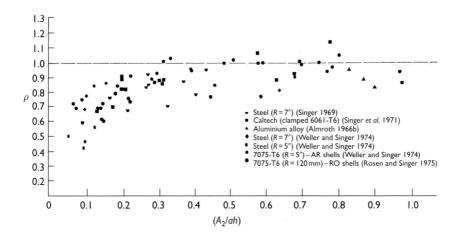

Figure 11.18 'Knock down factor' ρ (or 'linearity' = ratio of experimental buckling load to that predicted by linear theory) of ring-stiffened shells under axial compression as a function of the ring area parameter (from Singer and Rosen 1976; reproduced by kind permission of the Society for Experimental Mechanics, Inc.).

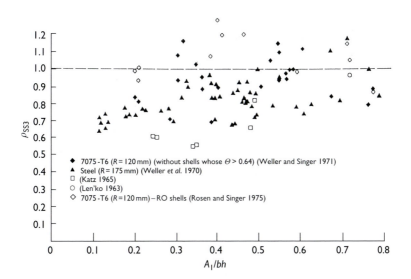

Figure 11.19 'Knock down factor' ρ (or 'linearity') of simply supported stringer-stiffened cylindrical shell under axial compression as a function of the stringer area parameter (from Singer and Rosen 1974).

of the stiffener area ratio (A_1/bh). In Fig. 11.19, ρ is shown for simply supported stringer-stiffened cylindrical shells (ρ_{SS3}) made of different materials and
tested by different investigators (Len'ko 1963; Katz 1965; Rosen and Singer
1975; Weller and Singer 1977). A similar plot for clamped ends can be found
in chapter 13 of Singer et al. (1997, 2001) or in Singer and Rosen (1974). The
scatter for the stringer-stiffened shells in Fig. 11.19, and in the similar plot for
clamped ends, is larger than that for the ring-stiffened ones in Fig. 11.18, but the
trend of the primary dependence on the stiffener area ratio is still evident, though
less clearly defined.

The knock down factor ('linearity') of all the simply supported shells
(Fig. 11.19), except those of Katz (1965), is above 0.65 even for weak stringers,
and for $(A_1/bh) > 0.5$ all shells have $\rho > 0.72$. In some of the shells tested by
Katz, which are the exceptions here, early local buckling was clearly observed
and local buckling may have occurred also in some of the other specimens of this
series. The scatter in Fig. 11.19 is usually about $\pm 20\%$, and even $\pm 30\%$ in regions
where many tests accumulate, about an average ρ that rises from $\rho \cong 0.7$ for
weak stringers to $\rho \cong 0.9$ for heavier ones. This scatter is partly due to incomplete
definition of the boundary conditions, and can be significantly reduced by a better determination of the actual boundary conditions with the aid of the Vibration
Correlation Technique, as shown in chapter 15 of Singer et al. (1997, 2001).

For clamped shells, lower values of knock down factor (or 'linearity'), down
to $\rho \cong 0.6$, were observed. Since clamping is usually incomplete, more realistic
boundary conditions, obtained by vibration correlation or a similar method, would
improve the knock down factors. However, clamped shells appear to be more
imperfect, probably due to additional imperfections introduced by the clamped
edges.

The influence of shell geometry and stiffener spacing in stringer-stiffened cylindrical shells was also studied (e.g. Singer 1972; Weller and Singer 1977), but
found to be less significant than the stringer area ratio in determining the bounds
of validity of linear smeared-stiffener theory.

Experimental results for other loading cases have provided additional support
for the adequacy of linear theory in the basic analysis of the buckling of stiffened
cylindrical shells. For example, in some NASA Langley bending tests (Peterson
1956; Anderson and Peterson 1971) and some US Navy hydrostatic pressure tests
(Reynolds and Blumenberg 1959) discussed in Singer (1972), $\rho > 0.9$ for all
cases. Tests at Technion on ring-stiffened conical shells (see Weller and Singer
1971) also support the conclusions arrived at for cylindrical shells.

It may be pointed out that the adequacy (or good approximation) of linear theory
for prediction of the buckling loads of closely stiffened shells stems from the
similarity of the actual *buckling modes* to those predicted. Often, this similarity
holds even under axial compression, contrary to isotropic cylindrical shells, where
the experimentally observed buckling modes differ substantially from the ones
predicted by linear theory.

Imperfection sensitivity

The influence of initial imperfections on the buckling of cylindrical shells has been widely studied in the last four decades, since initial imperfections have been identified as the main cause of the large discrepancies between experimental and theoretical buckling loads of unstiffened cylindrical shells under axial compression (e.g. Koiter 1963; Budiansky and Hutchinson 1966; Hutchinson and Amazigo 1967; Arbocz and Babcock 1969; Hutchinson and Koiter 1971; Arbocz 1976). For closely stiffened shells, the influence of the imperfections is less pronounced, but they remain, however, a major degrading factor.

The imperfection sensitivity concept introduced by Koiter has become a powerful tool, which has been extensively employed also for stiffened cylindrical shells by the Harvard and Caltech groups and many others (see, e.g. Hutchinson and Amazigo 1967; Brush 1968; Hutchinson and Frauenthal 1969; Singer *et al.* 1971; Arbocz 1974; Singer *et al.* 1997, 2001, chapter 3).

These imperfection sensitivity studies show that though stiffened cylindrical shells are less sensitive than isotropic shells, their sensitivity is still appreciable. The predicted sensitivities seem to depend strongly also on the assumptions regarding quantities that affect the classical buckling load only slightly (torsional stiffness of stiffeners and prebuckling deformations). The predicted regions of large imperfection sensitivity shift and the magnitude of the sensitivity coefficient 'b' changes accordingly from one study to another (see, e.g. fig. 11 of Singer 1972).

The lower imperfection sensitivity of closely stiffened cylindrical shells compared to isotropic ones can be physically explained by consideration of the difference in their buckling behaviour. Whereas, for example, in an isotropic cylindrical shell under axial compression (the most imperfection sensitive loading case) many buckling patterns and their initiation are possible, and hence practically any imperfection anywhere on the shell can trigger buckling; in a closely stringer-stiffened shell only one unique buckling pattern can appear (see Fig. 11.6), and hence only imperfections that somewhat resemble this unique wave shape can effectively start buckling.

One may point out here that in recent years new, different ideas have been employed to explain the imperfection-sensitive buckling behaviour of thin shells (e.g. Calladine 1995). It has been postulated in particular that, in addition to the usual stress-free geometric imperfections, other imperfections, in the form of locked-in stresses, may significantly contribute to the imperfection sensitivity of the shell. Such new ideas may eventually also further clarify the buckling behaviour of stiffened shells.

The development of imperfection measurement techniques and instrumentation has been traced and discussed in detail in Singer and Abramovich (1995) and in chapter 10 of Singer *et al.* (1997, 2001). Here, it will suffice to state that some type of geometric imperfection measurement has become an integral part of any properly carried out shell buckling test, be it on a laboratory scale or on a large scale (e.g. most of the papers in the special 1995 issue of *Thin Walled Structures*, edited by Galletly *et al.*).

In stiffened-cylindrical shells in particular, the geometric imperfections have been extensively measured and some of the data have been stored in the International Imperfection Data Bank, with branches at Delft University and at the Technion in Haifa (Arbocz and Abramovich 1979; Singer *et al.* 1979; Arbocz 1982, 1983; Abramovich *et al.* 1987; Arbocz and Hol 1995; Singer and Abramovich 1995).

As already pointed out in the section on 'Boundary effects' the influence of boundary conditions, and in particular the in-plane ones, in stiffened shells is very significant and usually as important as that of the initial imperfections. Hence, the definition of the effective boundary conditions by a non-destructive approach, like the vibration correlation technique VCT (e.g. Singer and Rosen 1976; Singer 1971, 1983; chapter 15 of Singer *et al.* 1997, 2001), is as essential a part of the buckling test of a stiffened cylindrical shell, as the measurement of initial imperfections.

Furthermore, the data acquired in the imperfection data banks are being evaluated to yield meaningful correlations with the relevant fabrication processes. As more data are recorded and evaluated, the imperfection-degrading characteristics of different fabrication processes will become available to designers.

One may expect poor quality fabrication to have a pronounced degrading effect on the buckling load of a stiffened shell, and this is indeed verified by tests. For example, in a test programme on ring-stiffened cylindrical shells under external pressure studying the effect of cut rings (Singer *et al.* 1972; Singer 1976), parallel test series were carried out on integrally stiffened aluminium alloy shells and on similar spot welded steel shells. The spot welded shells (Fig. 11.1) were of relatively poor fabrication quality and for both general instability and local inter-ring buckling yielded results 30–40% below those for the corresponding integrally stiffened ones (which, however, yielded rather high values even for external pressure).

These tests were concerned with the possible effect of elastic joints in the stiffening rings represented by cuts in the rings. This type of local imperfection, which may occur in practice, can have a very significant effect if the joints (or cuts) are situated along a generator, since it then induces a special type of buckling mode originating from the weakened generator. Figure 11.20 shows such a general instability failure under external pressure of an integrally ring-stiffened aluminium alloy shell (WA-3 of Singer *et al.* 1972), with three cuts in the ring along the generators. The reduction in buckling pressure caused by the cuts is 35% in this case, with reductions up to 45% with more cuts along the circumference. If the same cuts are, however, located along lines inclined at 30° or more to the vertical, the reduction is only half its original value. Similar results were also obtained in the case of inter-ring buckling.

Since, however, the degrading effect due to the imperfect fabrication of the spot welded shells was of similar magnitude to the reduction in buckling pressure caused by the cuts in the rings, the design information (about the possible alleviation of

Figure 11.20 General instability failure under external pressure of an integrally ring-stiffened 7075-T6 aluminium alloy shell (WA3 of Singer *et al.* 1972). The shell was tested at Technion in a programme studying the effect of cut rings. It had a radius of 120 mm, an $(R/h) \cong 480$ and closely spaced rings, which prevented local shell buckling.

the effect of the cuts by locating them along inclined lines) is rather qualitative and should be supported by additional tests and calculations.

When the coordinated international effort by industry and research laboratories, of evaluating imperfection and boundary condition data, has matured and yielded useful correlations with fabrication processes, a new, improved design method for stiffened cylindrical shells, as proposed by Singer and Abramovich (1995), will probably appear. It can be outlined as follows:

1   In the preliminary design stage, various shell construction methods will be evaluated with respect to their availability, cost and imperfection-degradation characteristics, taken from the International Imperfection Data Bank. A shell type and fabrication process will thus be decided upon, probably with the aid of some optimisation method.

2   The expected imperfection shape and magnitude of this designed shell, obtained from the imperfection bank, will be fed as input into one of the nonlinear codes for imperfect shells, which by then will also have been further improved. The required dimension of the shell will thus be computed. At this stage a stochastic approach could be introduced as an alternative.

3   A small-scale model of the shell will then be made by a fabrication process as close as possible to that to be used in the full-scale production. Its imperfection

will be measured and compared with that used in the design. The relevant scale factors for this comparison will be obtained from the imperfection data bank. The actual effective boundary conditions will also be determined by a non-destructive technique, like VCT, and compared with computed or assumed ones. Finally, the model will be tested under the scaled design loads.

4   The model test results will serve as a second iteration to the shell design, determined in Eq. (2), and if necessary the dimensions will be adjusted to bridge the gap between the two iterations. In extreme cases where this gap is large, a second model test could be carried out to serve as a third iteration.

5   The full-scale shell fabrication can then proceed. An imperfection scan will be made of the first full-scale shell, and the results of this scan will be compared with those of the model scan and also employed to recalculate the buckling load of the shell.

6   The full-scale imperfection scanning setup could eventually also become a quality control device for the production run, and an acceptable imperfection pattern will be one of the cardinal quality control parameters.

7   Eventually, if the production run justifies it, a full-scale buckling test will be performed and possibly utilized to relax the pass requirements in the quality control.

Such a design method is not simple, but yields less conservative buckling load predictions and, therefore, more efficient shell structures.

Inelastic effects

In the buckling of relatively thick-walled unstiffened and stiffened cylindrical shells inelastic effects are an important factor and have, therefore, motivated extensive studies of plastic buckling (e.g. Sewell 1972; Hutchinson 1974; Bushnell 1982; Bushnell and Meller 1984; Tvergaard 1987). A brief outline of methods of plastic buckling analysis for stiffened cylindrical shells is presented below.

However, it should be noted that even in shells of fairly high R/h ratios, considerable reductions in axial load carrying capability may occur due to a combination of imperfections and material nonlinearity, as was pointed out in Mayers and Wesenberg (1969). For example, in Technion tests on integrally stiffened cylindrical shells under axial compression (Weller *et al.* 1970, 1977) it was observed that one group of steel shells with a smaller R/h yielded consistently a lower knock down factor ρ than another group of steel shells with a larger R/h. Since all specimens were manufactured by a similar technique, an opposite trend would have been expected due to imperfections, and hence attention was drawn to the material properties of the small diameter shells. The material, AISI 4130 alloy steel, in a soft condition, indeed exhibited early nonlinear behaviour in its stress–strain curve, though its 0.1% yield stress was well above buckling stresses. Application of the maximum stress analysis of Mayers and his associates to a few of these steel specimens (Mayers and Wesenberg 1969; Wesenberg and Mayers 1969; Mayers and Meller 1972) showed that the correction for inelastic material effects indeed considerably raised the knock down factor ρ for the smaller diameter shells; whereas for the large diameter steel shells, whose material did not exhibit early nonlinear behaviour, no noticeable corrections in ρ appeared. The study of the inelastic material effects, therefore, explained the observed trends, and justified omission of the results of the smaller diameter steel shells from correlation studies for aerospace applications.

Returning to plastic buckling of thick-walled stiffened cylindrical shells, one should remember the 'plastic buckling paradox' (the flow theory versus deformation theory paradox), which has been comprehensively discussed by many researchers in recent years (e.g. Hutchinson 1974; Tvergaard and Needleman 1982; Galletly *et al.* 1990; Giezen *et al.* 1991; Singer *et al.* 1997, 2001, chapter 16), but has not yet been settled. Hence, for plastic bifurcation buckling analysis of stiffened shells, one can employ one of three alternative theories (Teng 1996): (1) flow (or incremental) theory; (2) deformation theory (which is actually not a physically acceptable plasticity theory) or (3) flow theory, but with the shear modulus taken from deformation theory.

Analyses employing the more rigorous flow theory (1) generally predict rather high buckling loads, while those using the 'less respectable' deformation theory (2), are usually conservative and predict lower buckling loads. They correlate, however, fairly well with experimental results. The third alternative (3), though not consistent, yields results that usually correlate quite well with tests. Some of the widely used computer codes, like BOSOR 5 (Bushnell 1976a) or ADINA, therefore, include all the three options.

One should remember that often, in particular in submarine and offshore applications, external pressure represents the primary loading case for stiffened cylindrical shells. Their buckling under external pressure, was, therefore, extensively studied since the 1950s in several research centres. In these studies, it was already realized in the 1960s that inelastic buckling is the usual ultimate mechanism of collapse (e.g. Krenzke and Reynolds 1967; Riks and Rankine 1997).

More recently, another alternative for estimation of the ultimate load carrying capacity of stiffened cylindrical shells was developed, the plastic mechanism approach. This method, which had evolved for the analysis of various steel structural elements, in particular of steel-plated structures (e.g. Murray 1973, 1983; Walker and Murray 1975; Murray and Khoo 1981; Singer *et al.* 1997, 2001, chapter 16), was further explored in the 1980s for ring-stiffened cylindrical shells, of the type used in offshore structures (Tsang and Harding 1984).

One may add here that several other methods, mostly numerical (using finite differences or finite elements) for the analysis of the collapse of stiffened cylindrical shells under axial compression or external pressure have also been developed (e.g. Harding and Dowling 1982; Walker *et al.* 1982a; Riks 1987; Ross 1990; Wunderlich *et al.* 1991; Bosman *et al.* 1993; Crisfield 1997). Another method represents a lower bound prediction, based on a reduced stiffness model, introduced by Croll at University College London for elastic buckling and extended later to elasto-plastic buckling (Croll 1981, 1985, 1995). These lower bound predictions provide in many cases reasonable agreement with results of the plastic mechanism approach and with experiments (Tsang and Harding 1984).

One should as well remember that residual stresses, primarily weld-induced residual stresses, may significantly affect inelastic buckling collapse. These effects may require additional measurements and further consideration (e.g. Dowling *et al.* 1982; Faulkner 1997; Galambos 1998, chapter 14).

Several multipurpose computer programs are available today for analysis of the elasto-plastic buckling of stiffened cylindrical shells, the most commonly used ones being BOSOR5 (Bushnell 1976a), STAGSC1 (Almroth and Brogan 1978, 1983), ADINA (1998), MSC/NASTRAN (1998) or ABAQUS (1997). These codes were validated by correlation with many tests, an essential requirement for their reliability.

For example, one of the many comparisons with experiments carried out with BOSOR5 was for 69 machined ring-stiffened aluminium alloy cylindrical shells, subject to external pressure, tested at the US Naval Ship Research Center (formerly the David Taylor Model Basin) in the 1960s (Boichot and Reynolds 1965). The paper (Bushnell 1976b) first presents the steps of the calculations with the BOSOR5 code. The determination of the best analytical model is discussed in detail. The geometry of all the test specimens is shown in Fig. 11.21. After some reasoning, Specimen 10-52F was chosen as the standard of comparison, since very good agreement between test and theory was expected for this quite thick shell with $R/h \cong 20$. Its other dimensions were $L_f = 0.300$ in., $b = 0.030$ in., $d = 0.097$ in. and $L_B = 1.921$ in.

The designation 'F' indicated the 24 (out of the 69) specimens that had fillets near the boundaries and where the rings join the shell wall. In the tests it was

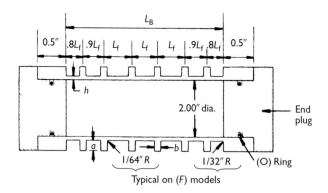

Figure 11.21 7075-T6 aluminium alloy ring-stiffened cylinders tested under external pressure by Boichot and Reynolds (1965) at the US Navy David Taylor Model Basin, Washington, DC (now the Naval Ship Research and Development Center), for which the comparisons with BOSOR5 predictions were carried out. The figure shows the geometry of all the 69 machined shells, and indicates the location of the fillets on the 'F' models.

observed that practically all the specimens without fillets fractured during failure and, therefore, the comparison was carried out in two groups: the models without fillets and those with fillets. In general, the test results and the BOSOR5 predictions were in reasonably good agreement. Test/prediction differences varied between $+10\%$ and -13% for the specimens without fillets and between $+7\%$ and -11% for those with fillets. If one restricts the comparison to those specimens which definitely did not fracture and were less sensitive to imperfections (i.e. excludes the thinnest shells whose $R/h \cong 50$), the maximum discrepancy between test and prediction is only 3.6%.

One should note, however, that in spite of many years of efforts by different approaches to develop good prediction techniques for plastic and elastic–plastic buckling of stiffened shells, comprehensive and reliable methods are still scarce. Hence, it is not surprising that even today new semi-empirical design methods are employed and proposed for inelastic buckling of stiffened cylindrical shells (e.g. Ross *et al.* 1995, 1999).

Nonlinear effects

In general, in order to characterise the buckling behaviour of a thin-walled structure, one has to investigate both its buckling (bifurcation) and its postbuckling behaviour. For stiffened cylindrical shells with many stiffeners, one should, therefore, employ a set of nonlinear bending equations (e.g. Brush and Almroth 1975), in which the stiffeners are 'distributed' over the entire shell, as outlined in the section on 'Smeared-stiffener theory', that is, a nonlinear smeared stiffener theory. However, it is often preferred to start with a first approximation, considering perfect shells and membrane prebuckling solutions, that is, to employ a linear smeared-stiffener theory, as detailed in the section on 'Linear smeared-stiffener theory', and then to correct for nonlinear effects.

The nonlinear effect that was apparently the first to be studied extensively in stiffened cylindrical shells was the effect of nonlinear prebuckling deformations, which can actually be considered a boundary effect (e.g. Almroth and Bushnell 1968; Block 1968; Singer 1972; Singer and Rosen 1974, 1976). Contrary to the case of isotropic cylindrical shells, in stiffened cylindrical shells the effect was generally found to be a minor factor in the calculation of the buckling load, except for short stringer-stiffened cylindrical shells. However, the effect is of sufficient significance to justify a design rule, whereby as a first approximation the linear theory is used for optimisation studies, and where the buckling load of the final design is always verified by a more rigorous buckling load computation, which includes the nonlinear prebuckling deformations resulting from the edge restraints. Almroth and Bushnell (1968) or Block (1968) indicate details of such typical calculations.

One possible approach to the computation of the buckling load of stiffened cylindrical shells, which includes nonlinear effects, is that proposed recently by Arbocz, Starnes and Nemeth (Arbocz *et al.* 1999). This hierarchical approach to buckling load calculations consists of three levels of analysis of increasing accuracy, applied successively. The approach evolved in the process of development of an interactive shell design code DISDECO (the Delft Interactive Shell DEsign Code) (Arbocz and Hol 1993) for shell stability analysis in a computer-aided engineering (CAE) environment, a code that is based on a modern graphic workstation.

In DISDECO three analyses levels are now available (Arbocz *et al.* 1999). *Level-1* represents the most approximate analyses, which, however, suffice to establish the main stability characteristics of the stiffened-shell design. *Level-2* analyses account for the primary nonlinear effects such as the edge restraint effects and satisfy the boundary conditions rigorously. The predictions of the Level-2 analyses for stiffened cylindrical shells are usually very accurate for axisymmetric loading. Only in cases of closely spaced eigenvalues and nearly simultaneous eigenmodes, where close to the limit or bifurcation point modal interaction becomes important, verification by computation with Level-3 analyses is necessary. The *Level-3* analyses are the most accurate, but usually require considerable analysis and computational efforts to model the various nonlinear effects accurately and reliably.

Essentially, the hierarchical approach to buckling load calculations represents a step-by-step approach from simple to more complex models and solution procedures, which are used only as necessary. Hence, the considerable computational efforts required in the direct application of Level-3 analyses, like currently available advanced capability two-dimensional finite element codes (e.g. STAGS, NASTRAN, ADINA or ABAQUS) can often be avoided. With interactive codes like DISDECO, the designer has a tool that enables him to use good engineering sense instead of direct 'brute force' computation, while leaving him the prerogative to verify his calculations with higher level analyses for selected cases.

Experimental investigations

Model fabrication

A broad survey of buckling experiments on stiffened cylindrical shells, their model fabrication and test procedures can be found in Chapter 13 of a recent book *Buckling Experiments* (Singer *et al.* 1997, 2001), which focuses on experimental methods in buckling of thin-walled structures. Partial surveys also appear in some earlier reviews, for example, Babcock (1974) and Singer (1982a,b). Here, the presentation, is therefore, limited to an outline and summary of these methods for stiffened shells.

The most common methods of model fabrication for shells are: machining, 'realistic' fabrication (using similar techniques to those employed on the production line), welding in special rigs, electrodeposition or manufacture from a low modulus material.

Integral machining is the most frequently employed fabrication method for small stiffened cylindrical shells and has been extensively used at the US Navy David Taylor Model Basin (e.g. Saunders and Windenburg 1932; Galletly *et al.* 1958;

Hom and Couch 1961; Krenzke and Reynolds 1967); at the Technion Aerospace Structures Laboratory (e.g. Singer 1969; Weller *et al.* 1970; Weller and Singer 1971; Rosen and Singer 1975); at the California Institute of Technology (Singer *et al.* 1971); at the Avco Corporation, Wilmington, MA (Midgley and Johnson 1967); at the UK Admiralty (Naval Construction) Research Establishment Dunfermline, Scotland (e.g. Kendrick 1953); at Det norske Veritas, Oslo (e.g. Valsgård and Foss 1982); at the Norwegian Institute of Technology, Trondheim (e.g. Odland 1981); at the University of Portsmouth (e.g. Ross *et al.* 1971, 1995, 1999); at Tokyo Denki University (Yamamoto *et al.* 1989); and many other research centers.

Models of integrally stiffened cylindrical shells are usually machined from thick-walled tubes, but for thin stiffened shells, representing, for example, the stringer-stiffened stages of the Saturn V launcher of the Apollo programme, with R/h of the order of 500, an internal support mandrel is necessary for accurate machining. The fabrication processes of the many integrally machined cylindrical shells tested at the Technion Aerospace Structures Laboratory are good examples of the employment of internal mandrels and of a thermal shrink fit by either cooling or heating.

The machining process for these steel specimens was divided into stages to yield higher precision. In the first stage, the internal and external surfaces of the tubes were roughly machined. Then, the internal surface was precisely turned to the dimension of the 'cooled mandrel', on which it was mounted later for machining of the stiffeners. The dimension of the inside diameter was chosen to give a medium press fit between the 'cooled mandrel' and the mounted blank. The steel blank was then mounted on the 'cooled mandrel', made of cast aluminium with a high silicon content and which has the shape of a reservoir with many fins around its inner surface. Liquid air poured into the reservoir of the mandrel cooled it appreciably and as a result its diameter contracted 0.4 mm, enabling the shell to slide onto the mandrel. After returning to room temperature the shell sat well on the mandrel and permitted accurate machining. After completion the shell was removed from the mandrel by another liquid air 'cooling' and a second shell was immediately mounted.

The machining process for the 7075-T6 aluminium alloy specimens was similar to that described above, except for the mounting of the blank on the mandrel and removing of the finished stiffened shell from it. The 'cooled' aluminum mandrel used for the steel shells was replaced here by a 'heated' steel mandrel. For details see section 13.2.1 of Singer *et al.* (1997, 2001).

Small-scale welded specimens are another extensively used fabrication method for small stiffened cylindrical shells. In relation to offshore structures and other marine structures, which are welded from high strength steel plates, the results obtained in tests of high precision integrally machined specimens were considered not reliable. This led not only to many tests on large shells, but also to the development of special welded model fabrication techniques.

In the late 1970s Walker and his associates at University College, London, developed a manufacturing and welding process for small-scale specimens with $R/t = 200–360$, which represented typical offshore elements on a 1/20 scale (Walker and Davies 1977; Walker *et al.* 1982a). The cylindrical shells were built up segmentally by rolling and machining thin steel sheets, having material characteristics closely similar to that employed in offshore structures, into curved panels and then welding these together with the stiffeners by a special technique. For details see Walker *et al.* (1982a) or chapter 13 in Singer *et al.* (1997, 2001).

The work was continued by Walker and his co-workers at the University of Surrey (Walker *et al.* 1981, 1982b) and small-scale ring-stiffened shells with $R/t = 150$ were made by a similar process.

As part of the same UK offshore element experimental programme, similar small-scale models were developed at Imperial College, London, by Dowling and his associates, using a different fabrication technique. Their models were fabricated by first TIG welding the longitudinal stiffeners to a flat plate, and then, after welding, this stiffener sheet was wrapped around a copper forming mandrel and the closing T-butt weld was completed. For details see Dowling *et al.* (1982), Scott *et al.* (1987) or chapter 13 in Singer *et al.* (1997, 2001).

Ring- and stringer-stiffened specimens were fabricated by both the University College and Imperial College techniques, and with the latter also orthogonally stiffened shells having internal stringers and rings.

Imperfection and residual stress measurements on 30 University College London and six Imperial College models yielded similar relative levels to those obtained on comparable large shells. The results of the small-scale tests could, therefore, be taken as a reliable complement to the large-scale tests performed in the same UK programme, carried out at Imperial College and the University of Glasgow (Dowling *et al.* 1982; Green and Nelson 1982). The test programme is also discussed in chapter 13 of Singer *et al.* (1997, 2001).

The *realistic specimens approach*, often used for unstiffened shells (for details see section 9.3.7 of chapter 9 in Singer *et al.* 1997, 2001) has also been employed for small-scale stiffened shells. For example, Miller (1977) fabricated his axially compressed small-ring-stiffened shells (with $R/t = 250–500$) by rolling and welding, similarly to his large-scale shells of $R/t = 500$.

Another example of 'realistically' fabricated small-scale stiffened cylindrical shells are the 20 hydrostatic pressure tests of Miller and Kinra (1981) on 16–48 in. outside diameter ring-stiffened shells, made from commonly used platform steels by routine platform fabrication procedures. Further examples are the cold-rolled and welded ring-stiffened shells, used by Odland (1981) and Yamamoto *et al.* (1989) for comparison with their integrally stiffened shells; or the ring- and stringer-stiffened model shells tested in the extensive joint CBI Industries, Plainfield, IL, and Glasgow University programme (Chen *et al.* 1985). Details of these and other models can be found in chapter 13 of Singer *et al.* (1997, 2001).

Low modulus materials, with high σ_y/E values (like Mylar or epoxy) have also often been employed in the fabrication of stiffened cylindrical shell specimens (e.g. Esslinger and Geier 1970; Tennyson *et al*. 1980).

Test rigs and procedures

When one embarks on a test programme for stiffened cylindrical shells, one should realise that the test rigs and test procedures for unstiffened shells can serve equally well for experiments on stiffened shells, except that for the latter more stiffness is usually required. Hence, there is no definite dividing line between test systems for stiffened and unstiffened shells, and many of the characteristics of isotropic shell testing also apply to stiffened cylindrical shells (e.g. chapter 9 of Singer *et al*. 1997, 2000).

Experiments on small-scale stiffened cylindrical shells subject to axial compression are usually carried out in the same test setups as similar unstiffened shells, except when very large stiffness is aimed at. An example of such a special test frame is that used at Imperial College, London (Fig. 11.22) for tests on their small-scale steel welded stiffened cylindrical shells (Dowling *et al*. 1982).

The circular frame, which was 10 m in diameter and 2.5 m high, consisted of four 125 mm thick platens acting as a 'multiple sandwich' for the shell model, load cells and screw jack. The 100 ton screw jack, which was driven by an electric motor, was located between the two bottom platens and the load was applied by reaction against the upper platen through six 50 mm diameter tension rods. The load was transferred directly through three spherically capped load cells that alone supported the third thick platen, which carried the model and the transducer frame. The load was then applied to the test shell by two smaller circular end plates, which were supported between the two upper platens by two thick-walled tubes. Columns on races ensured that the platens were kept parallel during loading.

Although self-reacting, the test rig was also anchored to the laboratory floor by means of a heavy cross-beam. This increased the stiffness of the test rig and facilitated better control of the postcollapse path. All the tests were carried out under displacement control.

A circular rotating framework, holding four vertical banks of transducers, was used for efficient measurement of the deflections of the shell under load, as well as of the initial geometrical imperfections. The remaining instrumentation, which also included extensive strain gaging, was similar to that used in unstiffened shells.

As pointed our earlier in this section, test procedures and instrumentation are indeed very similar for unstiffened and stiffened cylindrical shells. However, when the spacing of stiffeners is such that local buckling precedes general instability and collapse, additional measurements are usually required to investigate the influence of local buckling on the transition and on the final collapse.

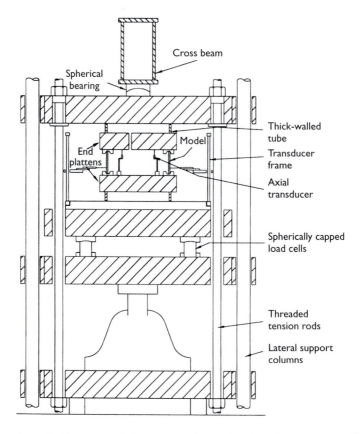

Figure 11.22 Imperial College test rig for small-scale welded stringer-stiffened shells subjected to axial compression (from Dowling *et al.* 1982; reproduced by kind permission of Springer-Verlag).

Large stiffened shells

The interaction between local buckling and general instability depends also on the shape of the stiffeners and the efficiency of the joints between stiffeners and shells. Since, however, stiffeners and their joints usually change somewhat when scaled down to a small model, several large, or even full-scale stiffened shells, have been tested to compare their buckling behaviour with that of the corresponding small-scale models.

Some earlier *axial compression tests* of large stiffened cylindrical aerospace shells were carried out in very large capacity testing machines, like the NASA Langley 1,200,000 lb. (≅545 ton) capacity machine (Peterson and Dow 1959; Dickson and Brolliar 1966), which was fitted in the late 1950s with a large capacity hydraulic jack that controlled the motion of the crosshead, and thus the load-shortening characteristics of the shells were obtained more accurately. Later NASA tests on large stiffened shells subject to axial compression were, however, mostly bending tests, which provided higher compressive stresses and larger rigidity (e.g. Card 1964; Peterson and Anderson 1966; Anderson and Peterson 1971).

Concurrently with the development of small-scale welded specimens, mentioned in the previous section, a special offshore industry related test rig for large stiffened cylindrical shells subject to axial compression was built at Imperial College and a series of large ring-and-stringer-stiffened shells (of 900, 1200 and 1500 mm diameter) were tested (Dowling and Harding 1982; Dowling *et al.* 1982).

In the large test rig the vertical loading was provided by 24 jacks, each of 20 ton rating, that were equally spaced beneath the cylinder's circumference and were all fed from a common pressure supply to ensure uniform concentric loading. Because the jacks were directly beneath the cylindrical shell, a relatively light bottom platen was sufficient to distribute the load. At the top, however, the reaction against the heavy overhead cross-frame was through a central spherical bearing and the top platen had to be substantial to minimize bending distortions. The cylinders were bedded onto the platens using an araldite and sand mixture to ensure uniform load application.

A rotating transducer frame was employed to measure the out-of-plane deflections during the test. The frame, which carried two vertical banks of transducers, rotated on an accurately machined steel ring bedded onto the top platen, and could be located at 20 predetermined positions using a spring loaded plunger. Using this method, a repeatability of readings to within 0.1 mm was achieved.

Further details of the fabrication of these typical large shells and of the test setup and test procedure can be found in Dowling and Harding (1982), Dowling *et al.* (1982) or in chapter 13 of Singer *et al.* (1997, 2001).

Many other similar tests on large stiffened cylindrical shells subject to axial compression were carried out in and for the offshore and marine industry, some of them also under combined loading. For example, one test setup, which was developed for large ring-stiffened cylindrical shells, subject to axial compression and external pressure, at the Chicago bridge and Iron Company (CBI) in the late 1970s and early 1980s, has been extensively used (Miller 1982). Other typical test setups were those of the DnV tests in Norway in the late 1970s (Grove and Didriksen 1977), or the extensive Mitsubishi Heavy Industries tests in Japan (e.g. Kawamoto and Yuhara 1986; Singer *et al.* 1997, 2001, chapter 13).

Experimental features of buckling of stiffened cylindrical shells

As pointed out earlier, the trend in modern stiffened circular cylindrical shells is towards close stiffening, and the design of these closely stiffened shells is mostly governed by general instability of the entire shell. Smeared-stiffener theory then predicts the buckling load and pattern fairly well, and even linear smeared theory presents a close approximation within the bounds of validity, as indicated in the section on 'Adequacy and bounds of validity of linear smeared-stiffener theory'.

A resulting experimental feature is well-defined unique buckling modes (e.g. Figs 11.5 or 11.6), hardly affected by initial imperfections. On the other hand, the influence of the boundary conditions, in particular the axial restraint on stringer-stiffened shells, is predominant, as are the eccentricity effects. The designer can estimate their influence if the boundary conditions are well defined, and experiments will usually verify his estimates with relatively little scatter.

However, because of the dominance of the effects of boundary conditions and eccentricity, non-destructive experimental determination of the actual boundary conditions has long been aimed at. One such promising non-destructive method is the Vibration Correlation Technique (VCT), developed at the Technion more than two decades ago (Singer and Rosen 1974, 1976; Singer 1979, 1983; Singer et al. 1997, 2001), which significantly reduces the experimental scatter.

Structural efficiency

The experimental evidence discussed earlier has shown fairly high values of knock-down factors for closely stiffened shells. The structural efficiency of stiffened shells, from a design point of view, can be evaluated by comparison with equal weight, or 'equivalent', unstiffened shells.

Since there are no reliable theoretical estimates for unstiffened cylindrical shells under axial compression, one has to rely on empirical formulae that show the primary dependence of the buckling load P_B on R/h. A very simple formula has been proposed by Pflüger (1963) for $R/h > 200$:

$$(P_B/P_{cl}) = 1/[1 + (R/100h)]^{1/2} \tag{9}$$

which in addition to its simplicity has the additional merit – for the purpose of comparison – of being unconservative for most test data. In Fig. 26 of Singer (1967), Pflüger's formula was superimposed on test results obtained by 14 investigators and found to be an upper bound for practically all the shells tested. Hence, P_B from Eq. (9) is a suitable standard for comparison, since – for the purpose of comparison – Eq. (9) is conservative, the structural efficiency obtained in reference to it is smaller than the actual efficiency of the stiffened shell!

For ring-stiffened cylindrical shells with external rings, axisymmetric buckling predominates and the simple formula for the general instability load, Eq. (8), makes the comparison with the 'equivalent' unstiffened shell very easy. The thickness of the equivalent shell (of identical weight) is

$$\bar{h} = [1 + (A_2/ah)]h \tag{10}$$

and, if Pflüger's empirical formula, Eq. (9), is employed, the buckling load of the equivalent shell is given by

$$(\bar{P}_B/P_{cl}) = (\bar{h}/h)^2[1 + (R/100\bar{h})]^{-1/2} \tag{11}$$

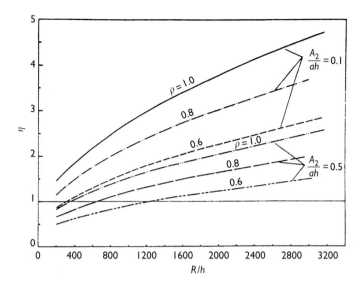

Figure 11.23 Structural efficiency of axially compressed ring-stiffened cylindrical shells – external rings (from Singer 1969; reproduced by kind permission of Springer-Verlag).

Hence, the efficiency of *externally ring-stiffened* cylindrical shells is given by

$$\eta = (\rho P_{\text{cr}}/\bar{P}_{\text{B}}) = \frac{\rho[\Delta_{\text{R}} + (R/100h)]^{1/2}}{\Delta_{\text{R}}^2} \tag{12}$$

where ρ is the knock down factor, or fraction of the 'linear' load achieved, and

$$\Delta_{\text{R}} = 1 + (A_2/ah) \tag{13}$$

With Eq. (12), design curves can readily be drawn that give η versus R/h for various values of ρ and (A_2/ah). In Fig. 11.23, reproduced from Singer (1969), a typical set of such curves is presented. It is immediately seen that even when a knock down factor ρ of only 60% is achieved, weak ring-stiffening is very efficient for thin shells (large R/h); or, in other words, thin shells with many closely spaced rings (to prevent local buckling) and external rings of small cross-sectional area carry axial compression very efficiently.

For *internal rings*, asymmetrical buckling occurs and Eq. (8) is no longer valid, unless the rings have very high torsional stiffness. The critical load parameter λ has, therefore, to be computed from Eq. (6) as detailed in Singer *et al.* (1967). The efficiency is given by Eq. (17) of Singer (1967). Internal rings will naturally be less efficient than external ones.

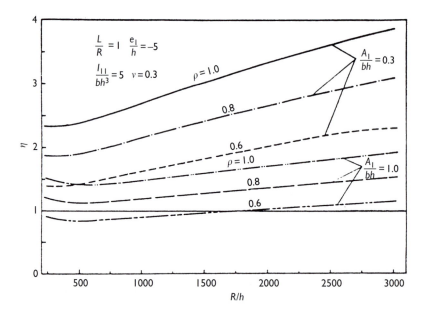

Figure 11.24 Structural efficiency of axially compressed stringer-stiffened cylindrical shells – external stringers (from Singer 1969; reproduced by kind permission of Springer-Verlag).

For *stringer-stiffened* cylindrical shells a similar expression can be obtained for the structural efficiency

$$\eta = \frac{\rho\lambda}{8[3(1-v^2)]^{1/2}} \cdot \frac{[\Delta_s + (R/100h)]^{1/2}}{(R/h)\Delta_s^{2.5}} \tag{14}$$

where

$$\Delta_s = 1 + (A_1/bh) \tag{15}$$

Some typical curves for *external stringers* are presented in Fig. 11.24. Again, the stiffening is seen to be more efficient for lighter stringers and a thinner shell, except for R/h below 500 where the efficiency may rise again slightly as R/h decreases.

For optimization of the stiffened shell, be it stringer- or ring-stiffened, one has to balance the likely knock down factor obtained for various stiffener area ratios with the η for the respective stiffener area and shell R/h.

From the discussion in the section on 'Adequacy and bounds of validity of linear smeared-stiffener theory', the range of $0.2 < (A_2/ah) < 0.5$ appears most promising for *rings*, and $0.3 < (A_1/bh) < 0.8$ for *stringers*, and with the ρ values obtained, stiffened shells are noticeably efficient. This can be clearly

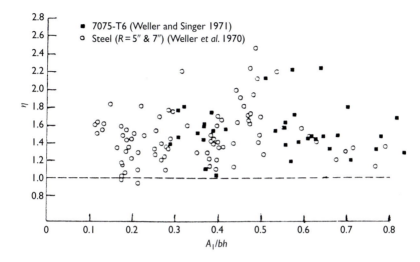

Figure 11.25 Structural efficiency of axially compressed stringer-stiffened cylindrical shells tested at Technion – external stringers (from Weller and Singer 1971).

seen in Fig. 11.25, reproduced from Weller and Singer (1971), which presents the structural efficiency η for the stringer-stiffened shells of Weller *et al.* (1970) and Weller and Singer (1971). It may be pointed out that Fig. 11.24 includes the 5 in. radius steel shells of Weller *et al.* (1970), discussed in the beginning of the section on 'Imperfection sensitivity', whose load carrying capacity was reduced by inelastic effects. If one now recalls, due to the conservative comparison, the actual structural efficiency will be higher than that presented in Fig. 11.24, and the pronounced advantage of stiffening is apparent.

Most minimum weight design studies of stiffened shells (e.g. Burns 1966; Burns and Almroth 1966; Singer and Baruch 1966; Harari *et al.* 1967; Block 1971) have employed linear theory with intuitive or qualitative justification of its applicability. The experimental evidence summarized earlier reinforces this justification and validates the conclusions arrived at in these studies. Usually, the studies with linear theory show plainly that optimization leads to stiffened shells. For example, Fig. 11.26, reproduced from Singer and Baruch (1966), in which the buckling loads for the most efficient distribution of added material in the case of combined axial compression and external pressure are presented, shows clearly the marked inferiority of the equivalent (equal weight) unstiffened shell.

Care has to be taken, however, in the interpretation of the results of linear theory. For example, in a study in the 1970s, aimed at design criteria (Tennyson *et al.* 1971), ring-stiffened shells under axial compression were stated to be always less efficient than equivalent isotropic shells. The analysis preceding the conclusion is a linear one, essentially identical to that of Singer (1969), summarized in that study. With the thickness of the equivalent shell given by Eq. (10), the classical

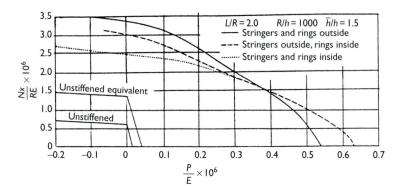

Figure 11.26 Theoretical interaction curves for most efficient distribution of stiffener material (from Singer and Baruch 1966).

buckling load of the equivalent shell is

$$P_{\text{cl.eq.}} = P_{\text{cl}}[1 + (A_2/ah)]^2 = P_{\text{cr}}[1 + (A_2/ah)]^{3/2} \qquad (16)$$

where P_{cr} is the buckling load for the ring-stiffened shell, given in Eq. (8). It is immediately obvious that, within the framework of linear theory, the equivalent isotropic shell will always appear to carry a higher buckling load.

If the imperfection sensitivity of isotropic and ring-stiffened shells is then assumed to be similar (Tennyson *et al.* 1971), the *incorrect conclusion*, that equivalent weight isotropic cylinders are actually superior from a design point of view, is obtained. In reality, however, the assumption of similar imperfection sensitivity is very unrealistic and experimental evidence (e.g. Singer 1969; Weller and Singer 1974) confirms that ring-stiffened shells are usually more efficient. As a matter of fact, fig. 6 of Tennyson *et al.* (1971) may be misleading, since a comparison with an empirical curve for the equivalent isotropic shells is missing. It is important, therefore, always to bear in mind the experimental evidence in the evaluation of the results of linear theory.

Today, more sophisticated optimization techniques are available, but these early elementary engineering type optimization studies clearly indicate to the designers the superior structural efficiency of stiffened shells and provide them with the trends, limitations and some guidance for preliminary design.

Combined loading

Buckling of stiffened cylindrical shells under combined loading

In practice, shells are usually subjected to a combination of loadings. Since the interactions of the buckling modes due to different types of load are usually nonlinear and may depend on the boundary conditions and on the imperfections, buckling of stiffened cylindrical shells under combined loading has been the focus of many studies. The load combinations commonly encountered in aerospace, marine or offshore engineering include axial compression and external pressure, axial compression and internal pressure, bending and external or internal pressure, as well as loadings with the addition of torsion or concentrated loads. One should recall that the combination of loadings may be *destabilizing*, as, for example, in the case of axial compression and external pressure, or *stabilizing* as in the case of the addition of internal pressure to any of the fundamental loads. Furthermore, one should note that in civil and offshore engineering also combinations of axial tension with external pressure and/or bending or torsion appear often, where in elastic buckling the tension stabilizes, but it destabilizes in plastic buckling (e.g. Giezen *et al.* 1991; Singer *et al.* 1997, 2001, chapter 16).

Interaction curves

For assessment of the buckling strength of a shell subjected to combinations of loading, the designer usually requires *interaction curves*, which represent the buckling loads corresponding to the different combinations of loading (e.g. Fig. 11.27).

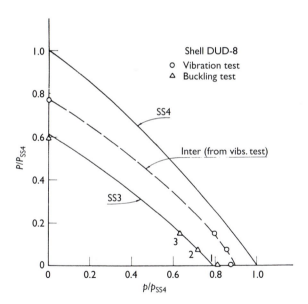

Figure 11.27 Interaction curves for a 7075-T6 aluminium alloy stringer-stiffened circular cylindrical shell, DUD-8 (with $R/h = 483$), tested at Technion (from Abramovich *et al.* 1988). Note the improved interaction curve (INTER) obtained from non-destructive vibration tests by VCT, as well as the repeated buckling tests that yield an experimental interaction curve, which is here very close to the theoretical SS3 curve (reproduced by kind permission of the Society for Experimental Mechanics, Inc.).

Each component of the load, which constitutes the combined-loading condition, is referred to that state under which it acts alone as a single loading.

As pointed out earlier, the buckling behaviour of stiffened cylindrical shells differs from that of unstiffened shells. The knockdown factors for the stiffened shells are higher, while the shapes of the interaction curves are also different. Clearly, the interaction curves depend on the single load reference buckling loads, which strongly depend on the boundary conditions. Hence, in order to ensure reliable interaction curves, the reference boundary conditions themselves should be well defined. The Vibration Correlation Technique (VCT), developed at the Technion (e.g. Singer and Rosen 1976; Singer 1983; Singer *et al.* 1997, 2001, chapter 15) presents a useful non-destructive experimental method for definition of the boundary conditions.

At the Technion, VCT was applied for non-destructive generation of more realistic buckling interaction curves for stringer-stiffened circular-cylindrical shells, subjected to combined loading of axial compression and external pressure (Abramovich and Singer 1978; Abramovich *et al.* 1988, 1991a,b). The details of the test setup and procedure as well as experimental results can be found in Abramovich and Singer (1978) and Abramovich *et al.* (1988, 1991a,b). It should be noted that the more recent Technion test rigs for combined loading of external pressure and axial compression have the ability of imperfection measurement *in situ* (including initial imperfections, growth of imperfections and imperfections after buckling under external pressure and combined loading) as well as vibration scanning as required for VCT.

The application of VCT, indeed, provided better defined boundary conditions and yielded non-destructively improved interaction curves. An example is Fig. 11.27, from Abramovich *et al.* (1988), which shows these curves for a 7075-T6 aluminium alloy stringer-stiffened shell DUD-8, subjected to combined axial compression and external pressure.

Repeated buckling tests

In the case of combined loading, there are two approaches to the construction of experimental interaction curves: use of separate, nominally identical, shells for each point on the interaction curve, or use of repeated buckling of the same specimen. Singer (1964) emphasised that with the use of separate shells the scatter in the experimental interaction curves is mainly due to minor differences among the specimens, whereas repeated buckling of the same specimen can provide a better experimental interaction curve. The earlier Technion studies on stiffened cylindrical shells under combined loading (e.g. Abramovich and Singer 1978; Abramovich *et al.* 1981) used both approaches for obtaining experimental interaction curves. It was concluded that for stiffened shells the two approaches could be applied successfully, though the repeated buckling method seemed preferable. The initial results for the experimental interaction curves in Abramovich and Singer (1978), obtained by repeated buckling, were not conclusive, but the later ones (Abramovich *et al.* 1981, 1988) were more consistent.

In Abramovich *et al.* (1988), it was also shown that the *sequence of loading*, constant axial compression first and then increasing the external pressure until buckling occurs, or the reverse order of loading, *does not influence* the buckling loads significantly. This was somewhat surprising in view of the nonlinear behaviour of the interaction curve. Since this observation was based on six specimens only, an additional test series was embarked upon to validate and reinforce this main conclusion (Abramovich *et al.* 1991b). For more precision and more reliable interpretation of the data, a new feature was incorporated in that test series, the measurement of geometrical imperfections of the shell *in situ* before, during and after each buckling load combination.

The later Technion experimental studies have extended the generality and applicability of the VCT as an adequate non-destructive tool for more reliable generation of buckling interaction curves of stiffened shells subjected to combined loading. The *repeated buckling method* to obtain reliable interaction curves appears to be very appropriate, provided the 'knock down' factor of the shell due to the induced geometrical imperfections (caused by the repeated buckling process) and the real boundary conditions can be assessed.

These studies showed again that measurement of the imperfections of the shell is a valuable tool. Using the measured geometrical imperfections as an input of two multimode analyses, CPIUTAM and CPIUTAMN, which are 1989 modifications of MIUTAM, originally developed by Arbocz and Babock (1976), one can calculate the 'knock down' factor ρ_{th} for shells under external pressure, axial compression and a combination of constant external pressure and buckling under axial compression, as well as a combination of constant axial compression and varying external pressure till buckling. This capability facilitates a complete evaluation of the effect of the growth of imperfections at all the intermediate points.

Acknowledgement

This work has been supported in part by the Jordan and Irene Tark Aerospace Structures Research Fund.

References

ABAQUS/EXPLICIT Manuals – Version 5.7 (1997). Hibbitt, Karlsson and Sorensen Inc., Providence, RI, USA.

ADINA (1998). User Interface, Users Guide, Report ARD 98-1, ADINA R&D Inc., Watertown, MA, USA.

Abramovich, H. and Singer, J. (1978). Correlation between vibration and buckling of stiffened cylindrical shells under external pressure and combined loading. *Israel Journal of Technology* **16**(1–2), 34–44.

Abramovich, H., Singer, J. and Grunwald, A. (1981). Nondestructive determination of interaction curves for buckling of stiffened shells. TAE Report 341, Technion – Israel Institute of Technology, Department of Aeronautical Engineering, Haifa, Israel.

Abramovich, H., Yaffe, R. and Singer, J. (1987). Evaluation of stiffened shell characteristics from imperfection measurements. *Journal of Strain Analysis* **22**(1), 17–23.

Abramovich, H., Weller, T. and Singer, J. (1988). Effect of sequence of combined loading on buckling of stiffened shells. *Experimental Mechanics* **28**(1), 1–13.

Abramovich, H., Singer, J. and Weller, T. (1991a). Buckling of imperfect stiffened cylindrical shells under combined loading. TAE Report 653, Technion – Israel Institute of Technology, Department of Aeronautical Engineering, Haifa, Israel.

Abramovich, H., Singer, J. and Weller, T. (1991b). The influence of initial imperfections on the buckling of stiffened cylindrical shells under combined loading. In *Buckling of Shell Structures, on Land, in the Sea and in the Air* (ed. J.-F. Jullien). Elsevier Applied Science, London, pp. 1–10.

Almroth, B.O. (1966a). Influence of edge conditions on the stability of axially compressed cylindrical shells. *AIAA Journal* **4**(1), 134–140.

Almroth, B.O. (1966b). Influence of imperfections and edge restraint on the buckling of axially compressed cylinders. NASA CR-432. Also presented at the *AIAA/ASME 7th Structures and Materials Conference*, Cocoa Beach, Florida, 18–20 April 1966.

Almroth, B.O. and Bushnell, D. (1968). Computer analysis of various shells of revolution. *AIAA Journal* **6**(10), 1848–1855.

Almroth, B.O. and Brogan, F.A. (1978). The STAGS Computer Codes, NASA CR 2950.

Almroth, B.O., Bushnell, D. and Sobel, L.H. (1968). Buckling of shells of revolution with various wall constructions. NASA CR-1049, Vol. 1: Numerical Results.

Almroth, B.O., Brogan, F.A. and Stanley, G.M. (1983). Structural analysis of general shells. Lockheed Missiles and Space Co., Palo Alto, CA, Vol. 2, Users Instructions for STAGSC-1, Report LMSC D633873.

Anderson, J.K. and Peterson, J.P. (1971). Buckling tests of two integrally stiffened cylinders subjected to bending. NASA TN D-6271.

Arbocz, J. (1974). The effect of initial imperfection on shell stability. In *Thin-Shell Structures, Theory, Experiment and Design* (eds Y.C. Fung and E.E. Sechler). Prentice-Hall, Englewood Cliffs, NJ, pp. 205–245.

Arbocz, J. (1976). Prediction of buckling loads based on experimentally measured imperfections. *Buckling of Structures, Proceedings of IUTAM Symposium*, Harvard University, Cambridge, USA, June 1974 (ed. B. Budiansky). Springer-Verlag, Berlin, pp. 291–311.

Arbocz, J. (1982). The imperfection data bank, a means to obtain realistic buckling loads. *Buckling of Shells, Proceedings of a State-of-the-Art Colloquium* (ed E. Ramm). Springer Verlag, Berlin, pp. 535–567.

Arbocz, J. (1983). Shell stability analysis: theory and practice. *Collapse: The Buckling of Structures in Theory and Practice, Proceedings of IUTAM Symposium* (eds J.M.T. Thompson and G.W. Hunt). Cambridge University Press, Cambridge, pp. 43–47.

Arbocz, J. and Babcock, D.C. (1969). The effect of general imperfections on the buckling of cylindrical shells. *Journal of Applied Mechanics, Series E* **36**(1), 28–38.

Arbocz, J. and Babcock, C.D. Jr (1976). Prediction of buckling loads based on experimentally measured initial imperfections. *Buckling of Structures, Proceedings of IUTAM Symposium*, Harvard University, June 1974 (ed. B. Budiansky). Springer 1976, pp. 291–311.

Arbocz, J. and Abramovich, H. (1979). The initial imperfection data bank at the Delft University of Technology – Part I. Report LR-290, Delft University of Technology, Department of Aerospace Engineering, The Netherlands.

Arbocz, J. and Hol, J. (1993). Shell stability analysis in a computer aided engineering (CAE) environment. AIAA Paper 93-133, *Proceedings 34th AIAA/ASME/ASCE/ AHS/ASC Structures, Structural Dynamics and Materials Conference*, La Jolla, CA, pp. 300–314.

Arbocz, J. and Hol, J.M.A.M. (1995). Collapse of axially compressed cylindrical shells with random imperfections. *Thin-Walled Structures* **23**(1–4), 131–158.

Arbocz, J., Starnes, J.H., Jr. and Nemeth, M.P. (1999). A hierarchical approach to buckling load calculations. AIAA Paper 99-1232. Presented at the *40th AIAA/ASME/ASCE/AHS/ASC Structures, Structural Dynamics, and Materials Conference*, St. Louis, Missouri.

Babcock, C.D. (1974). Experiments in shell buckling. In *Thin-Shell Structures, Theory, Experiment and Design* (eds Y.C. Fung and E.E. Sechler). Prentice-Hall, Englewood Cliffs, NJ, pp. 345–369.

Baruch, M. (1965). Equilibrium and stability equations for discretely stiffened shells. *Israel Journal of Technology* **3**(2), 138–146.

Baruch, M. and Singer, J. (1963). The effect of eccentricity of stiffeners on the general instability of stiffened cylindrical shells under hydrostatic pressure. *Journal of Mechanical Engineering Science* **5**(1), 23–27.

Baruch, M., Singer, J. and Weller, T. (1966). Effect of eccentricity of stiffeners on the general instability of cylindrical shells under torsion. *Proceedings of the 8th Israel Annual Conference on Aviation and Astronautics. Israel Journal of Technology* **4**(1), 144–154.

Batdorf, S.B. (1947). A simplified method of elastic stability analysis for thin cylindrical shells. NACA Report 874.

Becker, H. and Gerard, G. (1962). Elastic stability of orthotropic shells. *Journal of the Aerospace Sciences* **29**(5), 505–512.

Block, D.L. (1966). Buckling of eccentrically stiffened orthotropic cylinders under pure bending. NASA TND-3351.

Block, D.L. (1968). Influence of discrete ring stiffeners and prebuckling deformations on the buckling of eccentrically stiffened orthotropic cylinders. NASA TND-4283.

Block, D.L. (1971). Minimum weight design of axially compressed ring and stringer stiffened cylindrical shells. NASA CR-1766.

Block, D.L., Card, M.F. and Mikulas, M.M. (1965). Buckling of eccentrically stiffened orthotropic cylinders. NASA TND-2960.

Bodner, S.R. (1957). General instability of a ring-stiffened cylindrical shell under hydrostatic pressure. *Journal of Applied Mechanics* **24**(2), 269–277.

Boichot, L. and Reynolds, T.E. (1965). Inelastic buckling tests of ring-stiffened cylinders under hydrostatic pressure. David Taylor Model Basin, Washington DC, Rep. 1992.

Bosman, T.N., Pegg, N.G. and Kenning, P.J. (1993). Experimental and numerical determination of non-linear overall collapse of imperfect pressure hull compartments. In *Fourth International Symposium on Naval Submarines*, RINA, London.

Brush, D.O. (1968). Imperfection sensitivity of stringer stiffened cylinders. *AIAA Journal* **6**(12), 2445–2447.

Brush, D.O. and Almroth, B.O. (1975). *Buckling of Bars, Plates and Shells*. McGraw-Hill Book Company, New York, Chapter 5.

Budiansky, B. and Hutchinson, J.W. (1966). A survey of some buckling problems. *AIAA Journal* **4**, 1505–1510.

Burns, A.B. (1966). Structural optimization on axially compressed cylinders, considering ring-stringer eccentricity effects. *Journal of Spacecraft and Rockets* **3**(8), 1263–1268.

Burns, A.B. and Almroth, B.O. (1966). Structural optimization of axially compressed ring-stringer stiffened cylinders. *Journal of Spacecraft and Rockets* **3**(1), 19–25.

Bushnell, D. (1976a). BOSOR 5 – program for buckling of elastic–plastic shells of revolution including large deflections and creep. *Computers and Structures* **6**, 221–239.

Bushnell, D. (1976b). Buckling of elastic–plastic shells of revolution with discrete elastic–plastic ring stiffeners. *International Journal of Solids and Structures* **12**, 51–66.

Bushnell, D. (1982). Plastic buckling. In *Pressure Vessels and Piping: Design Technology* (eds S.Y. Zamrik and D. Dietrich). ASME, New York, pp. 47–117.

Bushnell, D. (1984). Computerized analysis of shells – governing equations. *Computers & Structures* **18**(3), 471–536.

Bushnell, D. (1985). *Computerized Bucking Analysis of Shells*. Martinus Nijhoff, Dordrecht/Boston.

Bushnell, D. and Meller, E. (1984). Elastic–plastic collapse of axially compressed cylindrical shells: a brief survey with particular application to ring-stiffened cylindrical shells with reinforced openings. *ASME Journal of Pressure Vessel Technology* **106**, 2–15.

Bushnell, D., Almroth, B.O. and Sobel, L.H. (1968). Buckling of shells of revolution with various wall constructions. NASA CR's 1049, 1050 and 1051, vols 1–3.

Calladine, C.R. (1995). Understanding imperfection-sensitivity in the buckling of thin-walled shells. *Thin-Walled Structures* **23**(1–4), 215–235.

Card, M.F. (1964). Bending tests of large-diameter stiffened cylinders susceptible to general instability. NASA TND-2200.

Card, M.F. and Jones, R.M. (1966). Experimental and theoretical results for buckling of eccentrically stiffened cylinders. NASA TND-3639.

Chang, L.K. and Card, M.F. (1971). Thermal buckling analysis for stiffened orthotropic cylindrical shells. NASA TN D-6332.

Chen, Y., Zimmer, R.A., de Oliveira, J.G. and Jan, H.Y. (1985). Buckling and ultimate strength of stiffened cylinders: model experiments and strength formulations. Presented at *17th Annual OTC*, Houston, TX, Paper OTC 4853.

Cox, H.L. (1933). Buckling of thin plates in compression. British Ministry of Aviation, Aeronautical Research Council, Reports and Memoranda, No. 1554.

Crisfield, M.A. (1997). *Nonlinear Finite Element Analysis of Solids and Structures*, Vols. 1 & 2. John Wiley and Sons, Chichester, UK.

Croll, J.G.A. (1981). Lower bound elasto-plastic buckling of cylinders. *Proceedings, Institution of Civil Engineers, Part 2* **71**, 235–261.

Croll, J.G.A. (1985). Stiffened cylindrical shells under axial and pressure loading. In *Shell Structures – Stability and Strength* (ed. R. Narayanan). Elsevier Applied Science Publishers, London, pp. 19–56.

Croll, J.G.A. (1995). Towards a rationally based elastic–plastic shell buckling design methodology. *Thin-Walled Structures* **23**(1–4), 67–84.

Dickson, J.N. and Brolliar, R.H. (1966). The general instability of ring-stiffened corrugated cylinders under axial compression. NASA TND-3089.

Donnell, L.H. (1933). Stability of thin-walled tubes under torsion. NACA Report 479.

Dowling, P.J. and Harding, J.E. (1982). Experimental behaviour of ring and stringer stiffened shells. In *Buckling of Shells in Offshore Structures* (eds J.E. Harding, P.J. Dowling and N. Agelidis). Granada, London, pp. 73–107.

Dowling, P.J., Harding, J.E., Agelidis, N. and Fahy, W. (1982). Buckling of orthogonally stiffened cylindrical shells used in offshore engineering. *Buckling of Shells, Proceedings of State-of-the-Art Colloquium* (ed. E. Ramm). Universität Stuttgart, Germany, 6–7 May. Springer-Verlag, Berlin, pp. 242–273.

Ellinas, C.P., Supple, W.J. and Walker, A.C. (1984). *Buckling of Offshore Structures.* Granada, London.

Esslinger, M. and Geier, B. (1970). Buckling and postbuckling behavior of discretely stiffened thin-walled circular cylinders. *Zeitschrift für Flugwissenschaften* **18**(7), 240–253.

Esslinger, M. and Geier, B. (1975). *Postbuckling Behavior of Structures.* CSIM Courses and Lectures No. 236, Springer-Verlag, Wien.

Faulkner, D. (1997). Effects of residual stresses on the ductile strength of plane welded grillages and of ring stiffened cylinders. *IMechE Journal of Strain Analysis* **12**(2), 130–139.

Flügge, W. (1932). Die Stabilität der Kreiszylinderschale. *Ingenieur Archiv* **3**, 463–506.

Galambos, T.V. (ed.) (1998). *Guide to Stability Design Criteria for Metal Structures*, 5th edn. John Wiley & Sons, New York.

Galletly, G.D., Slankard, R.C. and Wenk, E. (1958). General instability of ring-stiffened cylindrical shells subject to external hydrostatic pressure – a comparison of theory and experiment. *Journal of Applied Mechanics* **25**, 259–266.

Galletly, G.D., Blachut, J. and Moreton, D.N. (1990). Internally pressurised machined domed ends – a comparison of the plastic buckling predictions of the deformation and flow theories. *Proceedings of the Institution of Mechanical Engineers, Part C* **204**, 169–186.

Galletly, G.D., Rhodes, J. and Chong, K.P. (eds) (1995). Buckling strength of imperfection-sensitive shells. *Thin-Walled Structures* **23**(104), pp. 1–411.

Giezen, J.J., Babcock, C.D. and Singer J. (1991). Plastic buckling of cylindrical shells under biaxial loading. *Experimental Mechanics* **31**(4), 337–343.

Green, D.R. and Nelson, H.M. (1982). Compression tests on large-scale stringer-stiffened tubes. In *Buckling of Shells in Offshore Structures* (eds J.E. Harding, P.J. Dowling and N. Agelidis). Granada, London, pp. 25–43.

Grove, T. and Didriksen, T. (1971). Buckling experiments on 4 large ring-stiffened cylindrical shells subjected to axial compression and lateral pressure. Report No. 77-431, Det norske Veritas, Oslo.

Harari, O., Singer, J. and Baruch, M. (1967). General instability of cylindrical shells with non-uniform stiffeners. *Proceedings of the 9th Israel Annual Conference on Aviation and Astronautics. Israel Journal of Technology* **5**(1), 114–122.

Harding, J.E. and Dowling, P.J. (1982). Analytical results for the behaviour of ring and stringer stiffened shells. In *Buckling of Shells in Offshore Structures* (eds J.E. Harding, P.J. Dowling and N. Agelides). Granada, London, pp. 231–256.

Hedgepeth, J.M. and Hall, D.B. (1965). Stability of stiffened cylinders. *AIAA Journal* **3**(12), 2275–2286.

Hoff, N.J. (1967). Thin shells in aerospace structures. The 4th von Kármán Lecture of the American Institute of Aeronautics and Astronautics. *Aeronautics and Astronautics* **5**, 26–45.

Hom, K. and Couch, W.P. (1961). Hydrostatic tests of inelastic and elastic stability of ring-stiffened cylindrical shells machined from strain-hardening steel. US Navy Department David Taylor Model Basin, DTMB Report 1501.

Hutchinson, J.W. (1974). Plastic buckling. In *Advances in Applied Mechanics* (ed. C.S. Yih). Academic Press, New York, pp. 67–144.

Hutchinson, J.W. and Amazigo, J.C. (1967). Imperfection sensitivity of eccentrically stiffened cylindrical shells. *AIAA Journal* **5**(3), 392–401.

Hutchinson, J.W. and Frauenthal, J.C. (1969). Elastic postbuckling behavior of stiffened and barreled cylindrical shells. *Journal of Applied Mechanics, Series E* **36**(4), 784–790.

Hutchinson, J.W. and Koiter, W.T. (1971). Postbuckling theory. *Applied Mechanics Review* **24**, 1353–1366.

Katz, L. (1965). Compression tests on integraly stiffened cylinders, NASA TMX-55315, August.

Kawamoto, Y. and Yuhara, T. (1986). Buckling of fabricated ring-stiffened steel cylinders under axial compression. In *Advances in Marine Structures* (eds C.S. Smith and J.D. Clarke). Elsevier Applied Science, London, pp. 262–280.

Kendrick, S. (1953). The buckling under external pressure of circular cylindrical shells with evenly spaced equal strength circular ring frames – Part I. UK Naval Construction Research Establishment Report No. 211.

Kendrick, S. (1985). Ring-stiffened cylinders under external pressure. In *Shell Structures – Stability and Strength* (ed. R. Narayanan). Elsevier Applied Science, London, pp. 57–95.

Kobayashi, S. (1968). The influence of prebuckling deformation on the buckling load of orthotropic cylindrical shells under axial compression. *Transactions of the Japan Society for Aeronautical and Space Sciences* **11**(19), 60–68.

Koiter, W.T. (1956). Buckling and post-buckling behavior of a cylindrical panel under axial compression. Report S.476, National Luchtvaartlaboratorium (NLL) Amsterdam, Reports and Transactions, vol. 20.

Koiter, W.T. (1963). The effect of axisymmetric imperfections on the buckling of cylindrical shells under axial compression. *Proceedings of the Royal Netherlands Academy of Sciences, Amsterdam, Series B* **66**(5), 265–279.

Krenzke, M.A. and Reynolds, T.E. (1967). Structural research on submarine pressure hulls at the David Taylor Model Basin. *Journal of Hydronautics* **1**(1), 27–35.

Kuhn, P. (1933). A summary of design formulas for beams having thin webs in diagonal tension. NACA TN 469.

Len'ko, O.N. (1963). The stability of orthotropic cylindrical shells, Raschet Prostranstvennykh Konstrunktsii, Issue IV, pp. 499–524, Moscow 1958, Translation NASA TT F-9826, July.

Mayers, J. and Meller, E. (1972). Material nonlinearity effects in optimization considerations of stiffened cylinders and interpretation of test data scatter for compressive buckling. Stanford University USAAMRDL Technical Report 71-70, US Army Air Mobility Research & Development Lab, Fort Eustis, Virginia.

Mayers, J. and Wesenberg, D.L. (1969). The maximum strength of initially imperfect, axially compressed, circular cylindrical shells. Stanford University, USAAVLABS Technical Report 69-60, US Army Material Labs, Fort Eustis, Virginia.

Midgley, W.R. and Johnson, A.E., Jr. (1967). Experimental buckling of internal integral ring-stiffened cylinders. *Experimental Mechanics* **7**, 145–153.

Miller, C.D. (1977). Buckling of axially compressed cylinders. *ASCE, Journal of the Structural Division* **103**(ST3), 695–721.

Miller, C.D. (1982). Summary of buckling tests on fabricated steel cylindrical shells in USA. In *Buckling of Shells in Offshore Structures* (eds J.E. Harding, P.J. Dowling and N. Agelidis). Granada, London, pp. 429–471.

Miller, C.D. and Kinra, R.K. (1981). External pressure tests of ring-stiffened fabricated steel cylinders. *13th Annual Offshore Technology Conference*, Houston, May 1981, Paper No. OTC 4107, Proceedings OTC '81, Vol. 3, pp. 371–386.

Milligan R., Gerard, G., Lakshmikantham, C. and Becker, H. (1966). General instability of orthotropic stiffened cylinders under axial compression. *AIAA Journal* **4**(11), 1906–1913.

MSC/NASTRAN Users Manual – Version 70.5 (1988). The MacNeal Schwendler Corporation, Los Angeles, CA.

Murray, N.W. (1973). Buckling of stiffened panels loaded axially and in bending. *The Structural Engineer* **51**, 285–301.

Murray, N.W. (1983). Ultimate capacity of stiffened plates in compression. In *Plated Structures, Stability and Strength* (ed. R. Narayanan). Applied Science Publishers, London, pp. 135–163.

Murray, N.W. and Khoo, P.S. (1981). Some basic plastic mechanisms in the local buckling of thin-walled steel structures. *International Journal of Mechanical Sciences* **23**(12), 703–714.

Odland, J. (1981). An experimental investigation of the buckling strength of ring-stiffened cylindrical shells under axial compression. *Norwegian Maritime Research* **4**(9), 22–39.

Peterson, J.P. (1956). Bending tests on ring-stiffened circular cylinders. NACA TN 3735.

Peterson, J.P. (1969). Buckling of stiffened cylinders in axial compression and bending, a review of test data. NASA TN-D5561.

Peterson, J.P. and Anderson, J.K. (1966). Bending tests of large-diameter ring-stiffened corrugated cylinders. NASA TND-3336.

Peterson, J.P. and Dow, M.B. (1959). Compression tests on circular cylinders stiffened longitudinally by closely spaced Z-section stringers. NASA MEMO 2-12-59L.

Pflüger, A. (1963). Zur praktischen Berechnung der axial gedrückten Kreiszylinderschale. *Der Stahlbau* **43**(6), 161–165.

Reynolds, T.E. and Blumenberg, W.F. (1959). General instability of ring-stiffened cylindrical shells subject to hydrostatic pressure. David Taylor Model Basin Report 1324.

Riks, E. (1987). Progress in collapse analysis. *ASME Journal of Pressure Vessel Technology* **109**, 27–41.

Riks, E. and Rankine, C.C. (1997). Computational tools for stability analysis. AIAA Paper 97-1138. Presented at the *38th AIAA/ASME/ASCE/AHS/ASC Structures, Structural Dynamics, and Materials Conference*, Kissimmee, FL.

Rosen, A. and Singer, J. (1974). Vibrations of axially loaded stiffened cylindrical shells. *Journal of Sound and Vibration* **34**(3), 357–378.

Rosen, A. and Singer, J. (1975). Further experimental studies on the buckling of integrally stiffened cylindrical shells. TAE Report 207, Technion Israel Institute of Technology, Department of Aeronautical Engineering, Haifa, Israel.

Rosen, A. and Singer, J. (1976). Vibrations and buckling of eccentrically stiffened cylindrical shells. *Experimental Mechanics* **16**(3), 88–94.

Ross, C.T.F. (1990). *Pressure Vessels Under External Pressure: Statics and Dynamics*. Elsevier Applied Science, London.

Ross, C.T.F., Aylward, W.R. and Boltwood, D.T. (1971). General instability of ring-reinforced circular cylinders under external pressure. *Transactions of the Royal Institute of Naval Architects (RINA)* **113**, 73–92.

Ross, C.T.F., Haynes, P., Seers, A. and Johns, T. (1995). Inelastic buckling of ring-stiffened circular cylinders under uniform external pressure. In *ASME PD-Vol. 70 Structural Dynamics and Vibrations*. ASME, New York, pp. 207–215.

Ross, C.T.F., Coalter, B. and Johns, T. (1999). Design charts for the buckling of ring-stiffened cylinders and cones under external hydrostatic pressure. *Transactions of the Royal Institute of Naval Architects (RINA)* **141**(Part A), 15–31.

Saunders, H.E. and Windenburg, D.F. (1932). The use of models in determining the strength of thin-walled structures. *Transactions of the American Society of Mechanical Engineers* **54**(APM-54-25), 263–275.

Scott, N.D., Harding, J.E. and Dowling, P.J. (1987). Fabrication of small scale stiffened cylindrical shells. *IMechE Journal of Strain Analysis* **22**(2), 97–106.

Seggelke, P. and Geier, B. (1967). Das Beulverhalten Versteifter Zylinderschalen. *Zeitschrift für Flugwissenschaften* **15**(12), 477–490.

Sewall, J.L. and Naumann, E.C. (1968). An experimental and analytical vibration study of thin cylindrical shells with and without longitudinal stiffeners. NASA TN D-4705.

Sewell, M. (1972). Plastic buckling. In *Stability*. SM Study no. 6, University of Waterloo, Ontario, Canada, pp. 85–198.

Simitses, G.J. (1968). Buckling of eccentrically stiffened cylinders under torsion. *AIAA Journal* **6**(10), 1856–1860.

Singer, J. (1964). On experimental technique for interaction curves of buckling and shells. *Experimental Mechanics* **4**(9), 279–280.

Singer, J. (1967). The influence of stiffener geometry and spacing on the buckling of axially compressed cylindrical and conical shells (extended version). TAE Report 68, Department of Aeronautical Engineering, Technion, Haifa, Israel.

Singer, J. (1969). The influence of stiffener geometry and spacing on the buckling of axially compressed cylindrical and conical shells. *Theory of Thin Shells, Proceedings of the 2nd IUTAM Symposium*, Copenhagen, September 1967 (ed. F.E. Niordson). Springer-Verlag, Berlin, pp. 239–263.

Singer, J. (1972). Buckling of integrally stiffened cylindrical shells – a review of experiment and theory. In *Contributions to the Theory of Aircraft Structures*. Delft University Press, pp. 325–357.

Singer, J. (1976). Buckling, vibrations and postbuckling of stiffened metal cylindrical shells. *Proceedings of BOSS 1976 (1st International Conference on Behavior of Off-Shore Structures)*. Norwegian Institute of Technology, Trondheim, Norway, pp. 765–786.

Singer, J. (1979). Recent studies on the correlation between vibration and buckling of stiffened cylindrical shells. *Zeitschrift für Flugwissenschaften und Weltraum-forschung* **3**(6), 333–343.

Singer, J. (1982a). Buckling experiments on shells – a review of recent developments. *Solid Mechanics Archives* **7**, 213–313.

Singer, J. (1982b). The status of experimental buckling investigations of shells. *Buckling of Shells, Proceedings of a State-of-the-Art Colloquium* (ed. E. Ramm). Universität Stuttgart, Germany, 6–7 May, Springer-Verlag, Berlin, pp. 501–534.

Singer, J. (1983). Vibrations and buckling of imperfect stiffened shells – recent developments. In *Collapse: The Buckling of Structures in Theory and Practice* (eds J.M.T. Thompson and G.W. Hunt). Cambridge University Press, Cambridge, pp. 443–481.

Singer, J. and Abramovich, H. (1995). The development of shell imperfection measurement techniques. *Thin-Walled Structures* **23**(1–4), 379–398.

Singer, J. and Baruch, M. (1966). Recent studies on optimization for elastic stability of cylindrical and conical shells. *Aerospace Proceedings 1966, Proceedings of the 5th International Congress of the Aeronautical Sciences*, London. Macmillan, London, pp. 751–782.

Singer, J. and Haftka, R.T. (1968). Buckling of discretely ring-stiffened cylindrical shells. *Proceedings of the 10th Israel Annual Conference on Aviation and Astronautics. Israel Journal of Technology* **5**(1–2), 125–137.

Singer, J. and Haftka, R. (1975). Buckling of discretely stringer-stiffened cylindrical shells and elastically restrained panels. *AIAA Journal* **13**(7), 849–850.

Singer, J. and Rosen, A. (1976). Influence of boundary conditions on the buckling of stiffened cylindrical shells. *Buckling of Structures, Proceedings of IUTAM Symposium*, Harvard University, Cambridge, USA, 17–21 June (ed. B. Budiansky). Springer-Verlag, Berlin, pp. 227–250.

Singer, J. and Rosen, A. (1974). Design criteria for buckling in vibration of imperfect stiffened cylindrical shells. *ICAS Proceedings 1974, Proceedings of the 9th Congress of the International Council of the Aeronautical Sciences*, Haifa, August 1974 (eds R.R. Dexter and J. Singer). The Weizmann Science Press of Israel, Jerusalem, pp. 495–517.

Singer, J., Baruch, M. and Harari, O. (1966). Inversion of the eccentricity effect in stiffened cylindrical shells buckling under external pressure. *Journal of Mechanical Engineering Science* **8**(4), 363–373.

Singer, J., Baruch, M. and Harari, O. (1967). On the stability of eccentrically stiffened cylindrical shells under axial compression. *International Journal of Solids and Structures* **3**(4), 445–470.

Singer, J., Arbocz, J. and Babcock, C.D. (1971). Buckling of imperfect stiffened cylindrical shells under axial compression. *AIAA Journal* **9**(1), 68–75.

Singer, J., Libai, A., Weller, T. and Nachmani, S. (1972). The buckling under external pressure of ring-stiffened cylindrical shells with cut rings. Report ASL 48, Technion – Israel Institute of Technology, Haifa, Department of Aeronautical Engineering (in Hebrew).

Singer, J., Abramovich, H. and Yaffe, R. (1979). Initial imperfection measurements of stiffened shells and buckling predictions. *Proceedings, 21st Israel Annual Conference on Aviation and Astronautics. Israel Journal of Technology* **17**, 324–338.

Singer, J., Arbocz, J. and Weller, T. (1997, 2001). *Buckling Experiments – Experimental Methods in Buckling of Thin-Walled Structures* (two volumes). John Wiley & Sons, Chichester and New York.

Sobel, L.H. (1964). Effects of boundary conditions on the stability of cylinders subject to lateral and axial pressure. *AIAA Journal* **2**(8), 1437–1440.

Soong, T.C. (1967). Influence of boundary constraints on the buckling of eccentrically stiffened orthotropic cylinders. Presented at the *7th International Symposium on Space Technology and Science*, Tokyo.

Stein, M. (1968). Some recent advances in the investigation of shell buckling. *AIAA Journal* **6**, 2339–2345.

Stuhlman, C.E., De Luzio, A. and Almroth, B.O. (1966). Influence of stiffener eccentricity and end moment on stability of cylinders in compression. *AIAA Journal* **4**, 872–877.

Teng, J.G. (1996). Buckling of thin shells: recent advances and trends. *ASME Applied Mechanics Reviews* **49**(4), 263–274.

Tennyson, R.C., Muggeridge, D.B. and Caswell, R.D. (1971). New design criteria for predicting buckling of cylindrical shells under axial compression. *AIAA Journal of Spacecraft and Rockets* **8**(10), 1062–1067.

Tennyson, R.C., Booton, M. and Hui, D. (1980). Changes in cylindrical shell buckling behavior resulting from system parameter variations. *Theoretical and Applied Mechanics, Proceedings of the 15th IUTAM Congress*, Toronto, 1980 (eds F.P.J. Rimrott and B. Tabarrok). North Holland Publishing Co., Amsterdam and New York, pp. 417–430.

Thielemann, W.F. (1960). New developments in the nonlinear theories of the buckling of cylindrical shells. *Aeronautics and Astronautics, Proceedings of the Durand Centennial Conference 1959*. Pergamon Press, Oxford, pp. 76–119.

Thielemann, W. and Esslinger, M. (1964). Einfluss der Randbedingungen auf die Beullast von Kreiszylinderschalen. *Stahlbau* **33**(12), 353–361.

Thielemann, W. and Esslinger, M. (1965). Über den Einfluss der Exzentrizität von Längssteifen auf die axiale Beullast dünnwandiger Kreiszylinderschalen. *Stahlbau* **34**, 332–333.

Tokugawa, T. (1931). Model experiments on the elastic stability of closed and cross-stiffened circular cylinders under uniform external pressure. *Proceedings, World Engineering Congress*, Tokyo, 1929, vol. 29, pp. 219–279.

Tsang, S.K. and Harding, J.E. (1984). A mechanism approach for the prediction of the collapse strength of ring-stiffened cylinders under axial compression and external pressure. *Thin-Walled Structures* **2**, 325–353.

Tvergaard, V. (1987). Effect of plasticity on post-buckling behaviour. In *Buckling and Post-Buckling*. Lecture Notes in Physics 288. Springer-Verlag, Berlin, pp. 144–183.

Tvergaard, V. and Needleman, A. (1982). On the foundations of plastic buckling. In *Developments in Thin-Walled Structures* (eds J. Rhodes and A.C. Walker). Applied Science Publishers, London, pp. 205–233.

Valsgård, S. and Foss, G. (1982). Buckling research in Det Norske Veritas. In *Buckling of Shells in Offshore Structures* (eds J.E. Harding, P.J. Dowling and N. Agelidis). Granada, London, pp. 491–519.

van der Neut, A. (1947). The general instability of stiffened cylindrical shells under axial compression. Report S314, NLL Amsterdam, Reports and Transactions, vol. 13, pp. 57–84.

von Kármán, T., Sechler, E.E. and Donnell, L.H. (1932). The strength of thin plates in compression. *Transactions of the ASME* **54**(2), 53–57.

Wagner, H. (1929). Ebene Blechwandträger mit sehr dünnem Stegblech. *Zeitschrift für Flugtechnik und Motorluftschiffahrt* **20** (8, 9, 10, 11 and 12). (Translation appeared as Flat sheet metal girders with very thin metal webs. NACA TM 604, 605 and 606, 1931.)

Walker, A.C. and Davies, P. (1977). The collapse of stiffened cylinders. In *Steel Plated Structures* (eds P.J. Dowling, J.E. Harding and P.A. Frieze). Crosby Lockwood Staples, London, pp. 791–808.

Walker, A.C. and Kemp, K.O. (1976). Buckling of stringer-stiffened welded steel cylinders. *Proceedings of BOSS 1976 (1st International Conference on Behavior of Off-Shore Structures)*, Norwegian Institute of Technology, Trondheim, Norway, pp. 528–534.

Walker, A.C. and Murray, N.W. (1975). A plastic collapse mechanism for compressed plates. *International Association for Bridge and Structural Engineers* **35**, 127–134.

Walker, A.C., Andronicou, A. and Sridharan, S. (1981). Local plastic collapse of ring-stiffened cylinders. *Proceedings of the Institution of Civil Engineers, Part 2* **71**, 341–367.

Walker, A.C., Andronicou, A. and Sridharan, S. (1982a). Experimental investigation of the buckling of stiffened shells using small scale models. In *Buckling of Shells in Off-shore Structures* (eds J.E. Harding, P.J. Dowling and N. Agelidis). Granada, London, pp. 45–71.

Walker, A.C., Segal, Y. and McCall, S. (1982b). The buckling of thin-walled ring-stiffened steel shells. *Buckling of Shells, Proceedings of a State-of-the-Art Colloquium*, Universität Stuttgart, Germany, 6–7 May (ed. E. Ramm). Springer-Verlag, Berlin, Heidelberg, New York, pp. 275–304.

Weller, T. (1971). Further studies on the effect of in-plane boundary conditions on the buckling of stiffened cylindrical shells. TAE Report No. 120, Technion Research and Development Foundation, Haifa, Israel.

Weller, T. (1978). Combined stiffening and in-plane boundary condition effects on the buckling of circular cylindrical stiffened-shells. *Computers & Structures* **9**, 1–16.

Weller, T. and Singer, J. (1971). Experimental studies on buckling of 7075-T6 aluminum alloy integrally stringer-stiffened shells. TAE Report 135, Department of Aeronautical Engineering, Technion, Haifa, Israel.

Weller, T. and Singer, J. (1974). Further experimental studies on buckling of integrally ring-stiffened cylindrical shells under axial compression. *Experimental Mechanics* **14**(7), 268–273.

Weller, T. and Singer, J. (1977). Experimental studies on the buckling under axial compression of integrally stringer-stiffened circular cylindrical shells. *Journal of Applied Mechanics* **44**(4), 721–730.

Weller, T., Singer, J. and Nachmani, S. (1970). Recent experimental studies on the buckling of stringer-stiffened cylindrical shells. TAE Report 100, Department of Aeronautical Engineering, Technion, Haifa, Israel.

Weller, T., Singer, J. and Batterman, S.C. (1974). Influence of eccentricity of loading on buckling of stringer-stiffened shells. In *Thin Shell Structures, Theory, Experiment and Design* (eds Y.C. Fung and E.E. Sechler). Prentice-Hall, Englewood Cliffs, NJ, pp. 305–324.

Weller, T., Baruch, M. and Singer, J. (1979). Influence of in-plane boundary conditions on buckling under axial compression of ring-stiffened cylindrical shells. *Proceedings of the 5th Annual Conference on Mechanical Engineering*, Israel Journal of Technology **9**(4), 397–410.

Wenk, E., Jr., Slankard, R.C. and Nash, W.A. (1954). Experimental analysis of the buckling of cylindrical shells subjected to external hydrostatic pressure. *Proceedings of the Society for Experimental Stress Analysis* **12**(1), 163–180.

Wesenberg, D.L. and Mayers, J. (1969). Failure analysis of initially imperfect, axially compressed, orthotropic, sandwich, and eccentrically stiffened, circular cylindrical shells. Stanford University, USAAVLABS Technical Report 69-86, US Army Aviation Material Laboratories, Fort Eustis, Virginia.

Wunderlich, W., Lu, Z. and Obrecht, H. (1991). Elastic and inelastic buckling of ring-stiffened circular cylindrical shells subjected to external pressure. In *Buckling of Shell Structures, on Land, in the Sea and in the Air* (ed. J.-F. Jullien). Elsevier Applied Science, London, pp. 233–241.

Yamamoto, Y., Homma, Y., Oshima, K., Mishiro, Y., Terada, H., Yoshikawa, T., Morihana, H., Yamauchi, Y. and Takenaka, M. (1989). General instability of ring-stiffened cylindrical shells under external pressure. *Marine Structures* **2**, 133–149.

Chapter 12

Large diameter fabricated steel tubes

A.E. Elwi and G.L. Kulak

Introduction

Large diameter fabricated steel cylinders are often used as conveyor galleries, stacks or masts, or as members in offshore structures. As an illustration of proportions, cylinders used for conveyor galleries typically have diameters of 2.5–4.0 m and radius to thickness ratios, R/t, of 100–400. Spans can reach 60 m. As such, these members act as beams in flexure and they may also be subjected to axial forces. The cylinders are fabricated by cold-rolling steel plates to form short cans. The cans are then joined together by circumferential girth welds to produce long spans. Circumferential (or, ring) stiffeners at intervals of 0.5–2.0 times the diameter are almost always present to help maintain the circular shape of the tube. Longitudinal stiffeners may be used as well. If they are present, they will be welded to the cylinder surface, thereby improving the behavior under axial compression or beam bending. For cases where flexure predominates, it is usual to attach the longitudinal stiffeners over only part of the tube.

Because of the shape and size of the overall cross-section of these tubes, overall buckling modes such as lateral–torsional buckling and more localized modes such as Brazier buckling, are not design issues. Under flexure, the important mechanisms of failure involve local buckling of the shell in one of several possible modes. For example, the flexural capacity of a longitudinally stiffened cylinder can be governed by the instability of a local area that is highly stressed in compression. One possible mode of failure is so-called shell buckling, in which the shell plating buckles between longitudinal stiffeners, while the stiffeners themselves remain essentially straight. A second possible failure mode is that of general buckling, in which case the shell and longitudinal stiffeners buckle simultaneously. This is a possibility when the stiffeners are closely spaced. Lastly, failure may be initiated by stiffener buckling, in which the stiffeners are twisted and distorted before shell buckling occurs. This mode may take place when stiffeners with low torsional stiffness, such as thin-walled open cross-sections, are used. Of course, the three buckling modes may interact with one another.

When the cylinders do not have longitudinal stiffeners, of course the only form of local buckling that can take place is simply the local buckling of the shell in the compression portion of the cross-section.

For either unstiffened or longitudinally stiffened cylinders, shear buckling can take place as a consequence of the application of transverse loads. This happens when inclined buckles form as a result of the shear forces that arise from the transverse loads. These buckles are consistent with the principal compressive stresses in the shell.

A large part of the database available on large diameter tubes has been obtained from small-scale aerospace-quality specimens tested under axial compression. Well-known design guidelines such as API (1987), DnV (1995) and others are all based on such a database. However, in recent years large-scale tests on large diameter fabricated steel tubes subjected to axial compression, flexure and transverse shear have been carried out. These are much more representative of the kinds of tubes used in construction practice. The tests have generally been augmented with sophisticated nonlinear finite element analyses and parametric studies. The purpose of this chapter is to provide a design approach for large diameter fabricated steel tubes that is rationally based on large-scale testing and on the analysis of realistic specimens and experimental programmes. The proposed design approach covers the capacity of unstiffened and longitudinally stiffened large diameter tubes under beam action, that is, their flexural and transverse shear force capability.

Unstiffened tubes under flexure

Early test results on the buckling capacity of unstiffened tubes came from small specimens, and these specimens were generally of aerospace quality. Consequently, they did not reflect the initial imperfections and residual stress patterns of hot-rolled steel tubes fabricated by cold-bending and welding. It is only in about the last two decades that test results from large-scale fabricated steel specimens have become available. It is clear that testing of large-scale specimens is the most direct and reliable way to investigate the buckling behavior of such cylinders. In order that the material characteristics, initial imperfections and residual stresses be representative of the situation in practical structures, the specimens in such tests should be fabricated using industrial procedures.

Unstiffened cylinders with relatively small radius to thickness ratios, R/t, have been tested extensively under axial compression and flexure (Chen and Han 1985; Prion and Birkemoe 1987; SSRC 1998). However, the stiffness of these shells is high because of their large curvature. Consequently, local shell buckling behaviour cannot necessarily be expected to be the same as that for cylinders with large radius to thickness ratios. Stephens et al. (1982) and Bailey and Kulak (1984) tested a series of five unstiffened steel cylinders under axial compression followed by two series of matched large-scale specimens tested separately under axial compression and under bending. Each of these two series consisted of two large diameter tubes fabricated from hot-rolled steel plates (nominal yield and ultimate strengths of 300 and 450 MPa, respectively). The R/t ratios were 149 and 222. The results of the pure bending tests showed that both shells failed by local buckling in the extreme compression zone. Stevens et al. suggested that a critical stress expression developed from the abundant test data on specimens under axial compression was adequate to predict failure of otherwise similar tubes under bending. Bailey and Kulak (1984) drew similar conclusions.

This work on longitudinally unstiffened cylinders was used to develop a three-part equation to predict the critical shell buckling stress, σ_{cr} (Kulak *et al.* 1988). It is a function of the steel plate yield stress, σ_y, and a non-dimensional parameter, γ_s. The non-dimensional parameter reflects the material yield stress, the modulus of elasticity, E, and the thickness to radius ratio, t/R. The equations are:

$$\sigma_{cr} = 119.3\gamma_s\sigma_y \qquad\qquad \gamma_s \leq 0.0036 \qquad\qquad (1a)$$

$$\sigma_{cr} = \sigma_y(1.625 + 0.489\log\gamma_s) \qquad 0.0036 < \gamma_s \leq 0.0527 \qquad (1b)$$

$$\sigma_{cr} = \sigma_y \qquad\qquad\qquad \gamma_s > 0.0527 \qquad\qquad (1c)$$

in which γ_s is obtained as:

$$\gamma_s = (E/\sigma_y)^{1/2}(t/R)^{3/2} \qquad\qquad\qquad (2)$$

The non-dimensional buckling parameter, γ_s, is based on the panel strip theory proposed by Edlund (1977). It reflects the dependence of the elastic buckling strength on the modulus of elasticity, E, the thickness of the plate, t, the radius of curvature of the panel, R, and the static yield strength of the material, σ_y. Figure 12.1 shows a plot of Eq. (1) together with data that cover a large range of

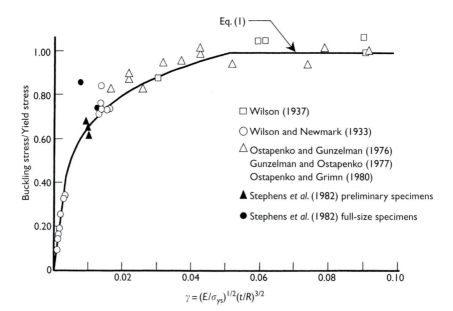

Figure 12.1 Local buckling capacity under axial compressive stresses.

the parameter γ_s. These data are taken from the Stevens *et al.* and Bailey and Kulak tests and from work by Wilson and Newmark (1933), Wilson (1937), Ostapenko and Gunzelman (1976), Gunzelman and Ostapenko (1977), and Ostapenko and Grimm (1980). Equation (1) clearly constitutes a lower bound on the strength of longitudinally unstiffened tubes. It is therefore recommended as a design base.

It should be noted that Eq. (1c), which applies to stocky tubes ($\gamma_s > 0.5$), is quite conservative. Kulak *et al.* (1988) have presented an alternative that overcomes this difficulty.

Longitudinally stiffened tubes

The American Petroleum Institute (API 1987), Det norske Veritas (DnV 1995), and others have all developed design guidelines for longitudinally stiffened tubes. The intended application for these rules is primarily for components of offshore tension leg platforms. For example, API provides a stability design bulletin (Bulletin 2U) for cylindrical shells. Design equations are based on the allowable stress method and have a factor of safety that is expected to range from 1.25 to 2. The general expression for the buckling stress, S, is given as:

$$S = \alpha \eta \sigma_{cr} \tag{3}$$

in which σ_{cr} is the critical elastic buckling stress for a perfect shell. The critical stress σ_{cr} is obtained from a buckling analysis of stiffened or unstiffened shells under various loading conditions. This is calibrated to elastic buckling test results using the knockdown factors α and η, which are intended to account for imperfections and inelasticity, respectively.

The value of the critical stress, σ_{cr}, depends on the type of buckling assumed. The API rules recognize two modes – a general buckling mode and a shell buckling mode. In the former, the shell and the stiffeners buckle together. In the latter, only the shell buckles, and this may be either between the stiffeners or outside the stiffened region when longitudinal stiffeners are present over only a portion of the cross-section. The philosophy of the API guidelines is that interaction of the different buckling modes must be avoided. This is assured by requiring that the load-carrying capacity be governed by local shell buckling between stiffeners. Specifically, the calculated buckling stress for general buckling must be at least 20% greater than that for shell buckling.

The general buckling mode in longitudinally stiffened tubes, called 'bay instability' by the API, is obtained from one of two approaches – an orthotropic shell treatment or an equivalent column equation. The orthotropic shell equation is based on the work of Block *et al.* (1965). It assumes that the longitudinal stiffeners are 'smeared' over the shell surface, and thus the shell has orthotropic stiffness properties in the longitudinal and circumferential directions. In this procedure, the key factors to be determined are the effective shell width between stiffeners and the number of buckling waves in each of the circumferential and longitudinal directions. These three factors are calculated iteratively so as to obtain the lowest buckling stress. According to Walker and Sridharan (1980) there should be at least two longitudinal stiffeners in a circumferential buckling halfwave length.

The other approach permitted by API uses the equivalent column buckling equation of Faulkner *et al.* (1983). The equivalent column consists of a longitudinal stiffener and a portion of the adjacent shell plating. The elastic general buckling stress is approximately the sum of the Euler buckling stress of the pseudo-column and the classical shell buckling stress of a shell panel. Compared with the orthotropic shell equation, the procedure is lengthy, even though no iterations are involved.

The elastic local buckling stress of the shell between stiffeners is given by the classical formula for unstiffened cylindrical shells (SSRC 1998):

$$\sigma_{cr} = CE(t/R) \tag{4}$$

in which E is Young's modulus and the parameter C varies according to the geometry of the cylinder. It is noted that the advantage of stiffening is not taken into account. However, the knockdown factor for imperfections, α (Eq. 3), is higher for densely stiffened shells than for unstiffened shells.

Experimental evidence

The assumption that buckling stress in flexure can be taken as equivalent to the buckling stress in uniform compression is widely employed in design guidelines. Although there have only been a few large-scale flexural tests of large diameter longitudinally stiffened steel cylinders subjected to pure bending, the experimental evidence shows that the flexural behavior of a longitudinally stiffened cylinder can differ significantly from axial compression behaviour. This is because the bending stress is not uniform through the cross-section and, unlike the uniform axial compression case, stresses can be redistributed after local failure as the neutral axis shifts away from the failed regions.

The experimental evidence from large-scale tests includes one specimen tested by Bailey and Kulak (1984), and four specimens tested by Chen *et al.* (1995a). The cylinders had a nominal diameter of 1.26 m, they were fabricated from hot-rolled plate, and they were longitudinally stiffened by hollow structural sections welded to the shell in the compression zone. The cold-forming and welding procedures used were typical of industrial practice. Although the number of specimens tested was limited, the experiment design considered the effects of the R/t ratio and the size and arrangement of the stiffeners. Initial imperfections and residual stresses were measured in some of these tests.

Evaluation of existing design provisions

In contrast to the case of large diameter unstiffened cylinders, which tend to fail in flexure at stress levels that are comparable to those achieved under uniform compression, Bailey and Kulak (1984) and Chen *et al.* (1995a) found that the ultimate moment of stiffened cylinders could be as high as 90% of their fully plastic moment. The API and DnV rules and other design criteria based on uniform compression do not predict such high capacities. The rules needed to be re-examined so that the effectiveness of longitudinal stiffeners can be realistically included.

In order to explore the role of longitudinal stiffeners, the experimental database was expanded significantly by means of a sophisticated finite element analysis procedure that included large deformation formulation, elasto-plastic material behaviour, and the effect of both initial imperfections and residual stresses (Chen *et al.* 1996). The parameters explored were the radius to thickness ratio, R/t, the ratio of stiffener spacing to shell thickness, s/t, and the angle covered by stiffeners, ϕ. Both the level of the residual stresses and the distribution of initial imperfections were included in the analysis. Figure 12.2 shows a typical cross-section of a longitudinally stiffened tube and Fig. 12.3 shows a typical distribution of residual stresses. Table 12.1 shows the parameters of the available database.

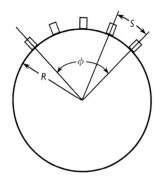

Figure 12.2 Typical cross-section of a longitudinally stiffened tube.

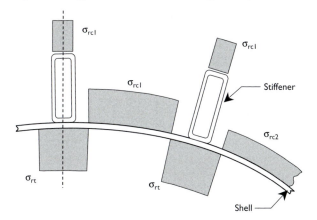

Figure 12.3 Typical distribution of residual stresses due to installation of stiffeners.

Table 12.1 Parametric study of longitudinally stiffened large diameter fabricated steel tubes

Specimen	Type	R/t	s/t	ϕ (degrees)	No. of stiffeners	M_u (kN m)[a]	Failure mode	M_{U} API (kN m)[b]	Error (%)[c]
SB1	Test	128	43	76.1	5	3068	General buckling	2248	−26.7
SB3b	Test	206	57	95.3	7	1333	General buckling	1026	−23.0
SB4	Test	189	42	89.0	8	1857	General buckling	1461	−21.3
S11	Model	120	54	77.3	4	2321	General buckling	1404	−39.5
S12	Model	120	54	51.6	3	2262	General buckling	1404	−37.9
S13	Model	120	81	77.3	3	2078	General buckling	928	−55.3
S21	Model	240	54	77.3	7	7960	General buckling	5807	−27.0
S22	Model	240	54	51.6	5	7680	General buckling	5807	−24.4
S31	Model	360	54	77.3	10	16,526	General buckling	12,955	−21.6
S32	Model	360	54	51.6	7	15,727	General buckling	12,955	−17.6
S14	Model	120	108	51.6	2	1852	Shell buckling	1573	−15.1
S23	Model	240	81	77.3	5	6499	Shell buckling	5181	−20.3
S24	Model	240	108	51.6	3	5846	Shell buckling	4868	−16.7
S33	Model	360	81	77.3	7	12,776	Shell buckling	10,447	−18.2
S34	Model	360	108	51.6	4	11,027	Shell buckling	9869	−10.5

Notes
a Ultimate moment obtained from experiments or finite element analysis.
b Ultimate moment predicted by API (1987).
c Error $= [M_{u\ API}/M_u) - 1]\ 100\%$.

In order to illustrate the inadequacy of current design rules for longitudinally stiffened cylinders, the API guideline is chosen for evaluation. The rules given by API are only applicable to cylinders with stiffeners covering the whole circumference, that is, shell buckling outside the stiffened region is not considered. Therefore, specimens that buckled outside the stiffened region are not included in the comparison. Table 12.1 shows the experimental and numerical results compared to predictions based on the API guideline. For the 10 cylinders that failed in general buckling, the API guideline predicts a capacity that is between 18 and 55% less than the experimental and numerical results. For the case of shell buckling between the stiffeners, the estimated capacity is 10–20% less than that predicted by the finite element study. On average, the API rules underestimate the case of general buckling by 29% and underestimate shell buckling by 16%. (The safety factor specified in the API guideline has not been applied in the calculation of $M_{u\ API}$ in Table 12.1.)

All of the current design procedures first provide an estimate of the shell buckling stress and then use this stress to predict the failure moments. A portion of the discrepancy between the predicted moments and the results generated herein arises from the process of converting the shell buckling stress into a bending moment. First, the ultimate moment is calculated according to linear beam theory:

$$M_u = \sigma_{cr} \pi R^2 t_e \tag{5}$$

in which the combined term $\pi R^2 t_e$ is the elastic section modulus and t_e is an equivalent thickness of a stiffened shell, defined by the codes as:

$$t_e = \frac{A_s + s_e t}{s} \qquad s_e \le s \tag{6}$$

In Eq. (6), A_s is the cross-sectional area of a stiffener, t the shell thickness, s the stiffener spacing, and s_e is an effective stiffener spacing. The latter takes the reduced stiffness of the buckled shell into account. Equations (5) and (6) imply that the stiffener spacing, s, should be reduced to s_e throughout the circumference. Such an approach is appropriate for the case of uniform axial compression, where the shell can be expected to buckle over the entire circumference. However, when subjected to flexure, the shell buckling in the cylinder takes place only at the location of maximum compressive stress: most of the cross-section remains fully effective.

Since the design specifications require that σ_{cr} not be greater than the yield stress, σ_y, the ultimate moment, M_u, can never exceed the yield moment. This is usually less than 80% of the plastic moment for a stiffened cylinder of large diameter. Equation (5) assumes an elastic failure and defines failure as the moment at initiation of buckling. However, a stiffened cylinder in bending behaves in a different way when it fails by general buckling. As has been observed in both the experimental and numerical studies, ultimate failure in flexure usually takes place well past the initiation of general buckling. This is in contrast to the case of

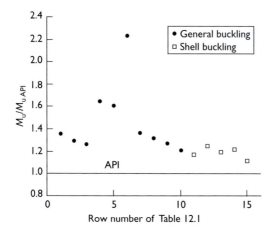

Figure 12.4 Comparison of API guidelines with tests and numerical analysis results of Chen *et al.*

axial compression. In the case of bending, stress redistribution allows the moment to continue to increase after yield and initial buckling. Inelastic behaviour is adequately developed over the cross-section for a general buckling case, and in typical cases the ultimate moment can be 90% or more of the fully plastic moment.

The above comparison is also shown graphically in Fig. 12.4, where the horizontal axis simply refers to the data in the corresponding row number of Table 12.1. It is apparent that the flexural capacity is significantly underestimated by the current design methods, especially for cylinders that fail by general buckling.

Rational design

It is obvious that a rational and successful design procedure for cylinders must take a number of factors into account. Clearly, the design will be influenced by the R/t and s/t ratios and by the extent of the stiffened circumference, which is represented by the angle ϕ. The method must identify the failure mode and account for variation of the circumferential residual stresses and the initial imperfections. An examination of the database put together by Chen *et al.* (1996) allows the construction of a design approach.

In the proposed approach, the designer will have to check a number of issues. It can be assumed that the radius is known: it is a function of the industrial parameters. Logically, the designer next will assume a plate thickness and stiffener distribution and then check the capacity. Iteration of the design process will be required in order to produce a cost-effective and safe design. A successful design should ensure that failure will take place inside, rather than outside, the stiffened region. (The desirability of failure within the stiffened region is discussed below.) This requires that a decision be made as to the extent of the stiffened region, ϕ. Having restricted the location of failure, the designer must next decide whether failure will be by shell buckling or by general buckling. This means that the size and distribution of the stiffeners must be selected. (It is assumed that buckling of the stiffeners will not be permitted, and they must be chosen with this in mind.) Other considerations such as the magnitude and distribution of initial imperfections and residual stresses are obviated since the database under question has already included these effects.

Failure outside versus failure inside the stiffened region

Since only part of the circumference in the compression zone is longitudinally stiffened, the shell outside the stiffened region may buckle under compression. Physical testing indicates that such buckling typically is an abrupt failure. Once the ultimate moment is reached, it is followed immediately by shell buckling – there is no subsequent increase of load and little subsequent deformation. Since this type of behaviour is generally undesirable, the goal should be that the failure mode is one of buckling within the stiffened region.

It is apparent that the type of failure, inside of or outside of the stiffened region, is primarily dependent on the extent of the stiffened region, ϕ, and on R/t. There are other parameters that affect the failure mode, however. The most important of these is the spacing of the stiffeners, s, because this controls the level of residual stresses in the stiffened region. This, in turn, affects the level of residual stresses outside the stiffened region. A high level of residual stresses outside the stiffened region will promote shell buckling in that zone.

The residual compressive stress between the stiffeners, $\sigma_{r\,c1}$, and the residual compressive stress outside the stiffened region, $\sigma_{r\,c2}$, are indirectly related because both are dependent on the shell thickness in the same manner and they act together to balance the tensile residual stress. The ratio $\sigma_{r\,c1}/\sigma_y$ measured in a 5 mm thick test cylinder (Chen *et al.* 1995a) that had a stiffener spacing of 180 mm, was 0.35. Masubuchi (1980) indicates that at large spacing this ratio would reduce to about 0.15. Thus, at small spacing the residual stress is high and the increased likelihood of failure outside the stiffened region can be reflected indirectly by notionally reducing the angle ϕ. At large spacing, the opposite is true. Chen *et al.* (1995b) showed that this can be accommodated by using an effective angle, ϕ_e:

$$\phi_e = \frac{0.35}{(0.15 + 36/s)}\phi \qquad (s \text{ in mm}) \tag{7}$$

The mode of failure for each of the cylinders investigated in this study is plotted in Fig. 12.5 in terms of the radius to thickness ratio, R/t, and the effective angle ϕ_e. A boundary is drawn (by judgement) between the two failure patterns, using a solid line, and the upper and lower bounds of the failure region limits are represented by dashed lines. For a cylinder of given R/t, the value of ϕ_e determines where buckling will occur. The cylinders to the left side of the line failed by shell buckling outside the stiffened region, while those on the right side had a failure in the stiffened region. The limit lines were placed close to SB3a and S32 because the deformed shapes showed that the two failure modes were nearly concurrent in these two cylinders. Buckles appeared both inside and outside the stiffened region in these cases.

The failure mode is also affected by the R/t ratio; the required ϕ_e increases for large R/t values if a failure in the stiffened region is desired. This reflects the fact that a larger R/t ratio results in a more uniform distribution of compressive stress in the top part of the circumference, which, in turn, promotes local shell buckling outside the stiffened region.

For failure to take place within the stiffened region, the effective angle ϕ_e has to lie on the right-hand side of the straight line indicated in Fig. 12.5 such that

$$\phi_e > 46 + 0.046(R/t) \qquad \text{(degrees)} \qquad (8)$$

When Eq. (7) is combined with Eq. (8), a lower bound for the angle ϕ is obtained:

$$\phi > (131 + 0.131(R/t))(0.15 + 36/s) \qquad (\phi \text{ in degrees and } s \text{ in mm}) \quad (9)$$

On the other hand, the experimental and analytical results shown in Fig. 12.5 indicate that shell buckling will take place outside the stiffened region if a small value of ϕ_e is used:

$$\phi_e < 26 + 0.074(R/t) \qquad (10)$$

Again, for an upper bound, Eqs (10) and (7) are combined to obtain

$$\phi < (74 + 0.21(R/t))(0.15 + 36/s) \qquad (\phi \text{ in degrees and } s \text{ in mm}) \quad (11)$$

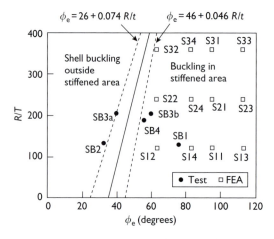

Figure 12.5 Influence of R/t on the failure mode of longitudinally stiffened tubes.

Both failure modes must be checked when the angle ϕ is between the values given by Eqs (9) and (11). The lowest value of M_u is then taken as the flexural capacity of the longitudinally stiffened cylinder.

General buckling and shell buckling inside the stiffened region

Failure within the stiffened region will be either by general buckling or shell buckling. Using the Chen *et al.* (1996) database, the cylinders that fall in this category are plotted in Fig. 12.6, where the parameters used are the stiffener spacing to shell thickness ratio, s/t, and the radius to thickness ratio, R/t. Figure 12.6 can be divided into two zones, corresponding to the general buckling and shell buckling modes. Some of the data points for cylinders in the general buckling zone are overlapped because they have identical values of s/t and R/t. For example, S21, S22 and S25 are shown by the same data point because they all have $s/t = 54$ and $R/t = 240$. They are different in other ways, however. For instance, both the angle ϕ and the size of stiffeners vary.

The straight line in Fig. 12.6 separates the two buckling modes in an approximate way. The key parameter that distinguishes the buckling modes is the spacing ratio, s/t. Whether general buckling or shell buckling will occur in this region depends largely on the value of this ratio. However, the R/t ratio indirectly influences such dependence. As R/t becomes large, the dividing line in Fig. 12.6 moves towards small s/t values. It indicates that close stiffener spacing is required for a cylinder with a large R/t ratio if general buckling is to be guaranteed.

The s/t ratio also affects the flexural capacity. This capacity can conveniently be expressed by the moment at failure normalised by the plastic moment, that is, M_u/M_P. Figure 12.7 displays the reduction of this normalised moment as the s/t

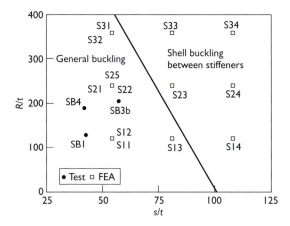

Figure 12.6 Combined influence of R/t and s/t on the failure mode of longitudinally stiffened tubes.

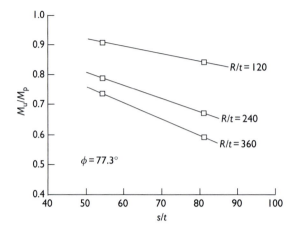

Figure 12.7 Influence of the s/t ratio on flexural capacity.

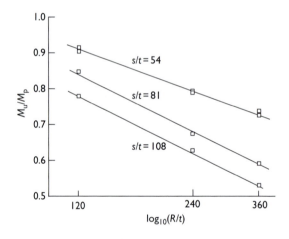

Figure 12.8 Influence of the R/t ratio on flexural capacity.

ratio increases for a certain value $\phi = 77.3°$. A similar trend was observed for other values of ϕ. A large s/t ratio may not only reduce the general buckling capacity, but it can also change the failure mode into shell buckling, which usually has a lower capacity than that of the general buckling.

Figure 12.8 demonstrates the effect of R/t on the capacity. The logarithm of the R/t ratio was found to be the best descriptor for this effect. As shown in the figure, the decrease in the moment ratio is linearly proportional to the increase in $\log_{10}(R/t)$. The slope of these straight lines is a function of the s/t ratio. Despite the fact that different buckling modes, general buckling or shell buckling, are contained within these data, the linear curves appear to be generally applicable.

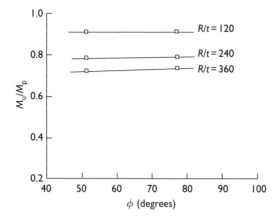

Figure 12.9 Influence of the stiffened region extent on the flexural capacity.

In order to examine the influence of the angle ϕ, cylinders identical in other aspects but with different values of ϕ are compared in Fig. 12.9 in three R/t groups. It is evident that the ultimate moment ratio M_u/M_P is practically independent of ϕ as long as the failure is clearly in the stiffened region. The decline of M_u for a small ϕ is offset by a reduction of the plastic section modulus used to calculate M_P.

Another parameter that affects the flexural capacity is the size of the stiffeners themselves. This can be observed by comparing cylinders S22 and S25. The only difference between the two is the size of the hollow structural shape (HSS) used as stiffeners (HSS 50.8 × 25.4 × 3.18 for S22 and HSS 25.4 × 25.4 × 2.54 for S25). The result was that $M_u/M_P = 0.79$ for S22 whereas $M_u/M_P = 0.70$ for S25. Since the shell buckling waves cause the stiffeners to twist, the development of shell instability relies on the torsional resistance of the stiffeners. It is apparent that the St. Venant torsional constant, J_s, is an influential factor.

Additional cases were examined to determine the effects of the stiffener size and the level of residual stresses. For the case of general buckling, the results showed that the sensitivity to changes in the level of residual stresses is negligible. For example, an increase in σ_{rc1}, the compressive residual stress in the shell between stiffeners, from $0.15\sigma_y$ to $0.35\sigma_y$ gave a maximum reduction of only 1.4% in the flexural capacity of cylinders S12 and S22. (The values $0.15\sigma_y$ and $0.35\sigma_y$ are the bounds of the measured compressive residual stresses in the shell between stiffeners in SB2.)

The effect of changes in the residual stress levels upon shell buckling outside the stiffened region, although more pronounced, is still within the tolerances usually associated with this kind of prediction. For example, Specimen SB2 showed a maximum change in ultimate load of about 6% when the tensile residual stress was increased by $0.10\sigma_y$ and the compressive residual stress reduced by $0.05\sigma_y$.

The eight cylinders in the groups of $R/t = 120$ and 240 were also re-analysed without any initial imperfections present. The increase in flexural capacity varied from 1 to 7% as compared to the models with initial imperfections. The sensitivity is much lower than that of elastic unstiffened shells. It was concluded that the effect of initial imperfections is not substantial here or, at the very least, it is already incorporated in the residual stress effect.

As a result of this examination, it is suggested that the designer attempt to establish the angle ϕ such that failure takes place inside the stiffened region. Failure outside the stiffened region can then be checked.

Capacity for the case of failure in the stiffened region

The flexural capacity that is reached when failure takes place within the stiffened region is a reflection of several key parameters. Chen *et al.* (1995b) showed that these parameters include the R/t and s/t ratios and the stiffness of the stiffeners. The 12 cylinders shown in Fig. 12.8 all have the same stiffeners and shell thickness. Differences in their capacity ratio, M_u/M_p, can be attributed to the variation in the R/t and s/t ratios. The figure shows that M_u/M_p can be expressed in terms of $\log_{10}(R/t)$ by using a family of straight lines. Such a family of curves was found to be

$$\frac{M^*_{u\ prd}}{M_P} = 0.527\left[2.30 - \left(\log\left(\frac{s}{t}\right) - 0.936\right)\left(\log\left(\frac{R}{t}\right) - 1.32\right)\right] \quad (12)$$

in which $M^*_{u\ prd}$ denotes the predicted ultimate moment for 5 mm thick cylinders with HSS stiffeners. (In this case, the stiffeners were HSS $50.8 \times 25.4 \times 3.18$.) Figure 12.10 shows that Eq. (12) provides a reasonable representation of the results obtained in the finite element analyses. However, so far the examination includes

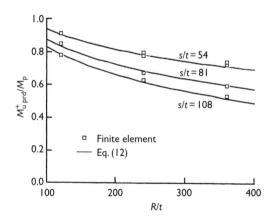

Figure 12.10 Predictions of Eq. (12) compared to the numerical analysis results.

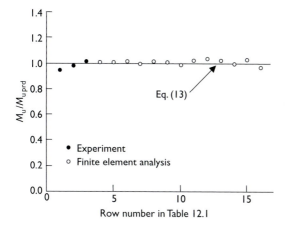

Figure 12.11 Predictions of Eq. (13) compared to the full Chen *et al.* database.

only the effects of R/t and s/t, and does not account for the influence of different stiffeners.

The pertinent geometric parameters of the HSS stiffeners that could affect the flexural capacity are the cross-sectional area and St. Venant torsional stiffness. It was found that the St. Venant torsional constant, J_s, provides the best description of the influence of the stiffeners. Here, the torsional constant is assumed to lie in the range of $200t^4 < J_s < 400t^4$. A complete design equation can then be written as:

$$\frac{M_{u\,prd}}{M_P} = 0.371 \left(1.25 + \frac{J_s/t^4}{1000} \right)$$
$$\times \left[2.30 - \left(\log\left(\frac{s}{t}\right) - 0.963 \right) \left(\log\left(\frac{R}{t}\right) - 1.32 \right) \right] \qquad (13)$$

Equation (13) defines the ultimate moment in a simple manner. Only three parameters, the ratios R/t, s/t and J_s/t^4, are involved. This, then, is a unified equation for the case of failure inside the stiffened region, whether it is by general buckling or by shell buckling. Predictions obtained using Eq. (13) are compared with the experimental and numerical results of Chen *et al.* (1996) in Fig. 12.11.

Flexural capacity for the case of shell buckling outside the stiffened region

The parameters that affect shell buckling outside the stiffened region are the R/t ratio and the angle ϕ that defines the stiffened portion of the shell. The capacity decreases for large R/t ratios and with smaller values of ϕ. In addition, the presence of compressive residual stresses must be considered.

A practical way to establish the flexural capacity when shell buckling occurs outside the stiffened region is to modify the formulas for the buckling of unstiffened cylindrical shells to account for the gradient of bending stress and the compressive residual stress. The stress gradient in the shell outside the stiffened region is more significant than it is near the apex of an unstiffened cylinder. In addition, the level of residual stresses in longitudinally stiffened cylinders is much higher than it is in unstiffened tubes.

Normally, the critical stress is provided in terms of the yield stress. However, the presence of compressive residual stresses will reduce the effective yield stress. The use of $\sigma_{rc2} = 0.15\sigma_y$ is consistent with the effective angle ϕ_e introduced in Eq. (7). Thus, an effective yield stress of $0.85\sigma_y$ is recommended here.

There are a number of formulas available for the calculation of the critical stress of unstiffened cylinders. The equations suggested by Stephens et al. (1982) earlier in this chapter, which are based on about 35 axial compression tests and include five large R/t specimens of their own experiments, are quite good. For the material and the R/t range of the cylinders in this study, the critical stress is obtained accordingly as:

$$\sigma_{cr} = \left(\frac{1+\chi\phi}{\cos(\phi/2)}\right) 0.85\sigma_y(1.625 + 0.489 \log \gamma_s) \tag{14}$$

in which ϕ is in degrees and γ_s is the non-dimensional constant defined as:

$$\gamma_s = (E/0.85\sigma_y)^{1/2}(t/R)^{3/2} \tag{15}$$

The factor χ is a linear expression of R/t, designed to account for the steep gradient of strains outside the stiffened region as compared to that at the apex where the Stephens et al. results were obtained. The term χ is obtained as follows (Chen et al. 1995b):

$$\chi = 0.10 + 0.0024\left(\frac{R}{t}\right) \tag{16}$$

The critical stress defined by Eq. (14) is assumed to occur at angle $\phi/2$, measured from the apex, and takes place under largely elastic conditions. This is conservative, but is justified. The relation between the stress and moment is assumed to follow elastic beam theory such that

$$M_u = \frac{\sigma_{cr}I_y}{R\cos(\phi/2) - z_0} \tag{17}$$

in which I_y is the moment of inertia of the cross-section, including the stiffeners, and z_0 is the shift of the neutral axis relative to the centroidal axis of the cylinder.

The suggested design procedure is as follows:

1   Choose a trial steel plate thickness and trial stiffener size and spacing.
2   Using Eqs (9) and (11), determine the extent of the stiffened area, ϕ, that will force shell failure in the stiffened region.
3   Determine the plastic moment capacity, M_P, assuming the entire cross-section is yielded.
4   Using Eq. (13), check the capacity for buckling within the stiffened region. Change one parameter at a time, as required, to achieve a design that is both safe and economical.
5   Calculate the critical stress for buckling outside the stiffened region and the corresponding moment using Eqs (14)–(17) to ensure that this moment is larger than the demand moment.

Transverse shear

Designers recognised that large diameter tubes subjected to transverse shear forces over a short shear span could fail following the initiation of diagonal buckles. Such buckles are reflective of compressive stresses in the diagonal direction as a result of transverse shear forces acting on the member. Early on, tests of this phenomenon were carried out on small specimens subjected to torsion. Consequently, most of the existing design approaches were established using a database of these torsional failures.

A review of the literature on tests of large diameter tubes fabricated from hot-rolled steel plate and subjected to transverse shear forces reveals that there are very few test results. Tests on small specimens fabricated from thin cold-rolled plate are more abundant, however. Galletly and Blachut (1985) tested a group of 14 small specimens ($R = 150$ mm) fabricated from cold-rolled steel sheet. The specimens had R/t ratios that ranged from 125 to 188 and R/L ratios ranging from 0.83 to 1.37, and were all tested as cantilevers. Baker and Bennett (1984) investigated the buckling interaction curves for both unstiffened and ring-stiffened cylinders. They used an apparatus that could apply different ratios of axial compression and shear loads on test specimens. The cylinders, which were made of 0.38 mm thick Lexan sheet and were used repeatedly, had proportions that are characteristic of nuclear steel containment vessels. They had a value of R/t equal to 460 and a value of R/L equal to 0.57. Both of these research groups reported that the unstiffened cylinders demonstrated a smooth transition from the classical diamond-shaped buckling mode when specimens were loaded in axial compression into the diagonal wave buckling mode when loaded in shear.

From their tests, Galletly and Blachut found that the inelastic critical shear stress can be accurately predicted by a quadratic interaction equation of the form:

$$\tau_p = \frac{\tau_y}{\sqrt{1 + (\tau_y/\tau_e)^2}} \tag{18}$$

in which τ_y is the yield shear stress and τ_e is the elastic critical shear stress of a perfect cylinder in torsion. The latter was developed by Batdorf *et al.* (1947) as a function of the t/R and R/L ratios:

$$\tau_e = k_s E \left(\frac{t}{R}\right)^{1.25} \left(\frac{R}{L}\right)^{0.5} \tag{19}$$

Here $k_s = 0.74$ and 0.825 for simply supported and fixed end conditions, respectively.

A European specification (ECCS 1988) uses an interaction equation similar to Eq. (18) to calculate the inelastic shear buckling stress. They suggest that the elastic shear buckling stress be reduced to 65% of the calculated value to allow for the effects of initial imperfections.

Bailey and Kulak (1984) reported the results of two specimens loaded under transverse shear as beams with fixed end supports, one fabricated from hot-rolled steel plate and the other fabricated from cold-rolled steel sheet. These specimens had R/t ratios of 75 and 250, respectively, and a radius to shear spans ratio (R/L) of 0.5. In each case, the cylinder was fixed at both ends and loaded in the middle by a concentrated transverse load. The more slender specimen, B1, was fabricated from 0.76 mm thick steel sheet. Considerable deformation was induced during the welding process in one of the two shear spans, but no significant differences in the behaviour between the two spans was observed during the test. Cylinder B1 failed as a consequence of a series of successive inclined buckles followed by compression buckling at the extreme fibre.

Obaia *et al.* (1992a) reported two large-scale tests on cylinders fabricated from hot-rolled steel plate. Those tubes had an R/t ratio of 185. One specimen, S1, was tested as a beam with fixed end supports and had an R/L ratio of 0.53, while the second specimen, S2, was tested as a cantilever and had an R/L ratio equal to 0.76. The nominal diameter of each of these specimens was 1270 mm. These tests and the Bailey and Kulak (1984) test piece B1 are considered to be the only available test results that reflect the material, proportions, and fabrication practice that are representative of full-size fabricated steel tubes. (The Bailey and Kulak specimen, which had $R/t = 75$, is somewhat stockier than usually encountered in practice, however.)

Roman and Elwi (1989) conducted a numerical analysis of the Bailey and Kulak tests. The analysis confirmed the suggestion by Bailey and Kulak that a tension field similar to that developed in plate girder can provide a significant load-carrying mechanism. However, they also noted that the tension field exists only in the postbuckling region and observed that it could be modelled as a truss. In that model, the diagonal element is aligned with the diagonal buckles and is composed primarily of the tension field. Figure 12.12 depicts the directions and magnitude of a typical principal membrane force on the sides of one of the Obaia *et al.* specimens that exhibited diagonal buckles.

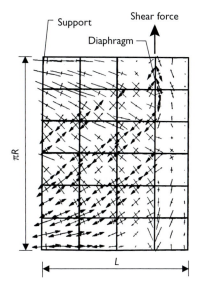

Figure 12.12 Postbuckling principal membrane forces.

In addition to carrying out the large-scale tests, Obaia *et al.* (1992b) also conducted a successful numerical simulation of those tests and used the same large-deformation elasto-plastic finite element analysis model to extend the database by another nine cylinders. This then covered the dimensions considered typical in practice. The model included initial imperfections and a simulation of residual stresses. Based on a nonlinear regression analysis of the Galletly and Blachut (1985) database, the large-scale tests of Bailey and Kulak, and the Obaia *et al.* results, a nonlinear regression design formula was developed. The formula takes into account the radius to thickness ratio, the radius to shear span ratio, the yield strength of the plate, and the elastic properties of the material:

$$V_{max} = 0.05V_y(e^{-0.0033(R/t)})(R/L)^{0.387}(E/\sigma_y)^{0.52} \tag{20}$$

This equation is bounded by the fully plastic yield shear strength V_y and by the elastic buckling capacity, which is

$$V_e = \tau_e \pi R t \tag{21}$$

Equation (20) is considered to be valid only within the ranges

$$100 < R/t < 300 \qquad 0.5 < R/L < 1.4 \qquad 450 < E/\sigma_y < 850$$

As an alternative, a modified K-truss model has been suggested that offers a rational mechanical model for the postbuckling shear capacity of these tubes

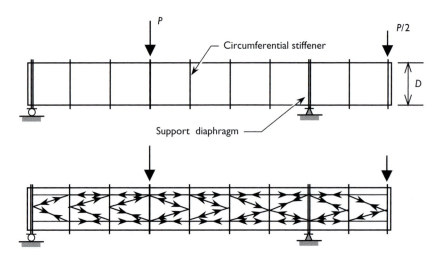

Figure 12.13 A possible K-truss depicting the postbuckling shear carrying mechanism.

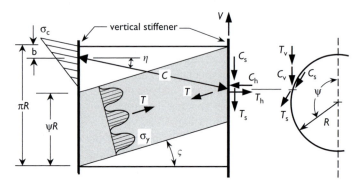

Figure 12.14 Details of the Obaia *et al.* tension field mechanism.

(Obaia *et al.* 1993). Figure 12.13 shows a possible K-truss projected on the side of a tube. Here the vertical elements are the circumferential stiffeners, the tension field in the diagonal buckles forms the tension diagonal elements, and the compression struts are the contribution of the compression field that crosses the buckles. Figure 12.14 shows a detail of the panel forces suggested above. Clearly, the extent of the tension field (angle ψ), is linked to the inclination of the tension field (angle ς). Both are thus a function of the R/L ratio so that:

$$\psi = \pi - (L/R)\tan\varsigma \qquad (22)$$

An analysis of the force diagram shown in Fig. 12.14 gives the postbuckling shear capacity as:

$$V = Rt\sigma_m\{0.5 \sin 2\varsigma(1 - \cos\psi) + (\psi \cos^2\varsigma)(\tan\eta)(\sin(\psi/2))\} \quad (23a)$$

Here σ_m is the maximum stress achieved by the tension field. Obaia *et al.* (1993) showed that $\sigma_m = 0.82\sigma_y$. Hence, the shear capacity can be written finally as:

$$V = Rt\sigma_y\{0.41 \sin 2\varsigma(1 - \cos\psi) + (\psi \cos^2\varsigma)(\tan\eta)(\sin(\psi/2))\} \quad (23b)$$

The compressive strut capacity is limited by the critical buckling stress of a shell. That capacity can be determined using the Stephens *et al.* formula (Eqs (1) and (2)). Determination of the angles ψ, ς and η is iterative.

This information was used to develop a series of nomographs. The one that is shown in Fig. 12.15 is for $E/\sigma_y = 600$. It will cover many practical applications. Two other nomographs, for E/σ_y values of 400 and 800, are available in Obaia *et al.* (1991).

The tension field will have to be anchored at the circumferential stiffener. The force required here is:

$$T_h = 2Rt\psi(1 + \cos 2\varsigma)(\sigma_m/2) \quad (24a)$$

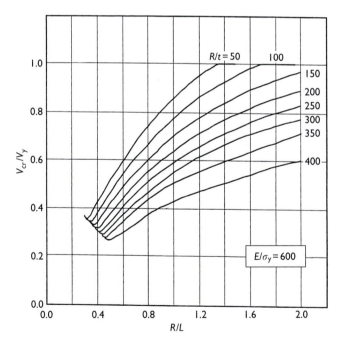

Figure 12.15 Design nomograph for the postbuckling shear capacity of thin-walled cylinders.

Using the substitution $\sigma_m = 0.82\sigma_y$, described earlier, this can finally be written as:

$$T_h = 0.82Rt\psi(1 + \cos 2\varsigma) \tag{24b}$$

This force is the last item required to complete the design of a large diameter fabricated tube. As has been already noted, these tubes are almost always stiffened circumferentially. The ostensible reason is to maintain the round shape of the section. When the tube is subjected to large shear forces coupled with flexure, such as would apply in the case of continuous tubes over several supports, transverse shear forces may be critical. It is imperative in these cases to provide circumferential stiffeners. The rational design for these stiffeners should be based on the force described by Eq. (24). It is also noted that the shear capacity rules described were developed on the basis of unstiffened tubes. There are no tests available that explore the shear capacity of longitudinally stiffened tubes. The solutions given here should be conservative for these cases.

Summary

Large diameter fabricated steel cylinders have many civil engineering applications. Such members almost always have circumferential stiffeners and may be longitudinally stiffened as well. Designers must proportion these members with consideration of their flexural and shear strength.

In recent years, large-scale tests on fabricated steel cylinders subjected to axial compression, flexure, and transverse shear have been conducted. The data obtained from such tests have been augmented by finite element analyses and parametric studies. Based on all of this work, recommendations for the calculation of the strength of longitudinally stiffened or unstiffened tubes have been presented. These recommendations can be employed with confidence by designers and can be used by those responsible for the development of design specifications and codes. Users are reminded that the strength descriptions presented herein are for ultimate capacity. Selection of a suitable resistance factor for each case is required.

References

API (1987). Stability design of cylindrical shells. *API Bulletin 2U*, 1st edn, American Petroleum Institute, Washington, DC (ANSI/API Bull 2U-1992).

Bailey, R.W. and Kulak, G.L. (1984). Flexural and shear behavior of large diameter steel tubes. Struct. Eng. Rep. No. 119, Dept. of Civil Engineering, University of Alberta, Edmonton, Alberta.

Baker, W.E. and Bennett, J.G. (1984). Experimental investigation of the buckling of nuclear containment-like cylindrical geometries under combined shear and bending. *Nuclear Engineering and Design* **79**(2), 211–216.

Batdorf, S.B., Stein, M. and Schildcrout, M. (1947). Critical stress of thin-walled cylinders in torsion. Technical Note No. 1344, NACA, Washington, DC.

Block, D.L., Card, M.F. and Mikulas, M.M. (1965). Buckling of eccentrically stiffened orthotropic cylinders. NASA TN D-2960.

Chen, Q., Elwi, A.E. and Kulak, G.L. (1995a). Flexural tests of longitudinally stiffened fabricated cylinders. *Journal of Constructional Steel Research* **34**(6), 1–25.

Chen, Q., Elwi, A.E. and Kulak, G.L. (1995b). Longitudinally stiffened large diameter fabricated steel tubes. *Journal of Constructional Steel Research* **35**(1), 19–48.

Chen, Q., Elwi, A.E. and Kulak, G.L. (1996). Finite element analysis of longitudinally stiffened cylinders in bending. *ASCE, Journal of Engineering Mechanics* **122**(11), 1060–1068.

Chen, W.F. and Han, D.J. (1985). Tubular members in offshore structures. *Surveys in Structural Engineering and Structural Mechanics*. Pitman, Boston, MA.

DnV (1995). Buckling strength analysis. Classification Notes, No. 30.1, Det Norske Veritas, Norway.

ECCS (1988). Buckling of shells. European Recommendations for Steel Construction, European Convention for Constructional Steelwork, Brussels, Belgium.

Edlund, Bo L.O. (1977). Buckling of axially compressed thin-walled cylindrical shells with asymmetric imperfections – preliminary report. *Second International Colloquium on the Stability of Steel Structures*, Liege, Belgium.

Faulkner, D., Chen, Y.N. and de Oliveira, J.G. (1983). Limit state design criteria for stiffened cylinders of offshore structures. *ASME, 4th National Congress of Pressure Vessels and Piping Technology*, Portland, Oregon.

Galletly, D.G. and Blachut, J. (1985). Plastic buckling of short cylindrical shells subjected to horizontal edge shear loads. *Journal of Pressure Vessel Technology, Transactions, ASME* **107**(5), 101–107.

Gunzelman, S.X. and Ostapenko, A. (1977). Local buckling tests on three large-diameter tubular columns. Report No. 393.8, Fritz Engineering Laboratory, Lehigh University, Bethlehem, PA.

Kulak, G.L., Stephens, M.J. and Bailey, R.W. (1988). Bending tests of large diameter fabricated steel cylinders. *Canadian Journal of Civil Engineering* **15**, 183–189.

Masubuchi, K. (1980). *Analysis of Welded Structures: Residual Stresses, Distortion, and their Consequences*. Pergamon Press, Oxford, UK.

Obaia, K.H., Elwi, A.E. and Kulak, G.L.(1991). Inelastic transverse shear capacity of large fabricated steel tubes. Struct. Eng. Rep. No. 170, Dept. of Civil Engineering, University of Alberta, Edmonton, Alberta.

Obaia, K.H., Elwi, A.E. and Kulak, G.L. (1992a). Tests of fabricated steel cylinders subjected to transverse loading. *Journal of Constructional Steel Research* **22**, 21–37.

Obaia, K.H., Elwi, A.E. and Kulak, G.L. (1992b). Ultimate shear strength of large diameter fabricated steel tubes. *Journal of Constructional Steel Research* **22**, 115–132.

Obaia, K.H., Elwi, A.E. and Kulak, G.L. (1993). Shear capacity of large diameter fabricated steel cylinders: a truss model. *Journal of Constructional Steel Research* **25**(3), 211–227.

Ostapenko, A. and Grimm, D.F. (1980). Local buckling of cylindrical tubular columns made of A36 steel. Rep. No. 450.7, Fritz Engineering Laboratory, Lehigh University, Bethlehem, PA.

Ostapenko, A. and Gunzelman, S.X. (1976). Local buckling of tubular steel columns. *Methods of Structural Analysis*, Vol. 2. ASCE, New York, NY.

Prion, H.G.L. and Birkemoe, P.C. (1987). Beam column behaviour of unstiffened fabricated steel tubes. *Proceedings of the SSRC Annual Technical Session*, Houston, TX.

Roman, V. and Elwi, A.E. (1989). Analysis of post-buckling behaviour of large diameter fabricated steel tubes. *Journal of Engineering Mechanics, ASCE* **115**(12), 2587–2600.

SSRC (1998). *Guide to Stability Design Criteria for Metal Structures* 5th edn. Structural Stability Research Council (ed. T.V. Galambos) Wiley, New York.

Stephens, M.J., Kulak, G.L. and Montgomery, C.J. (1982). Local buckling of thin-walled tubular steel members. Structural Engineering Report No. 103, Department of Civil Engineering, University of Alberta, Edmonton, Alberta, Canada.

Walker, A.C. and Sridharan, S. (1980). Analysis of the behavior of axial compressed stringer-stiffened cylindrical shells. *Proc. ICE, Part 2* **69**, 447–472.

Wilson, W.M. (1937). Tests on steel columns. University of Illinois Engineering Experimental Station Bulletin No. 292.

Wilson, W.M. and Newark, N.M. (1933). The strength of thin cylindrical shells as columns. University of Illinois Engineering Experimental Station Bulletin No. 255.

Chapter 13

Shell junctions

J.G. Teng

Introduction

Cylinders, cones, spheres and tori are some of the common basic shell elements. Steel shell structures such as silos, tanks, pressure vessels, offshore platforms, chimneys and tubular towers generally consist of two or more of these basic shell elements. Axisymmetric junctions featuring meridional slope mismatches between the connected elements are common features in steel shell structures. These junctions are a structural weakness, and their buckling and collapse strengths are a key design consideration.

Simple cone–cylinder junctions are the most common form of junctions and are found in steel silos and tanks with a conical roof, elevated conical water tanks with a cylindrical shell support, large tubular members and pipes with a transition cone between two cylinders of different diameters and pressure vessels with a conical end closure. Cone–cone junctions are also widely used in pressure vessels and piping components. A variety of cone–cylinder and cone–cone junctions are shown in Fig. 13.1. Complex junctions consisting of more than two shell segments include transition junctions in steel silos and tanks where the cylindrical vessel, the conical hopper (or end closure), and the supporting skirt meet (Fig. 13.2). Finally, examples of junctions containing curved shell segments include the junction between a cylindrical shell and a shallow spherical roof in a storage tank, and the junction between a cylindrical nozzle and a sphere in a pressure vessel (Fig. 13.3).

As the junction is a structural weakness, a ring is often provided to strengthen it. This ring stiffener may be in the form of an annular plate, a T-section or an angle section. Concerning buckling behaviour of junctions, it should be pointed out that if a ring is provided at a junction, the buckling behaviour of this ring often dominates the strength of the junction. Buckling of rings at shell junctions is thus separately treated in Chapter 14.

The most common load case for these junctions is a uniform or non-uniform pressure, although other load cases such as axial compression and bending can become important in some applications (e.g. Schmidt and Swadlo 1997). This chapter is concerned with shell junctions under a uniform internal pressure, on which there has been extensive recent research, although formulas for plastic limits loads are applicable to both internal or external uniform pressures.

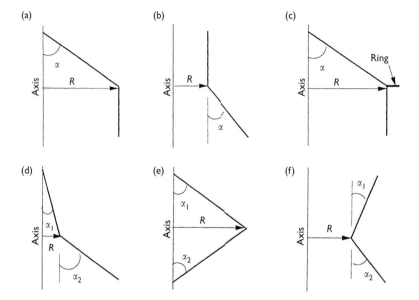

Figure 13.1 Cone–cylinder and cone–cone junctions: (a) cone large end-to-cylinder junction; (b) cone small end-to-cylinder junction; (c) ring-stiffened cone large end-to-cylinder junction; (d) cone large end-to-cone small end junction; (e) cone large end-to-cone large end junction; (f) cone small end-to-cone small end junction.

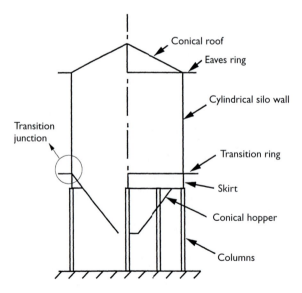

Figure 13.2 Typical structural form of an elevated steel silo.

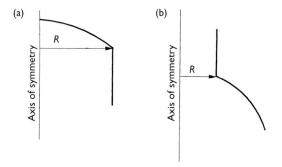

Figure 13.3 Sphere–cylinder junctions: (a) spherical roof-to-cylindrical wall junction; (b) spherical vessel-to-cylindrical nozzle junction.

Finite element collapse and buckling analysis

Most of the results presented in this chapter have been obtained from finite element plastic collapse or buckling analyses using an axisymmetric shell element to model the perfect geometry of the junction, so certain features of these analyses are briefly described here to assist in the interpretation of these results. Detailed descriptions of these analyses are available elsewhere (Teng and Rotter 1989a,b).

Axisymmetric shell junctions subject to internal pressure can fail by axisymmetric plastic collapse. This is the most obvious mode of failure and is due to yielding of the material. To determine the axisymmetric failure strength, one can carry out a small deflection elastic–plastic analysis. Such an analysis leads to a failure load which is equivalent to the classical plastic limit load obtainable from a classical limit analysis. The advantage of this analysis is that it provides a clearly defined failure load, while the disadvantage is that the effect of geometric changes (i.e. effect of geometric nonlinearity) is ignored in such an analysis. This effect can be very significant and is beneficial for internally-pressurised junctions. To take this effect into account, a large deflection elastic–plastic analysis has to be carried out. As this effect is strengthening, the load–deflection curve resulting from a large deflection analysis keeps rising, so failure has to be defined artificially. Further discussion of the failure criteria is given in the next section. The axisymmetric failure strength so determined is referred to as the large deflection collapse load. An elastic–perfectly plastic material is generally assumed in such analyses as the results are intended for application to structures made of ductile steel with a yield plateau.

When a junction is subject to circumferential membrane compression, non-symmetric buckling failure is possible. For the determination of the non-symmetric buckling failure load, one may carry out a linear elastic bifurcation buckling analysis, in which both geometric and material nonlinearities are ignored in the pre-buckling analysis. Such analyses lead to results which correspond to those from classical buckling analyses. There is, however, one difference between a linear elastic finite element bifurcation buckling analysis and a classical buckling analysis. The former invariably employs a linear elastic bending analysis for determining the prebuckling stress state, while the latter is generally based on a pure membrane state of prebuckling stresses. For shell junctions, the effect of geometric changes is so significant that such a linear elastic bifurcation buckling analysis is generally not very useful. All finite element buckling results presented in the present chapter are from a nonlinear bifurcation buckling analysis based on the perfect geometry, with the effect of geometric changes duly accounted for. In a nonlinear bifurcation buckling analysis, checks for the possibility of non-symmetric bifurcation buckling are carried out along the nonlinear axisymmetric load–deflection path. Using an axisymmetric shell element, the buckling mode can be described by a single harmonic term consisting of n waves. Many different circumferential buckling wave numbers need to be checked at a given load level, and many different load levels need to be checked to find the critical load level.

When material yielding occurs before buckling, the plastic behaviour of material needs to be properly modelled in an elastic–plastic large deflection prebuckling analysis using the J_2 flow theory of plasticity. For the bifurcation buckling analysis, this analytically more rigorous J_2 flow theory usually produces bifurcation buckling loads that agree less closely with experimental results and are higher than those from an analysis employing the J_2 deformation theory. This is the well-known paradox in the theory of plastic buckling yet to be properly resolved (Teng 1996). As a result, different plasticity theories have been used in plastic bifurcation buckling analyses. These include: J_2 flow theory, J_2 deformation theory and modified J_2 flow theory, which is the flow theory with its shear modulus replaced by that from the deformation theory. These three different options may lead to significantly different results for shell junctions (Teng and Rotter 1989b; Teng 1995; Gabriel 1996), with results from the deformation theory being the most conservative. All plastic bifurcation loads presented here were obtained using the deformation theory.

Uniform thickness cone large end-to-cylinder junctions

General

Cone large end-to-cylinder junctions (simply referred to as cone–cylinder junctions hereafter) (Fig. 13.1(a)) subject to uniform internal pressure are among the simplest junctions and are commonly found in pressure vessels and piping. As the junction is a structural weakness, reinforcement of these junctions by local thickening of the shell wall or by a ring is generally required. When local thickening of the shell wall is used, both the cylinder and the cone generally have the same increased thickness near the point of intersection, though away from this zone their thicknesses may differ. Provided that failure of the junction is confined to the thickened zone adjacent to the point of intersection and the shell thickness is constant in this zone, the behaviour of the junction reduces to that of a cone–cylinder junction of uniform thickness. The effect of more localised thickening of a cone–cylinder junction is considered in the section 'Sleeved cone–cylinder junctions'.

Existing design provisions for internally pressurised cone–cylinder junctions in the pressure vessel codes are limited. For example, the Australian code AS 1210 (1989) and the British code BS 5500 (1994) currently do not recommend the use of cone–cylinder junctions without a toroidal knuckle when $\alpha > 30°$, and provide simple rules for the reinforcement requirement of other junctions by either local thickening or a ring stiffener. The American code (ASME 1992) does allow the use of junctions with $\alpha > 30°$, but requires the designer to perform a stress analysis.

Failure modes

When a cone large end-to-cylinder junction is subject to uniform internal pressure, high bending and circumferential membrane compressive stresses arise near the point of intersection (Fig. 13.4). Under the large circumferential compression, two failure modes are possible: (a) axisymmetric collapse; and (b) non-symmetric buckling.

The axisymmetric failure mode predicted by a finite element analysis is shown in Fig. 13.5. Axisymmetric failure is characterised by the development of excessive radial deformations of the junction, with the formation of three plastic hinges, one at the point of intersection, and one each inside the cone and the cylinder, respectively. The failure mode may be idealised as a plastic mechanism with three plastic hinge circles as shown in Fig. 13.6.

The non-symmetric buckling mode involves deformations near the point of intersection which are, theoretically, periodical around the circumference if the shape of the junction is perfect. A three-dimensional plot of a finite element bifurcation buckling mode is shown in Fig. 13.7(a). A recent buckling failure of a real

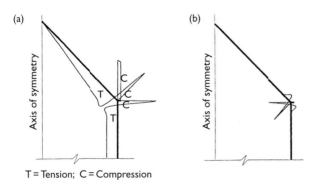

T = Tension; C = Compression

Figure 13.4 Stresses in a ring-stiffened cone–cylinder junction under internal pressure: (a) circumferential membrane stresses; (b) meridional bending stresses.

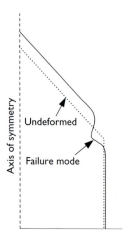

Figure 13.5 Axisymmetric failure mode of a cone–cylinder junction.

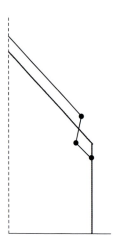

Figure 13.6 Idealised plastic collapse mechanism for a cone–cylinder junction.

cone–cylinder junction was reported and analysed by Jones (1994) and further studied in detail by Teng and Zhao (2000). In terms of the meridional shape, the non-symmetric bifurcation buckling mode is in one of the two forms (Teng 1995): the symmetric mode as shown in Fig. 13.7(b) and the anti-symmetric mode as shown in Fig. 13.7(c). Evidently, the modes are neither symmetric nor anti-symmetric in a strict sense, so these terms should not be taken to imply their exact shapes. Of course, the meridional buckling mode of each junction is different in details. In the symmetric mode, the buckling deformations involve a radial

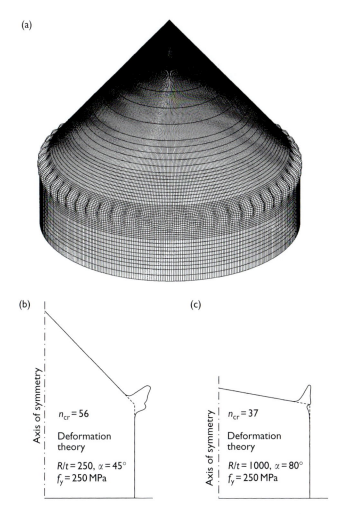

(a)

(b)

Axis of symmetry

$n_{cr} = 56$

Deformation theory

$R/t = 250$, $\alpha = 45°$
$f_y = 250\,\mathrm{MPa}$

(c)

Axis of symmetry

$n_{cr} = 37$

Deformation theory

$R/t = 1000$, $\alpha = 80°$
$f_y = 250\,\mathrm{MPa}$

Figure 13.7 Non-symmetric bifurcation buckling modes of cone–cylinder junctions: (a) three-dimensional plot of finite element bifurcation buckling mode; (b) symmetric mode; (c) anti-symmetric mode.

translation of the point of intersection which leads to membrane deformations in the cone. In the anti-symmetric mode, bending deformations are dominant. Junctions which are thick (i.e. the radius-to-thickness ratio is small) and/or have a steep cone are likely to buckle in the symmetric mode as the ratio of the cone membrane stiffness in the radial direction to the shell bending stiffness is comparatively small for these junctions. This ratio increases as the junction becomes thinner and the cone becomes shallower, making the anti-symmetric mode critical.

In experiments, roughly uniform short-wave buckles were observed (Fig. 13.16). Of course, the waves are not exactly periodical due to fabrication imperfections and material non-uniformity.

Axisymmetric failure strength

Limit load

Axisymmetric failure of cone–cylinder junctions is caused by the formation of plastic hinges, so plastic limit analyses (Save and Massonnet 1972) can be carried out to define their plastic limit loads. Such analyses have been carried out using a numerical optimisation procedure by Davie *et al.* (1978) and Myler and Robinson (1985) who produced graphs showing lower bound values of classical limit pressures for cone–cylinder junctions of uniform thickness for a limited range of geometry. Plastic limit loads can also be calculated using a small deflection elastic–plastic finite element analysis. Teng (1994) used the finite element method to obtain numerical results for a wide range of geometry ($10° \leq \alpha \leq 80°$ and $10 \leq R/t \leq 1000$). Based on his numerical results and the concept of effective section analysis first proposed by Rotter (1987), the following accurate and simple formula for the plastic limit load was derived by Teng (1994):

$$\frac{p_l R}{f_y t} = \frac{3}{\tan \alpha} \left(\frac{t}{R} \right)^{0.55} \left(1 + \frac{1}{\sqrt{\cos \alpha}} \right) \tag{1}$$

where p_l is the limit pressure, f_y the yield stress, R the cylinder radius, t the shell thickness and α the cone apex half angle. The above rationally based and simple equation fits the finite element limit loads almost precisely (Fig. 13.8).

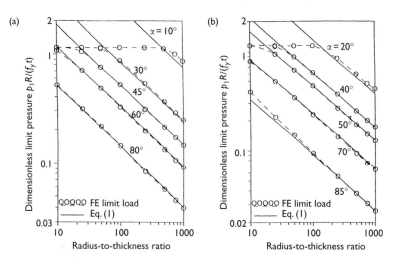

Figure 13.8 Approximation to limit loads of cone–cylinder junctions.

Large deflection axisymmetric collapse load

Equation (1) provides only a lower bound to the axisymmetric failure strength of these junctions, as the strengthening effect of geometric nonlinearity (i.e. geometric changes) available to these junctions is ignored in a small deflection finite element analysis (Teng 1994). This is illustrated in Fig. 13.9, where the load deflection curve from a small deflection analysis is compared with that from a large defection analysis which considers the effect of geometric changes. This comparison shows that loads much higher than the plastic limit load may be sustained, though at the expense of increased deformations.

Load–deflection curves from a large deflection analysis keep rising, which means that no definite point of failure can be identified on such curves. The final failure of a real junction, likely to be by rupture at welded joints, occurs after

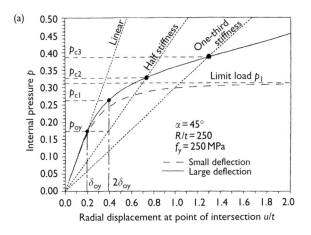

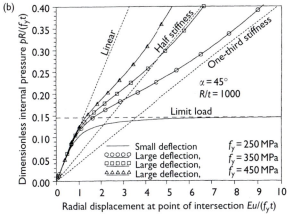

Figure 13.9 Effect of large deflections and definition of failure: (a) $R/t = 250$; (b) $R/t = 1000$.

substantial deformations and cannot be predicted by such an analysis based on the assumption of an elastic–perfectly plastic material. As a result, an artificial definition of failure based on limiting displacements or stiffness losses (Gerdeen 1979) has to be adopted to give a failure load. All or some of the following definitions of failure were employed by Gerdeen (1979), Galletly and Blachut (1985) and Hayakawa et al. (1977): (a) overall yield load p_{oy}, (b) twice yield deflection load p_{c1}, (c) half stiffness load p_{c2} and (d) one third stiffness load p_{c3}. Failure loads based on the above four definitions are shown in Fig. 13.9 and were considered by Teng (1994) in his study on cone–cylinder junctions. It should be noted that whilst the first two loads, p_{oy} and p_{c1}, can generally be obtained from the load–deflection curve as long as the junction fails by excessive inward radial deformations, the last two failure loads, p_{c2} and p_{c3}, may not exist if the secant stiffness of the structure never drops below half its initial stiffness or one third its initial stiffness (Fig. 13.9(b)).

Teng (1994) determined all four loads for a large range of geometry and established the following equation to predict the large deflection collapse load:

$$\frac{p_c R}{f_y t} = \left[1 + 400 \left(\frac{f_y}{E} \right)^2 \frac{R}{t} \right] \left[\frac{3}{\tan \alpha} \left(\frac{t}{R} \right)^{0.55} \left(1 + \frac{1}{\sqrt{\cos \alpha}} \right) \right] \qquad (2)$$

Equation (2) is a simple modification of Eq. (1) by a large deflection factor which is related to the yield stress-to-elastic modulus ratio f_y/E and the radius-to-thickness ratio R/t. This factor was derived based on the one-third stiffness loads for junctions with an apex half angle of 60° only, and is therefore independent of cone apex half angles (Fig. 13.10). Equation (2) was compared with finite element

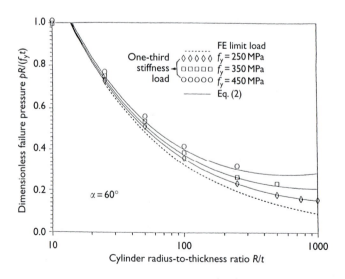

Figure 13.10 Approximation to the large deflection effect.

results over a wide range of geometries for the three yield stresses of 250, 350 and 450 MPa (Teng 1994). This comparison showed that Eq. (2) provides a close approximation for junctions with α between 40 and 70°, and a first, conservative approximation for other junctions (Teng 1994). The strength predicted by Eq. (2) is approximately three times that predicted by Eq. (1) for a yield stress of 450 MPa and an R/t ratio of 1000. The use of Eq. (2) instead of Eq. (1) in design can lead to much lighter structures. However, junctions with a large R/t ratio and/or a high yield stress are likely to fail by non-symmetric buckling before reaching the axisymmetric strength of Eq. (2) or Eq. (1).

Non-symmetric buckling strength

Elastic buckling strength

Elastic buckling may govern the strength of very thin junctions especially when the cone is very shallow and/or the yield stress is high. Teng (1996a) carried out elastic bifurcation buckling analyses for internally-pressurised cone–cylinder junctions with $30° \le \alpha \le 85°$ and $250 \le R/t \le 1000$ as part of a larger study into the elastic buckling strength of cone–cylinder junctions under localised circumferential compression. Figure 13.11 shows that the dimensionless buckling pressure $p_{cr}R/(Et)$ and the R/t ratio observe a power relationship (i.e. straight line in a log–log plot). The finite element elastic buckling loads were found to be closely approximated by the following equation (Teng 1996a):

$$\frac{p_{cr}R}{Et} = \frac{320}{\tan^{2.5}\alpha} \left(\frac{t}{R}\right)^{2.1} \left(1 + \frac{1}{\cos^{1.5}\alpha}\right) \tag{3}$$

Equation (3) was devised to provide a better approximation to the finite element results for junctions which are thin and/or have a shallow cone (Fig. 13.11), as elastic buckling may be the critical mode of failure for these junctions. For other junctions which are unlikely to be governed by elastic buckling, Eq. (3) is less accurate but conservative. It should be noted that for some junctions, no buckling was predicted by the finite element analysis (Teng 1996a) although Eq. (3) predicts a buckling pressure. For example, for $\alpha = 30°$, elastic buckling was predicted to be possible only when $R/t = 750$ and 1000 for the four different R/t values examined by Teng (1996a) (Fig. 13.11). This is not a real drawback of Eq. (3), as the strength of such junctions is unlikely to be controlled by buckling failures and the prediction of Eq. (3) is in any case a conservative answer.

Plastic buckling strength

Buckling of relatively thick cone–cylinder junctions occurs after the material has yielded. If this yielding is very limited, then the buckling behaviour is little affected. However, when significant through-thickness yielding has occurred, the buckling behaviour of cone–cylinder junctions is dominated by plastic material behaviour.

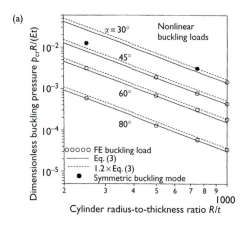

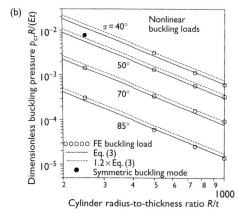

Figure 13.11 Approximation to elastic buckling loads of cone–cylinder junctions.

Finite element buckling loads were obtained by Teng (1995) for shells with $100 \leq R/t \leq 1000$, $30° \leq \alpha \leq 85°$ and $f_y = 250$, 350 or 450 MPa using the deformation theory of plasticity. Again, no buckling was predicted for some of the geometries investigated (Teng 1995). When the buckling loads are plotted in the dimensionless form as shown in Fig. 13.12, it is seen that unless elastic (or nearly elastic) buckling is critical, the variation of the dimensionless buckling pressure $p_{cr}R/(f_y t)$ with R/t for different yield stresses falls almost on a single straight line for each α value. Based on this observation, Teng (1995) arrived at the following equation which approximates the finite element results closely (Fig. 13.12):

$$\frac{p_{cr}R}{f_y t} = \frac{6}{\tan^{1.6}\alpha}\left(\frac{t}{R}\right)^{0.75}\left(3 + \frac{1}{\cos^{1.5}\alpha}\right) \tag{4}$$

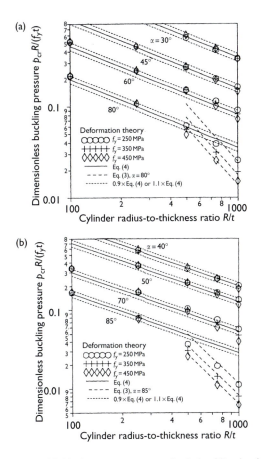

Figure 13.12 Approximation to plastic buckling loads of cone–cylinder junctions.

Experimental behaviour

General

The preceding buckling strength equations were developed based on finite element bifurcation buckling loads obtained using the perfect geometry of the junction. Since buckling of shells is generally known to be sensitive to geometric imperfections, it is necessary to quantify the effect of fabrication imperfections on the buckling strength before these equations can be applied with confidence in practical design. The most direct way to assess the applicability of Teng's (1995, 1996a) buckling strength equations to real shells with geometric imperfections is to compare their predictions with experimental results.

Up to now, there have been seven buckling tests on internally pressurised cone–cylinder junctions (Gabriel 1996; Teng and Zhao 2000; Zhao and Teng 2001).

The first three tests conducted by Gabriel (1996) were on specimens which had large geometric defects that are unrepresentative of full-scale structures, with two of them featuring large localised defects due to the matching/welding of the cone to the cylinder. The large localised defects in these two specimens led to the development of localised buckles and substantial discrepancies between the test results and finite element results based on the perfect geometry. The buckling load of the third specimen agreed well with a finite element prediction based on the perfect geometry. Teng and Zhao (2000) described one buckling test on a more perfect specimen which appears to have had a level of imperfections more representative of real junctions. The experimental buckling load of this specimen was found to be close to the finite element bifurcation load of the perfect geometry (Teng and Zhao 2000). The test described in Teng and Zhao (2000) was a trial test, prior to the three careful recent tests (Zhao and Teng 2001). Results from these three tests are briefly summarised below.

Zhao and Teng's experiments

Figure 13.13 shows the test set-up employed by Zhao and Teng (2001). In this test set-up, the initial shape of a test specimen was surveyed by a laser displacement meter which could revolve around and move up and down the specimen with the control of a computer. Loading was applied by filled water pressurised using a hydraulic pump. All specimens had a radius of around 500 mm, with the thicknesses of the cone and the cylinder being either 1 or 2 mm. For such thin shell model junctions, a careful fabrication procedure was followed to achieve good quality specimens (Zhao and Teng 2001). Figure 13.14 shows the initial shape of one of the three model junctions (Junction CC-1) plotted using half of the survey points to avoid congestion. Only part of the cone was surveyed due to the limited measurement range of the laser displacement meter, but the critical part of the cone was included (Fig. 13.14). The model junctions were fabricated from steel sheets by rolling and welding, with the four meridional welds in the cone and the two meridional welds in the cylinder causing localised ridges (Fig. 13.14). In addition, short-wave imperfections of a smaller amplitude near the point of intersection and longer-wave imperfections of a greater amplitude away from the point of intersection were found. The mean and maximum geometric deviations of all grid points from the nominal geometry in the vicinity of the joint (within 2λ from the joint where λ is the linear elastic meridional bending half wavelength) were determined for each specimen. The mean deviations for all specimens are less than one wall thickness. The maximum deviations are much larger but they always occurred at the meridional welds as shown in Fig. 13.14. The imperfections in these specimens are expected to be similar to those in real junctions, particularly in the critical region near the point of intersection. However, there appear to have been no measurements of imperfections in real cone–cylinder junctions.

Figure 13.13 Set-up for buckling experiments on cone–cylinder junctions.

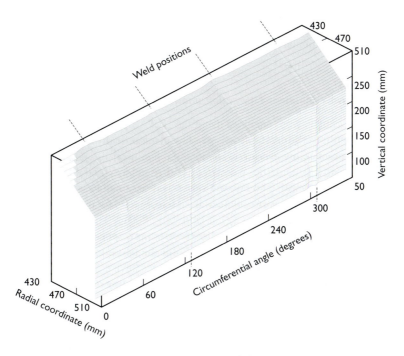

Figure 13.14 Initial surface of model junction CC-1.

For each specimen, a number of strain gauges were installed on the cone around a circumference at a meridional distance of 20 mm from the point of intersection. Fig. 13.15 shows the variations of strains with increasing pressures for all three specimens. In the initial stage of loading, the strains were similar (Fig. 13.15), indicating dominantly axisymmetric behaviour. As the pressure reached a certain value, the strains at different locations started to diverge from each other (Fig. 13.15). The divergence of the strain readings is an indication of the growth of non-symmetric deformations and is more obvious in Junctions CC-1 and CC-2 which are thinner than Junction CC-3. With further increases of the pressure, roughly uniform short-wave buckles centred around the circumferential weld could be observed clearly. The buckles continued to grow with the applied pressure until rupture and/or severe leaking occurred at one or more of the welded joints, leading to the release of pressure and signifying the end of the test. The main weld failures occurred at the circumferential weld. Junction CC-1 failed explosively by the rupture of the circumferential weld while failures of the other two specimens were less drastic. Figure 13.16 shows the three specimens after failure with the pressure already released.

Obviously, the postbuckling behaviour of these junctions is stable, so their experimental buckling loads cannot be determined in a straightforward manner. An approximate but rational approach is to take the pressure at which the strain readings started to diverge as the buckling pressure, although this does not allow a precise definition of the buckling pressure as the strain readings had small differences right from the beginning of loading due to the presence of initial imperfections, particularly for Junction CC-3 which is relatively thick.

Correlation with nonlinear bifurcation analysis

All three specimens were analysed using the NEPAS program (Teng and Rotter 1989b) for nonlinear bifurcation analysis of axisymmetrically loaded shells of revolution with prebuckling geometric and material nonlinearities taken into account. The nominal perfect geometries were used and the steel was modelled as an elastic–perfectly plastic material. The Poisson's ratio was assumed to be 0.3. The elastic moduli and yield stresses were obtained from tensile tests for the steel sheets used to fabricate the model junctions (Zhao and Teng 2001). The nonlinear bifurcation loads are shown in Fig. 13.15 and are seen to correspond well to the loads at which strains started to diverge, particularly for the two thinner model junctions (Junctions CC-1 and CC-2). These bifurcation loads have not included the consequences of welding including residual stresses and welding deposits. Both factors are not believed to have an effect which is significant enough to change this overall picture of close agreement, given that the residual stresses are self-equilibrating and the amount of welding deposit very small. It may therefore be concluded that due to a stable postbuckling behaviour, cone–cylinder junctions show only limited sensitivity to initial imperfections and that the bifurcation loads, on which

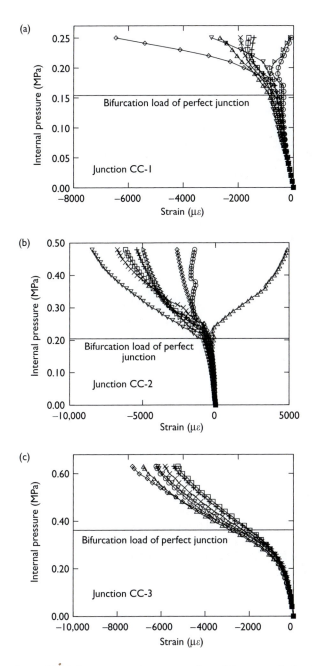

Figure 13.15 Pressure–strain curves of model cone–cylinder junctions: (a) CC-1; (b) CC-2; (c) CC-3.

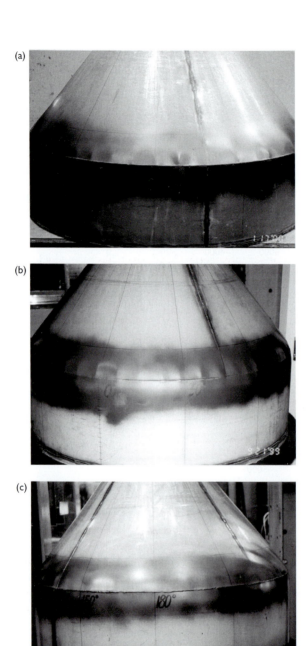

Figure 13.16 Postbuckled model cone–cylinder junctions: (a) CC-1; (b) CC-2; (c) CC-3.

Teng's (1995, 1996a) design approximations are based, are a good measure of the buckling strength of real imperfect junctions. Therefore, Teng's equations can be used with confidence in design.

Consequence of a buckling failure

Since the postbuckling path is stable, one may then question whether buckling constitutes the ultimate limit state of the junction since buckling is not associated with the peak load carrying capacity. The stable postbuckling behaviour does not diminish the importance of the buckling phenomenon, as the postbuckling growth of short-wave buckles can lead to large strains at the circumferential weld and consequent weld rupture failure. Because of this risk of rupture failure, buckling in a cone–cylinder junction has a more serious consequence than that in an internally pressurised torispherical head which has been widely studied (Galletly 1985, Galletly and Blachut 1985). This is because for a torispherical head, the buckles appear on the toroidal section away from the welds, while in a cone–cylinder junction, the buckling deformations are centred around the circumferential weld. Given this possibility of subsequent weld rupture, the buckling load should be taken as the ultimate load. With further research, the buckling load may be taken as a serviceability limit state if the ultimate rupture load can be determined.

In cases where the prediction of the weld rupture load is required for a safe design, as in the design of a fragile cone-to-cylinder joint in tanks (Yoshida 1999) when subject to internal explosion, it is essential that the growth of postbuckling deformations and strains be carefully studied and a suitable failure criterion developed. Experiments such as those by Zhao and Teng (2001) should be useful for future research on this important issue.

Controlling failure mode

In the design of internally pressurised cone–cylinder junctions, one is interested to know when buckling is a possible failure mode that he has to design against. This question can be answered easily by comparing Eqs (1)–(4). Based on these equations, Teng (1995) produced charts which clarified the range of geometry for junctions with R/t up to 1000 and α up to 85° controlled by either plastic collapse, plastic buckling or elastic buckling. Different charts were obtained by Teng (1995) depending on whether the plastic limit load or the large deflection collapse load was taken to be the axisymmetric failure strength. One of such charts is shown in Fig. 13.17 for a yield stress of 450 MPa. Figure 13.17 was developed by taking the large deflection collapse load of Eq. (2) as the axisymmetric failure strength. Buckling governs the strength of a wide range of geometries and is found to be critical for a junction with an R/t ratio being as low as about 117 when

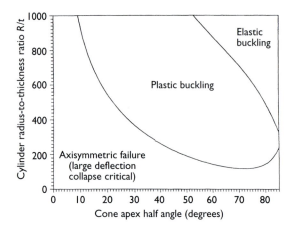

Figure 13.17 Controlling failure mode for cone–cylinder junctions with $f_y = 450\,\text{MPa}$.

$f_y = 450\,\text{MPa}$ and $\alpha = 72°$. A lower yield stress leads to a smaller range of buckling-critical junction geometries (Teng 1995).

Sleeved cone–cylinder junctions

General

When local thickening of the shell wall is used, both the cylinder and the cone generally have the same increased thickness near the point of intersection, though away from this area their thicknesses may differ. Local thickening is also referred to as sleeving or sleeve reinforcement (i.e. adding a sleeve to the original shell wall), and the length of the thickened portion is thus called the sleeve length. When the sleeve length is large enough that failure is confined to the sleeved part of the junction, the junction can be treated as one of uniform thickness. The buckling and collapse strengths of uniform thickness cone–cylinder junctions are discussed in the previous section.

Sleeve reinforcement can be conveniently achieved by the use of a thicker plate in the region near the point of intersection. As steel plates come only in a small number of different thicknesses, the choice for the thickness ratio between the thicker portion and the thinner portion is restricted by the thicknesses of steel plates available. Given a thickness ratio, it is often sufficient to prevent the junction from failure by a sleeve length shorter than one for failure confinement mentioned earlier. Reducing the thickness ratio to the next lower value may lead to inadequate junction strength even if the sleeve is made very long. It is therefore desirable and economical to allow designers the freedom to use a sleeve shorter than the minimum sleeve length for failure confinement.

It is easy to see that in the design of sleeve reinforcement, the designer needs to know: (a) how long the sleeve should be for failure confinement; and (b) if a sleeve shorter than is required for failure confinement is provided, how the strength can be assessed. Existing information on these two issues are summarised in the rest of this section. It should be noted that no study has examined the buckling strength of junctions with short sleeves.

Minimum sleeve length for failure confinement

Guidance on the minimum sleeve length for confinement of junction local failure is provided in various pressure vessel codes. Among them, the ASME (1992) code recommends a minimum sleeve length of $2(Rt_s/\cos\phi)^{0.5}$. Here, R is the radius of the cylinder, ϕ = the apex half angle α for the cone and 0 for the cylinder, and t_s is the thickness of the sleeved shell wall (Fig. 13.18). On the other hand, the British code (BS 5500 1994) recommends a minimum sleeve length of $(2Rt_s/\cos\phi)^{0.5}$.

No rigorous and comprehensive study appears to have been conducted on this minimum length requirement until the recent study of Teng and Gabriel (1998), although the issue was briefly discussed by Myler and Robinson (1985). Teng and Gabriel (1998) showed that for a sleeved junction to be treated as a junction of uniform thickness t_s for junction failure considerations, the minimum required sleeve length is about 0.8 times the linear elastic meridional bending half wavelength λ. This minimum sleeve length is also the maximum effective sleeve length, as any further increase in the sleeve length does not lead to an increase in the junction failure strength. This minimum length can thus be written as

$$l_{\min} = 0.8\lambda = 0.8 \times 2.44\sqrt{\frac{Rt_s}{\cos\phi}} = 1.95\sqrt{\frac{Rt_s}{\cos\phi}} \approx 2\sqrt{\frac{Rt_s}{\cos\phi}} \qquad (5)$$

which matches exactly that recommended in the ASME (1992) pressure vessel code. On the other hand, the minimum length required by the British pressure

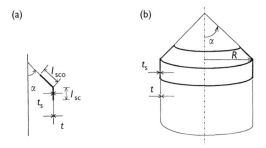

(a)   (b)

Figure 13.18 Sleeved cone–cylinder junction.

vessel code (BS 5500 1994) is too short for confining junction failure to the sleeved region.

Limit loads of locally sleeved junctions

Junctions with a sleeve length shorter than the minimum length specified in Eq. (5) is called locally sleeved junctions here. Two different expressions for their limit loads have been developed by Teng and Gabriel (1998) to cover junctions with different sleeve lengths and thicknesses based on finite element results for a wide range of geometry. For junctions with a short sleeve, the following equation approximates the finite element limit loads closely:

$$p_{ls} = p_l \left[1 + 7\kappa \left(\frac{t_s}{t} - 1 \right) \right] \tag{6}$$

where κ is the dimensionless sleeve length given by

$$\kappa = l_{sco}/\lambda_{sco} = l_{sc}/\lambda_{sc} \tag{7}$$

$$\lambda_{sco} = 2.44\sqrt{\frac{Rt_s}{\cos \alpha}}, \qquad \lambda_{sc} = 2.44\sqrt{Rt_s} \tag{8}$$

In Eq. (6), p_l is the limit pressure of the corresponding unsleeved junction and can be found using Eq. (1). Equation (6) was found by Teng and Gabriel (1998) to provide very close predictions of the finite element results overall. There are some errors on the safe side for extremely light sleeving which is not expected to be common in practice. For junctions with a longer sleeve, the following simple equation was proposed by Teng and Gabriel (1998) to provide a close lower bound approximation:

$$p_{ls} = p_l \left[1 + (0.75\kappa + 0.4) \left(\left(\frac{t_s}{t} \right)^{1.55} - 1 \right) \right] \qquad \text{for } \kappa \leq 0.8 \tag{9}$$

In practical design, Eqs (6) and (9) should be used together with p_l from Eq. (1) and the lower value from the two equations gives the limit pressure of the junction.

Approximation to the large deflection effect

Teng and Gabriel (1998) suggested that the effect of large deflections on the axisymmetric failure strength of locally sleeved cone–cylinder junctions be conservatively approximated by adopting the modification factor in Eq. (2) and replacing R/t by R/t_s. That is, the large deflection collapse load of a locally sleeved cone–cylinder junction p_{cs} may be related to the corresponding limit pressure p_{ls} through

$$p_{cs} = \left[1 + 400 \left(\frac{f_y}{E} \right)^2 \frac{R}{t_s} \right] p_{ls} \tag{10}$$

Limited comparisons between the predictions from Eq. (10) and the one-third stiffness loads from finite element large deflection analyses supported their proposition (Teng and Gabriel 1998).

Ring-stiffened cone–cylinder junctions

General

As an alternative to local thickening of shell walls, a ring stiffener may be provided at the cone–cylinder junction (Fig. 13.1(c)). This ring is likely to suffer out-of-plane buckling, a topic which is extensively treated in Chapter 14. In this section, the limited work carried out specifically on ring-stiffened cone–cylinder junctions under internal pressure is summarised.

Axisymmetric failure strength

The following simple equation proposed by Teng and Gabriel (1994) provides accurate predictions of limit loads of ring-stiffened cone–cylinder junctions subject to internal pressure:

$$\frac{p_{\mathrm{l}}R}{f_{\mathrm{y}}t} = \frac{3}{\tan\alpha}\left(\frac{t}{R}\right)^{0.55}\left(1+\frac{1}{\sqrt{\cos\alpha}}\right) + \frac{2A_{\mathrm{r}}}{Rt\tan\alpha} \tag{11}$$

where A_{r} is the cross-sectional area of the ring stiffener.

It is easy to recognise that the first term on the right-hand side of Eq. (11) represents the resistance offered by the shell segments (Eq. 1) and the second term that from the ring stiffener. If both local thickening and ring stiffening are used, the limit load may be estimated by replacing the first term in Eq. (11) by Eq. (6) or (9) as appropriate, although no rigorous study has been conducted to verify this proposal.

No rigorous study has examined the large deflection axisymmetric collapse behaviour of ring stiffened cone–cylinder junctions. It is, however, not unreasonable to assume that the contribution of the ring to the axisymmetric failure strength is little affected by junction deformations, so the second term on the right hand side of Eq. (11) may be added to Eqs (2) and (10) to provide good first estimates of the large deflection collapse loads of uniform thickness cone–cylinder junctions with a ring stiffener and locally sleeved junctions with a ring stiffener.

Buckling strength

The elastic buckling strength of cone–cylinder junctions with an annular plate ring stiffener under internal pressure has recently been studied by Teng and Ma (1999). No study has examined the plastic buckling strength of ring-stiffened cone–cylinder junctions under internal pressure.

Two types of buckling modes were identified by Teng and Ma (1999): the shell buckling mode and the ring buckling mode. The following approximate buckling strength formula was developed by Teng and Ma (1999) for the shell buckling mode:

$$p_{cr,r} = p_{cr}\left(1 + \frac{B_p T_p}{0.778\sqrt{Rt}\,(1 + 1/\sqrt{\cos\alpha})}\right) \tag{12}$$

where B_p and T_p are the width and thickness of the annular plate ring, respectively, and p_{cr} is the elastic buckling pressure of a cone–cylinder junction without a ring stiffener as given by Eq. (3). Equation (12) was derived by determining the amount of internal pressure taken up by a ring deforming with the shell, with the amount of pressure resisted by the shell assumed to be the same as that of a cone–cylinder junction without a ring stiffener. Equation (12) was developed for a cone–cylinder junction with an annular plate ring, but if a T-section or angle section ring is present instead, it is expected that close predictions can be obtained by replacing $B_p T_p$ in Eq. (12) with the ring cross-sectional area A_r.

The strength of the ring buckling mode can be conservatively predicted by the elastic buckling strength formula of Jumikis and Rotter (1983) developed for annular plate rings in steel silos (Teng and Ma 1999). The application of Jumikis and Rotter's formula for transition rings in steel silos to ring-stiffened cone–cylinder junctions was facilitated by an effective section analysis (Rotter 1983, 1985) so that the maximum circumferential compressive stresses at the inner edge of the annular plate at buckling could be compared for the two cases. This is explained in detail in Chapter 14.

Junctions of curved shell segments

Typical junctions with one or more curved shell segments include the junction between a spherical roof and the vertical wall of a tank (Fig. 13.3(a)), the junction between a spherical vessel and a cylindrical nozzle (Fig. 13.3(b)), and the junction formed by a toroidal segment and a supporting skirt in an egg-shaped digester. Limited recent work has been done on junctions containing curved shell segments.

Mahrenholtz (1998) studied the elastic and plastic buckling behaviour of internally pressurised spherical roof-to-wall junctions with large radius-to-thickness ratios typically found in storage tanks. The following two equations were proposed to approximate the elastic and plastic buckling pressures, respectively:

$$\frac{p_{cr}^e R_{wall}}{E t_{wall}} = 100\left(\frac{t_{wall}}{R_{wall}}\right)^{1.8}\left(\frac{t_{roof}}{t_{wall}}\right)^{2.2}\frac{\cos^{0.45}\alpha}{\tan\alpha} \quad \text{(elastic buckling)}$$

$$\tag{13}$$

and

$$\frac{p_{cr}^{p} R_{wall}}{f_y t_{wall}} = 60 \left(\frac{t_{wall}}{R_{wall}}\right)^{0.9} \left(\frac{t_{roof}}{t_{wall}}\right)^{1.8} \frac{\cos^{0.1}\alpha}{\tan\alpha} \qquad \text{(plastic buckling)} \quad (14)$$

for tank geometries in the following range

$$600 < \frac{R_{wall}}{t_{wall}} < 3500, \qquad 50° < \alpha < 88°, \qquad \frac{t_{roof}}{t_{wall}} < 1.25 \qquad (15)$$

where R_{roof} and t_{roof} are the radius and thickness of the spherical roof, respectively, and R_{wall} and t_{wall} are the radius and thickness of the cylindrical wall, respectively, and α is the change in slope between the cylindrical wall and the spherical roof at the point of intersection. For a given wall-to-roof junction, predictions from the two equations need to be compared and the lower prediction should be the critical pressure (Mahrenholtz 1998). Mahrenholtz (1998) also proposed a design recommendation in the Eurocode format, but the accuracy of the design recommendation depends on whether the assumed knockdown factors for fabrication imperfections are valid. Since for these junctions, the effect of imperfections is expected to be similar to that of cone–cylinder junctions, it may be the case that the direct use of Eqs (13) and (14) in design leads to more realistic predictions.

The influence of a ring-stiffener was also considered by Mahrenholtz (1998). In addition, he briefly compared the buckling behaviour of tanks with a conical roof and that of tanks with a spherical roof, and stated that the difference in buckling strength between the two is not great.

A possible way for a simple extrapolation of the existing knowledge on junctions consisting of conical or cylindrical segments to those with curved shell segments is to approximate a curved shell segment by an equivalent conical shell segment with a slope equal to that of the curved shell segment at the point of intersection in the determination of the effective section. This approximation has been shown to work well by Au (1999) for the plastic limit load of an egg-shaped digester with the skirt represented by a vertical restraint and is also supported by evidence generated by Mahrenholtz (1998). It is desirable that the validity of this approximation be further investigated in the future.

Plate–cylinder junctions

Circular plates are often used as end closures of small pressure vessels, leading to plate–cylinder junctions (Fig. 13.19). Limit analysis was considered in Hodge (1964), Save and Massonet (1972) and Save (1984) for such plate–cylinder junctions, but a limit analysis is not useful unless the junction is very thick. The effect of geometric changes is very strong in these junctions, so a large deflection analysis is necessary for accurate predictions of load–deflection responses. The elastic large deflection behaviour of these junctions was studied by Uddin (1987). He showed

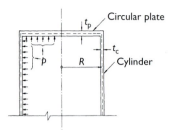

Figure 13.19 Plate–cylinder junction.

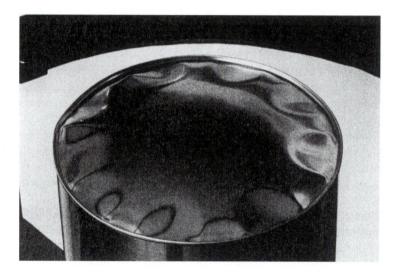

Figure 13.20 Buckles on a plate–cylinder junction.

that the nonlinear effect is very strong and the linear elastic bending theory is completely unsatisfactory in predicting the form of stress distributions.

Teng and Rotter (1989c) presented a comprehensive study of the elastic–plastic large deflection and bifurcation buckling behaviour of these plate–cylinder junctions. They showed for the first time that non-symmetric buckling can occur after large nonlinear deformations and presented buckling loads for junctions of uniform and non-uniform wall thickness. A further study of the buckling problem was carried out by Gabriel (1996) in which both experiments and numerical analyses were carried out. Figure 13.20 shows well-developed periodical postbuckling waves on a plate–cylinder junction in a simple exploratory experiment. Finite element buckling modes for three uniform thickness junctions (Teng and Rotter 1989c) are shown in Fig. 13.21, indicating that the buckling deformations are concentrated in

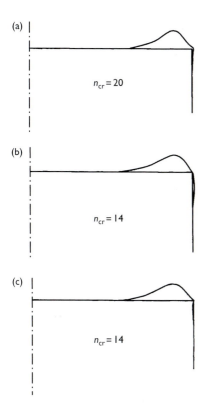

(a) $n_{cr} = 20$

(b) $n_{cr} = 14$

(c) $n_{cr} = 14$

Figure 13.21 Buckling modes of uniform thickness plate–cylinder junctions: (a) $R/t = 200$; (b) $R/t = 300$; (c) $R/t = 1000$.

the end plate. If, however, the plate end is made thicker, buckling deformations in the cylinder become more significant. Figure 13.22 shows the buckling mode of a junction with $R/t_p = 300$ and the plate being five times as thick as the cylinder with a yield stress of 250 MPa. This mode appears strange at first sight as both the end plate and the cylinder deform in the same direction. Compatibility is maintained at the corner, but strong local bending deformations occur near the yielded corner during buckling.

Figure 13.23 is plotted using results given in Gabriel (1996) for plate–cylinder junctions of uniform thickness obtained using the deformation theory of plasticity. Elastic or nearly elastic buckling controls for R/t ratios below about 500, above which plastic buckling controls. Gabriel proposed two design approximations for these results, one in the form of a fourth order polynomial and one in the form of two linear equations. One problem with these equations is that they were based on the results for a single yield stress of 250 MPa, so although the buckling pressure was normalised by the yield stress, the validity of these expressions for other yield

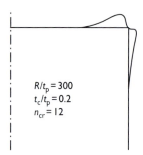

$R/t_p = 300$
$t_c/t_p = 0.2$
$n_{cr} = 12$

Figure 13.22 Buckling mode of a non-uniform thickness plate–cylinder junction.

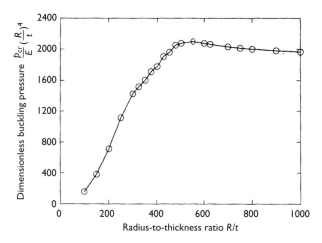

Figure 13.23 Buckling pressures of uniform thickness plate–cylinder junctions.

stresses has not been established. Furthermore, it is simple enough to read from Fig. 13.23 instead of using these equations if one is limited to a yield stress similar to 250 MPa.

Three tests were carried out by Gabriel (1996) on plate–cylinder junctions subject to internal pressure. Despite large geometric imperfections in the end plate in the form of inward dishing in the test specimens, Gabriel found the experimental buckling loads to agree well with nonlinear bifurcation buckling loads of the perfect geometry (1996), although there was some uncertainty in the definition of experimental buckling loads. Therefore, results such as those shown in Fig. 13.23 can be used to design against buckling. The tests also showed a stable postbuckling path, like that of internally pressurised cone–cylinder junctions. Teng and Luo (1997) showed theoretically that the buckling strength of the plate end is not sensitive to an axisymmetric inward dishing through a postcollapse bifurcation analysis, although inward dishing will cause snapping of the end plate at a low pressure.

Junctions in tanks subject to explosion

A number of recent studies (Yoshida and Miyoshi 1992; Yoshida 1995, 1999; Yoshida and Tomiya 1998) conducted in Japan have examined the failure of tanks subject to internal explosion caused by the ignition of hydrocarbon gas in the tank (Fig. 13.24). These studies are concentrated on geometries specific to oil storage tanks in terms of roof shapes (shallow cones or spherical caps) and wall thicknesses. The explosion-induced internal pressure can lead to buckling of the roof–cylinder junction and the bottom plate–cylinder junction. The general buckling behaviour of these junctions is similar to that discussed earlier for cone–cylinder junctions and plate–cylinder junctions under internal pressure.

For safety considerations, the roof–cylinder junction should be designed to fail before the bottom plate–cylinder junction during an explosion as failure of the latter can cause severe damage to both the tank itself and surrounding facilities. This is referred to as the fragile joint design concept. An interesting point to note is that in the design of a fragile roof–cylinder junction, a conservative design rule is not a good rule, contrary to normal cases. Yoshida and Miyoshi (1992) showed that the current API rule (1993) for the design of the roof–cylinder junction is too conservative for relatively thick-walled small diameter tanks. The API rule (1993) is based on the effective section concept but with the effective lengths of the shell walls underestimated in comparison with accurate values, and the assumption that failure occurs when the circumferential stress from such an effective section analysis reaches the yield stress. No account is taken of the possibility of non-symmetric buckling. Consequently, it is conservative for relatively thick-walled tanks and unconservative for thin tanks.

Studies have also been carried out on the buckling of the bottom plate–cylinder junctions in empty tanks (Yoshida 1995) and partially filled tanks (Yoshida and Tomiya 1998) subject to uplifting from internal pressure for a number of tank

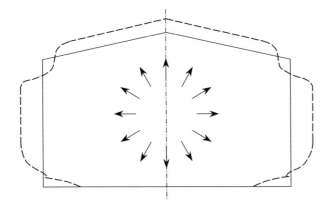

Figure 13.24 Tank subject to internal explosion.

geometries. In addition, a related study by Yoshia *et al*. (1998) examined peak strains in the bottom plate–cylinder junctions subject to seismic uplifting where the seismic uplifting is treated as being axisymmetric for simplicity.

All the above studies have been based on the premise that if the roof–cylinder junction has a lower buckling pressure than the bottom plate–cylinder junction, then the roof–cylinder junction will be the fragile link in the structure when subjected to explosion. This is only true provided a lower buckling pressure will lead to a lower final failure pressure. Since the postbuckling paths of these junctions are stable and their final failure is likely by rupture of the welded connection (Teng and Zhao 2000; Zhao and Teng 2001), there is no guarantee that a lower buckling load will lead to a lower final rupture load. A precise assessment of the rupture failure loads of these junctions requires accurate knowledge of the postbuckling behaviour and strain-based rupture failure criteria. Much further work is thus required on this topic.

Complex junctions

General

The preceding sections deal with cone large end-to-cylinder junctions under internal pressure and to a much more limited extent plate–cylinder junctions and spherical cap–cylinder junctions under internal pressure. There are many other forms of shell junctions (Fig. 13.1), for which only limited information exists on their behaviour and strength.

No study has been found on the buckling of junctions other than those covered in the preceding sections. However, several studies exist on the plastic collapse loads of some of these junctions. The earlier studies rely on the application of the plastic limit analysis and are concentrated on junctions in pressure vessels. This approach leads to upper and lower bound solutions of the classical limit load, and requires the use of simplifying assumptions. Such studies include those by Jones (1969), Davie *et al*. (1978), Taylor and Polychroni (1983), Myler and Robinson (1985) on cone–cylinder and cone–cone junctions and those by Gill (1964), Ellyin and Sherbourne (1965a, 1965b), Calladine (1969), Biron (1977) on cylinder–sphere and other forms of junctions. Experimental results (e.g. Dinno and Gill 1965; Ellyin 1969; Hayakawa *et al*. 1977) and limit loads from finite element elasto-plastic analyses (e.g. Hayakawa *et al*. 1977) have also been reported. These studies have typically produced numerical results presented in graphs or tables for simple cone–cylinder, cone–cone and sphere–cylinder junctions in pressure vessels, but no simple and general algebraic equation suitable for codification purposes has been developed. Furthermore, more complex shell junctions were not considered.

More recently, a simple general method for determining the plastic limit loads of shell junctions was developed by Teng (1998). The method (Teng 1998) is an extension of the effective area method for the plastic limit loads of silo transition

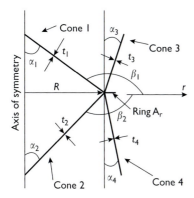

Figure 13.25 Complex shell junction.

junctions initially proposed by Rotter (1987) and subsequently improved and further elaborated by Teng and Rotter (1991a,b), and can be applied to any complex axisymmetric junctions formed form conical and cylindrical segments subject to uniform internal or external pressure (Fig. 13.25). The basic idea of the effective area method is that the radial components of the meridional forces from different shell segments forming the junction lead to an effective radial ring load which has to be resisted by circumferential compressive or tensile membrane stresses in the shell segments adjacent to the point of intersection. The plastic limit load can be found if at the point of plastic collapse, the contribution of each segment to both the resistance and the equivalent radial ring load can be defined. Details of this method are presented below.

Effective area

In this effective area method, the contribution of each shell segment to the resistance of the junction is defined in terms of an effective area (= effective length of the shell segment l_{pi} × shell thickness t_i) which is deemed to carry a circumferential membrane stress equal to the yield stress at plastic collapse. The effective area concept for plastic collapse analysis initially proposed by Rotter (1987) came from a re-interpretation of the solution of a classical limit analysis of a ring-loaded cylinder (Sawczuk and Hodge 1960; Calladine 1983). A complex junction consisting of four shell components and an annular plate ring stiffener is shown in Fig. 13.25. To define the effective area of each shell segment, the shell segments forming the junction need to be divided into two groups, those above and those below the ring stiffener. For each group with K (K is generally different for the two groups) shell segments, the equivalent thickness t_{eq} is found through

$$t_{eq} = \sqrt{\sum_{i=1}^{K} t_i^2} \qquad (16)$$

The group with a smaller equivalent thickness (t_{eqA}) is referred to as group A and the one with a larger equivalent thickness (t_{eqB}) is referred to as group B. The effective area of each shell segment for resisting the circumferential force at the junction is given by the following general expression

$$A_{pi} = l_{pi} t_i = \left(0.975 \gamma_{pi} \sqrt{\frac{Rt_i}{\cos \alpha_i}} \right) t_i \qquad (17)$$

where R is the radius of the point of intersection and α_i is the cone apex half angle of the ith shell segment (= 0 for a cylindrical segment) (Fig. 13.25), $\gamma_{pi} = 1.0$ for shell segments of the thinner group and $= 0.7 + 0.6\zeta^2 - 0.3\zeta^3$, with $\zeta = t_{eqA}/t_{eqB}$ for shell segments of the thicker group. The total circumferential force F that can be resisted by the junction is then given by

$$F = f_y \sum A_{pi} \qquad (18)$$

where f_y is the yield stress of the steel.

Equivalent radial ring load

The equivalent radial ring load at the junction contributed by each conical segment (segment i) is derived from two sources: the radial component of the meridional tension in the cone at the point of intersection and the radial component of the internal pressure on the effective compression ring (Fig. 13.26). The local pressure effect was ignored in Rotter (1987) and Teng and Rotter (1991a) when deriving

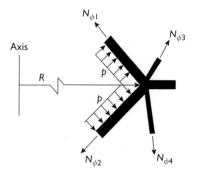

Figure 13.26 Equilibrium of an effective compression ring.

their limit load formulas for silo transition junctions, but was included by Teng (1998) in the generalised method for better accuracy. The contribution of the ith conical segment with an apex half angle α_i to the equivalent ring load is

$$q_i = N_{\phi i} \cos \beta_i + c l_{pi} p \cos \alpha_i = \frac{pR \cos \beta_i}{2 \cos \alpha_i} + c l_{pi} p \cos \alpha_i \qquad (19)$$

The meridional membrane stress resultant $N_{\phi i}$ at the point of intersection in the ith conical shell is

$$N_{\phi i} = \frac{pR}{2 \cos \alpha_i} \qquad (20)$$

The stress resultant $N_{\phi i}$ is positive when it is tensile. The internal pressure p is defined as pressure acting on the shell wall surface nearer to the axis of symmetry. The definition of β_i is illustrated in Fig. 13.25. Equilibrium consideration of the isolated effective compression ring (Fig. 13.26) requires the coefficient c in the above equation to be 1, but a different c value may lead to better results in some cases. Based on the numerical comparisons in Teng (1998), a value of 0.5 should be used for c when the local pressure effect is beneficial, and a value of 1 should be used when it is detrimental.

The total equivalent radial ring load q is the sum of the contributions from all conical segments loaded by an internal pressure p. If the total equivalent radial ring load is positive, the junction is subject to a circumferential tensile force. On the other hand, a negative q indicates that the resultant radial ring load is inward and the junction is subject to a circumferential compressive force.

Plastic limit pressure

At plastic collapse, this equivalent radial ring load should be balanced by the circumferential membrane stresses at the value of yield stress in the effective compression ring (Fig. 13.26) (assuming the ring stiffener and the shell components all have the same yield stress):

$$\sum_{i=1}^{n} pR \left(\frac{R \cos \beta_i}{2 \cos \alpha_i} + 0.975 c \gamma_{pi} \sqrt{Rt_i \cos \alpha_i} \right) = \pm f_y \left(\sum_{1}^{N} 0.975 \gamma_{pi} t_i \sqrt{\frac{Rt_i}{\cos \alpha_i}} + A_r \right) \qquad (21)$$

where n is the number of loaded shell segments, and N is the total number of shell segments forming the junction. The minus sign on the right-hand side of Eq. (21) should be used if the junction is subject to circumferential compression. The limit pressure is thus given by

$$\frac{p_1}{f_y} = \frac{\sum_{i=1}^{N} 0.975 \gamma_{pi} (t_i/R)^{3/2} \left(1/\sqrt{\cos \alpha_i} \right) + \frac{A_r}{R^2}}{\pm \sum_{i=1}^{n} \left((\cos \beta_i / 2 \cos \alpha_i) + 0.975 c \gamma_{pi} \sqrt{t_i \cos \alpha_i / R} \right)} \qquad (22)$$

Simpler formulas can be deduced for specific cases from the above general equation. For example, for an arbitrary cone–cone junction of uniform thickness with a ring stiffener, the limit pressure is given by

$$\frac{p_1 R}{f_y t}$$

$$= \frac{0.975\sqrt{\frac{t}{R}}\left(\left(1/\sqrt{\cos\alpha_1}\right) + \left(1/\sqrt{\cos\alpha_2}\right)\right) + A_r/Rt}{\frac{1}{2}\left(\left(\cos\beta_1/\cos\alpha_1\right) + \left(\cos\beta_2/\cos\alpha_2\right)\right) + 0.975c\left(\sqrt{\cos\alpha_1} + \sqrt{\cos\alpha_2}\right)\sqrt{t/R}} \tag{23}$$

Accuracy of the effective area method

The accuracy of the effective area method was verified by Teng (1998) through extensive numerical comparisons between its predictions and plastic limit loads from finite element small deflection elastic–plastic analysis. As an example, Fig. 13.27 shows the comparison of limit loads for arbitrary cone–cone junctions with $R/t = 500$ subject to internal pressure. The top cone has an apex half angle of $20°$. The angle β_2 for the bottom cone varies from $10°$ (cone large end–cone small end junction) to $170°$ (cone large end–cone large end junction). The effective area method is seen to predict limit loads close to finite element results (Fig. 13.27). The effective area method with a coefficient $c = 0.5$ predicts the finite element results more accurately for junctions between two large ends of cones than when $c = 1$, because for these junctions the effect of the local pressure is beneficial. The finite element result for $\beta_2 = 80°$ is significantly below that from the effective area method because junction collapse is not the critical mode of collapse for this geometry. Instead, plastic collapse occurs in the bottom cone.

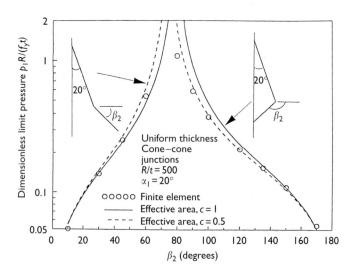

Figure 13.27 Limit loads of cone–cone junctions.

Additional remarks

The effective area method presented above can provide accurate predictions of plastic limit loads of all sorts of junctions formed from conical/cylindrical segments subject to internal pressure, including cone large end–cylinder junctions with or without a ring stiffener treated in detail in sections titled 'Uniform thickness cone large end-to-cylinder junctions' and 'Ring-stiffened cone–cylinder junctions', although Eqs (1) and (11) specifically developed for them are simpler and more accurate.

   In applying the effective area method, two things must be remembered. First, the prediction of this method is an approximation to the plastic limit load, so the effect of geometric nonlinearity has not been taken into account. For internally pressurised junctions which are the concern of this chapter, this effect is strengthening, so the method provides a safe prediction of the axisymmetric failure strength. Caution should be exercised if the effective area method is applied to junctions which are weakened by geometrically nonlinearity, for which the prediction of the effective area method is unsafe and caution should be exercised in using this prediction in design. Second, plastic collapse does not always govern the strength of junctions as they may fail by buckling when subject to circumferential compression. The possibility of buckling before the axisymmetric collapse strength is reached shall be checked by appropriate means.

Conclusions

This chapter has provided a summary of the existing knowledge of the buckling and collapse of shell junctions subject to internal pressure. The behaviour and strength of uniform thickness cone large end-to-cylinder junctions have been treated comprehensively in this chapter, based on extensive studies conducted in recent years. Other junctions covered in the chapter include sleeved or ring-stiffened cone–cylinder junctions, flat plate–cylinder junctions and junctions of curved shell segments, on which there has been only limited work. A general method for determining the plastic limit loads of complex shell junctions has also been presented.

   Axisymmetric failure controls the strength of shell junctions which are either reactively thick and/or are subject to circumferential tension. The plastic limit load provides a conservative measure of the axisymmetric failure strength of internally pressurised shell junctions, as the beneficial effect of geometric nonlinearity is not accounted for in this plastic limit load. The plastic limit load can be accurately determined for all forms of shell junctions formed from cylindrical/conical segments, using either the generalised effective area method or more accurate equations where these have been developed. For more efficient designs, the large deflection collapse load based on a limit to the stiffness loss or displacement should be used. Expressions for large deflection collapse loads have been proposed for uniform-thickness cone–cylinder junctions with or without sleeve reinforcement but not for other junctions.

Thinner shell junctions under internal pressure which leads to circumferential compression in the junction are likely to fail by non-symmetric buckling featuring periodical waves around the circumference. Such buckling has a stable postbuckling path, so buckling is not synonymous with catastrophic failure. Nevertheless, postbuckling deformations can lead to large strains at welded joints, causing weld rupture. The importance of buckling cannot thus be underestimated in thinner shells built with higher strength metals. Simple design equations are available only for uniform thickness cone–cylinder junctions.

Existing research has shown that the behaviour of shell junctions subject to internal pressure shows only limited sensitivity to geometric imperfections. Existing design equations, all based on the perfect geometry, can thus be used with confidence in design. For junctions for which simple design equations are not yet available, direct use of nonlinear and bifurcation buckling analyses based on the perfect geometry is a feasible alternative.

Acknowledgements

The author is grateful to the Hong Kong Research Grants Council and The Hong Kong Polytechnic University for their financial support in the last few years for some of the work covered in this chapter. He also wishes to thank Dr Y. Zhao for his assistance in the preparation of this chapter.

Notation

A_{pi}	plastic effective area of ith shell segment
A_r	cross-sectional area of ring
B_p	width of annular plate
E	elastic modulus
F	total circumferential compressive force in a junction
f_y	yield stress
K	number of shell segments in a given group in effective section analysis
l_{pi}	plastic effective length of ith shell segment
l_{min}	minimum sleeve length for failure confinement
l_{sc}	sleeve length in the cylinder
l_{sco}	sleeve length in the cone
N	total number of shell segments in a junction
$N_{\phi i}$	meridional tension from the ith shell segment at the point of intersection
n_{cr}	number of critical buckling waves
p	internal pressure
p_c	large deflection collapse pressure
p_{cr}	bifurcation buckling pressure
$p_{cr,r}$	bifurcation buckling pressure of ring-stiffened cone–cylinder junctions

p_{cs}	large deflection collapse pressure of locally sleeved cone–cylinder junctions
p_l	limit pressure
p_{ls}	limit pressure of locally sleeved cone–cylinder junctions
q_i	contribution to the equivalent ring load by the ith shell segment
R	radius of cylinder middle surface
R_{roof}	radius of spherical roof of a storage tank
R_{wall}	radius of cylindrical wall of a storage tank
T_p	thickness of annular plate
t	shell wall thickness
t_c	thickness of cylinder in a plate–cylinder junction
t_{eq}	equivalent thickness of shell wall
t_i	thickness of ith shell segment
t_p	thickness of plate in a plate–cylinder junction
t_{roof}	thickness of spherical roof of a storage tank
t_s	thickness of sleeved shell wall
t_{wall}	thickness of cylindrical wall of a storage tank
α	cone apex half angle; change in slope at the point of intersection between the cylindrical wall and the spherical roof of a storage tank
α_i	apex half angle of ith shell segment
β_i	angle between the shell meridian and the radial coordinate
γ_{pi}	modification coefficient for plastic effective length of ith shell segment
ζ	equivalent thickness ratio
κ	dimensionless sleeve length
λ_{sco}	linear elastic bending half-wavelength of the sleeved portion of a cone
λ_{sc}	linear elastic bending half-wavelength of the sleeved portion of a cylinder
ϕ	α for the cone; 0 for the cylinder

References

API (1993). *Welded Steel Tanks for Oil Storage*, API Standard 650, 9th edn. American Petroleum Institute, USA.

AS 1210 (1989). *Unfired Pressure Vessel Code*. Standards Australia, Sydney.

ASME (1992). *ASME Boiler and Pressure Vessel Code*. American Society of Mechanical Engineers, New York, USA.

Au, C. (1999). Analysis and design of steel egg-shaped digester. BEng Dissertation, Department of Civil and Structural Engineering, The Hong Kong Polytechnic University, Hong Kong, China.

Biron, A. (1977). Review of lower-bound limit analysis for pressure vessel junctions. *Journal of Pressure Vessel Technology, ASME*, 413–418.

BS 5500 (1994). *Specification of Unfired Fusion Welded Pressure Vessels*. British Standards Institution, London, UK.

Calladine, C.R. (1969). Lower bound analysis of symmetrically-loaded shells of revolution. *Pressure Vessel Technology, Proceedings of the First International Conference on Pressure Vessel Technology*, ASME (ed. I. Berman), Vol. 1, pp. 335–343, New York, USA.

Calladine, C.R. (1983). *Theory of Shell Structures*. Cambridge University Press, Cambridge, UK.

Davie, J., Elsharkawi, K. and Taylor, T.E. (1978). Plastic collapse pressures for conical heads of cylindrical pressure vessels and their relation to design rules in two British Standard Specifications. *International Journal of Pressure Vessels and Piping* **6**, 131–145.

Dinno, K.S. and Gill, S.S. (1965). An experimental investigation into the plastic behaviour of flush nozzles in spherical pressure vessels. *International Journal of Mechanical Sciences* **7**, 817–839.

Ellyin, F. (1969). Elastic–plastic behaviour of intersecting shells. *Journal of the Engineering Mechanics Division, ASCE* **95**(EM1), 69–94.

Ellyin, F. and Sherbourne, A.N. (1965a). Limit analysis of axisymmetric intersecting shells of revolution. *Nuclear Structural Engineering* **2**, 86–91.

Ellyin, F. and Sherbourne, A.N. (1965b). The collapse of cylinder/sphere intersecting pressure vessels. *Nuclear Structural Engineering* **2**, 169–180.

Gabriel, B. (1996). Behaviour and strength of plate-end and cone-end pressure vessels. MEngSc Thesis, Department of Civil and Systems Engineering, James Cook University, Australia.

Galletly, G.D. (1985). Torispherical shells. In *Shell Structures: Stability and Strength* (ed. R. Narayanan). Elsevier Applied Science Publishers, London, UK.

Galletly, G.D. and Blachut, J. (1985). Torispherical shells under internal pressure-failure due to asymmetric buckling or axisymmetric yielding. *Proceedings, Institution of Mechanical Engineers* **199**(C3), 225–238.

Gerdeen, J.C. (1979). A critical evaluation of plastic behaviour data, and a united definition of plastic loads for pressure components. WRC Bulletin 254, Welding Research Council, 64pp., New York, USA.

Gill, S.S. (1964). The limit pressure for a flush cylindrical nozzle in spherical pressure vessel. *International Journal of Mechanical Sciences* **6**, 105–115.

Hayakawa, T., Yoshida, T. and Mii, T. (1977). Collapse pressure for the small end of a cone-cylinder junction based on elastic–plastic analysis. *Proceedings, 3rd International Conference on Pressure Vessel Technology*, Tokyo, Japan, pp. 149–156.

Hodge, P.G. (1964). Plastic design of closed cylindrical structure. *Journal of the Mechanics and Physics of Solids* **12**, 1–10.

Jones, D.R.H. (1994). Buckling failures of pressurized vessels – two case studies. *Engineering Failure Analysis* **1**(2), 155–167.

Jones, N. (1969). The collapse pressure of a flush cylindrical nozzle intersecting a conical pressure vessel axisymmetrically. *International Journal of Mechanical Sciences* **11**, 401–415.

Jumikis, P.T. and Rotter, J.M. (1983). Buckling of simple ringbeams for bins and tanks. In *Proceedings, International Conference on Bulk Materials Storage, Handling and Transportation*, IEAust, Newcastle, Australia, August, pp. 323–328.

Mahrenholtz, J.C. (1998). Stability response of the roof to wall transition of tanks. Diploma Thesis, University of Karlsruhe, Germany.

Myler, P. and Robinson, M. (1985). Limit analysis of intersecting conical vessels. *International Journal of Pressure Vessels and Piping* **18**, 209–240.

Rotter, J.M. (1983). Effective cross-sections of ringbeams and stiffeners for bins. *Proceedings, International Conference on Bulk Materials Storage, Handling and Transportation*, IEAust, Newcastle, Australia, August, pp. 329–334.

Rotter, J.M. (1985). Analysis and design of ringbeams. In *Design of Steel Bins for the Storage of Bulk Solids*, School of Civil and Mining Engineering, University of Sydney, Sydney, Australia, pp. 164–183.

Rotter, J.M. (1987). The buckling and plastic collapse of ring stiffeners at cone/cylinder junctions. *Proceedings, International Colloquium on Stability of Plate and Shell Structures*, Gent, Belgium, April, pp. 449–456.

Save, M.A. (1984). Limit analysis and design of containment vessels. *Nuclear Engineering and Design* **79**, 343–361.

Save, M.A. and Massonnet, C.E. (1972). *Plastic Analysis and Design of Plates, Shells and Disks*, North Holland, Amsterdam, The Netherlands.

Sawczuk, A. and Hodge, P.G. (1960). Comparison of yield conditions for circular cylindrical shells. *Journal of the Franklin Institute* **269**(5), 362–374.

Schmidt, H. and Swadlo, P. (1997). Strength and stability design of unstiffened cylinder/cone/cylinder and cone/cone shell assemblies under axial compression. *Proceedings, International Conference on Carrying Capacity of Steel Shell Structures*, Brno, Czech Republic, October, pp. 361–367.

Taylor, T.E. and Polychroni, G.Y. (1983). Optimum reinforcement of uniform thickness cone–cylinder intersections in vessels subject to internal pressure. *International Journal of Pressure Vessels and Piping* **11**, 33–46.

Teng, J.G. (1994). Cone-cylinder intersection under pressure: axisymmetric failure. *Journal of Engineering Mechanics, ASCE* **120**(9), 1896–1912.

Teng, J.G. (1995). Cone–cylinder intersection under internal pressure: non-symmetric buckling. *Journal of Engineering Mechanics, ASCE* **121**(12), 1298–1305.

Teng, J.G. (1996). Buckling of thin shells: recent advances and trends. *Applied Mechanics Reviews, ASME* **49**(4), 263–274.

Teng, J.G. (1996a). Elastic buckling of cone–cylinder intersection under localized circumferential compression. *Engineering Structures* **18**(1), 41–48.

Teng, J.G. (1998). Collapse strength of complex metal shell intersections by the effective area method. *Journal of Pressure Vessel Technology, ASME* **120**, 217–222.

Teng, J.G. and Gabriel, B. (1994). Collapse and buckling of internally-pressurized cone–cylinder intersections. *Proceedings, Australasian Structural Engineering Conference-94*, Sydney, Australia, September, pp. 989–997.

Teng, J.G. and Gabriel, B. (1998). Sleeved cone–cylinder intersection under internal pressure. *Journal of Engineering Mechanics, ASCE* **124**(9), 971–980.

Teng, J.G. and Luo, Y.F. (1997). Post-collapse bifurcation analysis of shells of revolution by the accumulated arc-length method. *International Journal for Numerical Methods in Engineering* **40**(13), 2369–2383.

Teng, J.G. and Ma, H.W. (1999). Elastic buckling of ring-stiffened cone–cylinder intersections under internal pressure. *International Journal of Mechanical Sciences* **41**, 1357–1383.

Teng, J.G. and Rotter, J.M. (1989a). Elastic–plastic large deflection analysis of axisymmetric shells. *Computers and Structures* **31**(2), 211–235.

Teng, J.G. and Rotter, J.M. (1989b). Non-symmetric bifurcation of geometrically non-linear elastic–plastic axisymmetric shells subject to combined loads including torsion. *Computers and Structures* **32**(2), 453–477.

Teng, J.G. and Rotter, J.M. (1989c). Non-symmetric buckling of plate-end pressure vessels. *Journal of Pressure Vessel Technology, ASME* **111**(3), 304–311.

Teng, J.G. and Rotter, J.M. (1991a). Collapse behaviour and strength of steel silo transition junctions – Part I. Collapse mechanics. *Journal of Structural Engineering, ASCE* **117**(12), 3587–3604.

Teng, J.G. and Rotter, J.M. (1991b). Collapse behaviour and strength of steel silo transition junctions – Part II. Parametric study. *Journal of Structural Engineering, ASCE* **117**(12), 3605–3622.

Teng, J.G. and Zhao, Y. (2000). On the buckling failure of a pressure vessel with a conical end. *Engineering Failure Analysis* **7**(4), 261–280.

Uddin, M.W. (1987). Large deflection analysis of plate-end pressure vessels. *International Journal of Pressure Vessels and Piping* **29**, 47–65.

Yoshida, S. (1995). Elastic–plastic buckling of uplifted shell-to-bottom joint of oil storage tanks under internal pressure. *Proceedings, 1995 Joint ASME/JSME Pressure Vessel and Piping Conference*, PVP, Vol. 305, pp. 41–47.

Yoshida, S. (1999). Bifurcation buckling of aboveground oil storage tanks under internal pressure. *Advances in Steel Structures, Proceedings of the Second International Conference on Advances in Steel Structures*, (eds S.L. Chan and J.G. Teng). Elsevier Science Ltd, London, UK, pp. 671–678.

Yoshida, S. and Miyoshi, T. (1992). Bifurcation buckling of the top end closure of oil storage tanks under internal pressure. *Proceedings, 1992 ASME Pressure Vessel and Piping Conference*, PVP, Vol. 230, pp. 111–115.

Yoshida, S. and Tomiya, M. (1998). Elastic–plastic buckling analysis of the uplifted shell-to-bottom joint of internally pressurized oil storage tanks using axisymmetric shell finite element method. *Proceedings, 1998 Joint ASME/JSME Pressure Vessel and Piping Conference*, PVP, Vol. 370, pp. 121–128.

Yoshida, S., Kawano, K. and Tomiya, M. (1998). Analysis of uplift behaviour of the shell-to-bottom joint of a seismically-loaded oil storage tank using axisymmetric solid finite element method. *Proceedings, 1998 Joint ASME/JSME Pressure Vessel and Piping Conference*, PVP, Vol. 370, pp. 181–186.

Zhao, Y. and Teng, J.G. (2001). Buckling experiments on cone–cylinder intersections under internal pressure. *Journal of Engineering Mechanics, ASCE* **127**(12), 1231–1239.

Chapter 14

Rings at shell junctions

J.G. Teng and Y. Zhao

Introduction

Axisymmetric shell junctions are subject to a large circumferential force due to slope mismatches between the adjacent shell segments. A ring is often provided at these junctions to help resist this circumferential force. Such ring-stiffened shell junctions may fail by buckling if this circumferential force is compressive. This buckling behaviour is likely to be dominated by the ring, with the adjacent shell segments playing the role of restraining the buckling deformation of the ring. This behaviour is distinctly different from that of shell junctions without a ring extensively treated in Chapter 13.

Figure 14.1 shows a typical elevated steel silo which consists of a cylindrical vessel above a conical discharge hopper supported on a cylindrical skirt. The point of intersection between the cone and the cylinder is commonly referred to as the transition, and the intersection formed by the shell segments the transition junction. The ring at the transition junction of an elevated steel silo is a typical example of rings at shell junctions. The structural form of an elevated steel liquid storage tank is similar. In Japan, research has been carried out on the design of

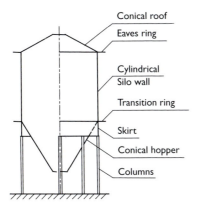

Figure 14.1 Typical steel silo.

double-decker tanks for the storage of oil products (Yoshida *et al.* 1998), and the proposed structural form features exactly the same transition junction, but with a shallower cone.

This chapter deals with the buckling of junction rings, based primarily on existing research conducted for transition rings in elevated steel silos uniformly supported around the circumference. For simplicity of presentation, the material of the chapter is generally presented with explicit reference to steel silos only, except in the penultimate section, where the application of the knowledge obtained from research on steel silo transition rings to the design of rings of other shell junctions is examined.

Cross-sectional forms of rings

The simplest ring is an annular plate (Fig. 14.2(a)). Simple annular plate rings are susceptible to out-of-plane buckling failure. To increase the out-of-plane resistance, a stiffened annular plate in the form of a T-section (Fig. 14.2(b)) or an angle section (Fig. 14.2(c)) can be used instead. In the extreme case, an eccentrically placed vertical stiffener may be added to an annular plate (Fig. 14.2(d)). The vertical stiffener increases both the out-of-plane bending and the torsional resistance of the annular plate ring, resulting in enhancement in buckling strength.

An alternative to the use of an edge stiffener is to employ an annular plate of small width-to-thickness ratio. For large elevated steel silos and tanks, this may lead to the use of very thick plates, which can cause fabrication inconvenience or difficulty. Therefore, the use of thinner plate elements is often required. Indeed, the use of a ring instead of thickening the shell segments of the junction is generally due to this same consideration. The buckling behaviour of shell junctions without a ring is dealt with in Chapter 13.

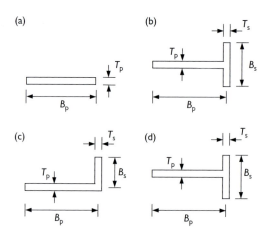

Figure 14.2 Ring cross-sectional forms: (a) annular plate ring; (b) T-section ring; (c) angle section ring; (d) eccentrically stiffened ring.

Failure modes

General

For an isolated ring of arbitrary cross-section, buckling involves simultaneous in-plane flexure, out-of-plane flexure and torsion. However, if the ring section possesses an axis of symmetry lying in the plane of the ring, then in-plane buckling of the ring, involving only periodical flexural deformations within the plane of the ring, is uncoupled from the out-of-plane bucking of the ring involving out-of-plane flexure and torsion. Classical stability solutions for both in-plane buckling (Timoshenko and Gere 1961; Smith and Simitses 1969) and out-of-plane buckling (Timoshenko and Gere 1961; Goldberg and Bogdanoff 1962; Wah 1967; Trahair and Papangelis 1987; Teng and Rotter 1987) of isolated rings are available, but these cannot generally be applied to predict the buckling load of a ring at a steel silo transition junction, as the adjacent shell walls provide strong restraints to the ring during buckling.

For rings at steel silo transition junctions, both in-plane and out-of-plane bucklings are again possible, with the buckling deformations being dependent on the restraints provided by the adjacent shell walls. Buckling, however, may not be the critical failure mode when a stocky ring is used. Instead, the ring together with the adjacent shell segments may fail by plastic collapse. In addition to the above three main failure modes, two secondary failure modes are also possible: local buckling of the ring when the ring section is made of slender plate/shell elements and buckling of the adjacent shell segments when the ring is stocky but the shell segments are thin.

In-plane buckling of ring

In-plane buckling is the only form of buckling recognised in earlier design guides for steel silos and tanks (Wozniak 1979; Gaylord and Gaylord 1984). However, in-plane buckling of rings at steel silo transition junctions appears to be almost always prevented by the membrane stiffness of the cone and the cylinder (Jumikis and Rotter 1983) as far as elastic buckling is concerned. In some situations, in-plane buckling may become important. These include rings on junctions consisting of shell segments with steep meridians and junctions with extensive yielding. In these situations, the risk of in-plane buckling is increased by the low radial resistance the shell segments offer to the ring.

Out-of-plane buckling of ring

For rings at steel silo transition junctions, out-of-plane buckling is generally the critical failure mode, and has been the subject of extensive recent research. This chapter is thus concerned only with the out-of-plane buckling mode of transition rings. Out-of-plane buckling usually involves twisting deformations of the ring

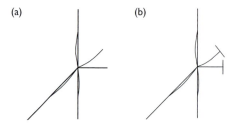

Figure 14.3 Out-of-plane buckling mode: (a) annular plate ring; (b) T-section ring.

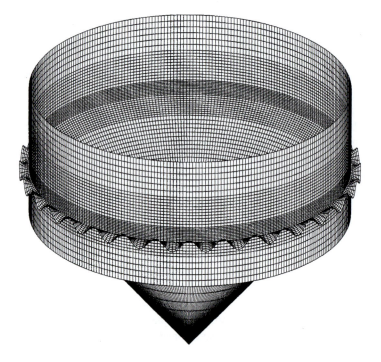

Figure 14.4 Three-dimensional plot of transition ring buckling.

about the point of attachment (Fig. 14.3) and the buckling deformations assume many waves around the circumference (Fig. 14.4). Significant cross-sectional distortion may be involved (Fig. 14.3).

Plastic collapse of junction

If a stocky ring is used so that its buckling load is high, then the junction will fail by plastic collapse. The failure mode is characterised by excessive axisymmetric

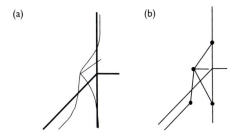

Figure 14.5 Plastic collapse of silo transition junction: (a) plastic collapse mode; (b) idealized plastic collapse mechanism.

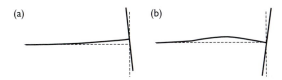

Figure 14.6 Buckling modes of clamped T-section rings: (a) distortional buckling; (b) local buckling.

deformations of the junction, with the formation of plastic hinge circles, one at the transition, and one each inside each adjacent shell segment (Fig. 14.5).

Local buckling of ring section

If the ring is made of slender plate/shell elements, it is possible for the plate/shell elements to suffer local buckling before the out-of-plane buckling load of the ring is reached. Figure 14.6(b) shows a local buckling mode of a T-section ring. Until further research is available to allow for a more liberal approach, the best current advice is to limit the slenderness of the plate/shell elements to avoid local buckling. Teng and Chan (1999a) has developed a chart which gives these limits for a T-section ring.

Adjacent shell wall buckling

The shell walls adjacent to the ring may buckle instead of the ring if the ring is stocky but the shell segments are thin. This mode generally involves buck-ling deformations in the cone, and can only occur if the cone is rather shallow. Figure 14.7 shows such a shell buckling mode for a rather shallow cone briefly discussed in Teng and Barbagallo (1997). As a result, it is an unlikely failure mode in steel silos for which the hopper is generally not shallow enough for this mode to become important but can be important in tanks which may have a shallow conical

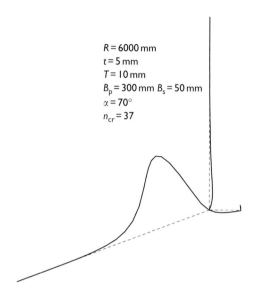

$R = 6000\,mm$
$t = 5\,mm$
$T = 10\,mm$
$B_p = 300\,mm$ $B_s = 50\,mm$
$\alpha = 70°$
$n_{cr} = 37$

Figure 14.7 Shell buckling mode.

enclosure. An exception is steel silos with an expanded flow hopper which starts from the transition with a shallow truncated cone followed by a steeper cone.

Circumferential compression in a junction

A steel silo transition junction is subjected to non-uniform internal pressures and corresponding downward frictional tractions (Fig. 14.8). For the development of strength rules for the transition junction, a simple way of characterizing this complex loading has to be used. As far as ring buckling or junction collapse failures are concerned, the controlling force is the circumferential compressive force carried by the junction. This compressive force is derived chiefly from the radial component of the meridional tension at the top of the hopper. Denoting the meridional tensile force at the top of hopper as $N_{\phi h}$ (Fig. 14.8), the circumferential compressive force F_0 due to the hopper meridional tension is given by

$$F_0 = N_{\phi h} R \sin \alpha \qquad (1)$$

in which R is the radius of the cylinder and α is the cone apex half angle. This value is only approximate because the effect of the local internal pressures on the cylindrical and conical shell walls adjacent to the transition has been conservatively ignored (Fig. 14.8).

This compressive force is resisted by the ring acting with a short segment of each shell wall forming the transition junction. The section formed by the ring and short

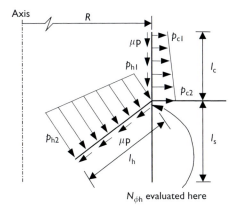

Figure 14.8 Effective section and local pressures.

segments of the shell walls is called the effective section of the ring (Fig. 14.8)
(Rotter 1985). Details of how this effective section can be defined is presented in the
next section. Once the effective section is defined, a more accurate assessment of
the circumferential compressive force F taking into account the effect of pressures
on the effective ring is given by (Rotter 1985)

$$F = N_{\phi h} R \sin \alpha - 0.5(p_{c1} + p_{c2})l_c R - 0.5(p_{h1} + p_{h2})(\cos \alpha - \mu \sin \alpha)l_h R$$

$$(2)$$

in which l_c and l_h are the effective lengths of the cylinder and the hopper, respectively, and the pressure values p_{c1}, p_{c2}, p_{h1} and p_{h2} are as defined in Fig. 14.8.
Although the variation of pressure p is generally nonlinear, a linear approximation over the effective lengths can be used without much loss of accuracy as the
effective lengths are generally small compared to the total lengths of the shell
walls. It should be noted that the effective lengths l_c and l_h are different for elastic and plastic collapse analyses, but for design use, the elastic effective lengths
can be used in Eq. (2) for evaluating the circumferential compressive force F
in both elastic and plastic collapse analyses as this only leads to small errors
(Teng 1997).

The circumferential compressive force F is an important parameter, as for a
given geometry, the failure strength in terms of F is largely independent of the
specific load distributions on the silo walls (Rotter 1987; Greiner and Ofner 1991;
Teng and Rotter 1991a–c; Teng 1997). It may be noted that the definition of
the circumferential compressive force above follows the same principle as that
used in the plastic effective section analysis in the section 'Complex junctions' of
Chapter 13.

Elastic effective section analysis

Circumferential membrane stress in the ring

Figure 14.9 shows the distribution of prebuckling circumferential compressive membrane stress in a typical transition junction with a T-section ring for two cases: one without a skirt (Fig. 14.9(a)) and the other with a skirt (Fig. 14.9(b)) (Teng and Chan 2000). Figure 14.9(a) shows that the T-section ring is under nearly uniform circumferential compression, with the maximum compressive stress being at the inner edge of the ring. Figure 14.9(b) shows a somewhat different situation: while the annular plate of the ring is under nearly uniform compression with its maximum circumferential membrane stress at the inner edge, the maximum circumferential membrane stress for the entire ring cross-section occurs at the top of the edge stiffener. The reason for this difference lies in the unbalanced meridional bending stiffnesses of the shell walls above and below the ring. For the case of Fig. 14.9(a), the cylinder and the hopper have the same thickness, so the meridional bending stiffness of the cylinder is similar to that of the hopper, requiring the T-section to carry little torsional loading. For the case of Fig. 14.9(b), since the combined bending stiffness of the hopper and the skirt exceeds that of the cylinder considerably, the junction has to accommodate this imbalance by anti-clockwise rotation during radial deformations. As a result, the ring is loaded by an anti-clockwise axisymmetric torque, leading to bending actions of the ring about the radial axis. These bending actions are responsible for the linear variation of the circumferential membrane stress over the height of the stiffener. As the annular plate lies on the horizontal centroidal axis of the T-section ring, the membrane stresses in the annular plate are not affected by this torsional deformation of the ring.

In all existing studies on elastic buckling of transition rings using the finite element method, the elastic buckling strength of rings is characterised by the inner

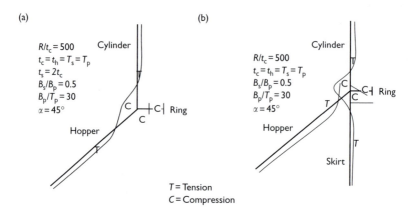

(a)

$R/t_c = 500$
$t_c = t_h = T_s = T_p$
$t_s = 2t_c$
$B_s/B_p = 0.5$
$B_p/T_p = 30$
$\alpha = 45°$

Cylinder

Hopper

C⊣ Ring

(b)

$R/t_c = 500$
$t_c = t_h = T_s = T_p$
$B_s/B_p = 0.5$
$B_p/T_p = 30$
$\alpha = 45°$

Cylinder

Hopper

C⊣ Ring

Skirt

T = Tension
C = Compression

Figure 14.9 Typical distributions of prebuckling circumferential compressive stress: (a) junction without a skirt; (b) junction with a skirt.

edge circumferential compressive stress in the annular plate (e.g. Jumikis and Rotter 1983; Rotter and Jumikis 1985). This stress can be related to the applied loading using the elastic effective section analysis developed by Rotter (1985), as explained below for both annular plate and other rings.

Annular plate rings

In an annular plate transition ring, the circumferential compressive stress varies across the annular plate and reaches its maximum value at the inner edge. The value of this maximum stress under elastic conditions may be determined accurately using the simple elastic effective section analysis proposed by Rotter (1985). The maximum circumferential compressive stress is given by

$$\sigma_\theta = \frac{F}{(1 + 0.3 B_p/R) A_e} \tag{3}$$

where the total effective area A_e is defined by

$$A_e = \frac{B_p T_p}{1 + 0.8 B_p/R} + \sum_{i=1}^{N} l_{ei} t_i \tag{4}$$

where N is the number of shell segments present at the junction, B_p and T_p are the width and thickness of the annular plate, respectively, and l_{ei} and t_i are the elastic effective length and thickness of the ith shell segment, respectively. To determine the effective length l_{ei} for each shell segment, it is necessary to first separate the shell segments into two groups, those above and those below the annular plate. For each group with K shell segments, the equivalent thickness of the group is defined as

$$t_{eq} = \sqrt{\sum_{i=1}^{K} t_i^2} \tag{5}$$

Denoting the thinner group as group A (the group with a smaller equivalent thickness) and the thicker one group B, the equivalent thickness ratio ζ is then found as

$$\zeta = t_{eqA}/t_{eqB} \tag{6}$$

The effective length of each shell segment is given by

$$l_{ei} = 0.778 \gamma_{ei} \sqrt{R t_i / \cos \phi} \tag{7}$$

where $\phi = 0$ for each cylindrical component and $\phi = \alpha$ for the hopper, and the appropriate value of γ_{ei} for the ith shell segment is

$$\gamma_{ei} = 1 \qquad \qquad \text{for the thinner group} \tag{8}$$

$$\gamma_{ei} = \gamma_e = 0.5(1 + 3\zeta^2 - 2\zeta^3) \qquad \text{for the thicker group} \tag{9}$$

It may be noted that the definition of the elastic effective section above is similar to the definition of the plastic effective section given in the section 'Complex junctions' of Chapter 13, although the actual values of the effective lengths differ under the two conditions.

Other ring sections

The effective section method initially proposed by Rotter (1985) for junctions with an annular plate ring as described above has also been applied by Teng and Chan (2000) with some minor modifications to evaluate the inner edge compressive stress of a T-section transition ring. The modifications involve the omission of terms of the order of B_p/R in comparison to 1. With such modifications, the inner edge circumferential compressive stress in a T-section transition ring may be found as

$$\sigma_\theta = \frac{F}{A_e} \tag{10}$$

where the total effective area A_e for elastic stress analysis is defined by

$$A_e = A_r + \sum_{i=1}^{N} l_{ei} t_i \tag{11}$$

where A_r is the cross-sectional area of the ring ($=B_p T_p + B_s T_s$). The same approach as that for annular plates is used to determine the effective length l_{ei} for each shell segment.

This modified approach was found by Teng and Chan (2000) to predict the inner edge compressive stress closely by comparison with finite element predictions. The accuracy of this approach has not been examined for other ring sections, although the same formula might be expected to provide reasonable predictions for angle and eccentrically stiffened rings. If the variation of the circumferential membrane stress over the entire ring cross-section is required, the more sophisticated method developed by Chen and Rotter (1998) should be used.

Classical buckling analysis

General

Classical solutions for out-of-plane buckling of rings based on the thin-walled member theory have been developed by a number of researchers. Most existing solutions of this type consider only unrestrained rings of monosymmetric section with the axis of symmetry lying in the plane of the ring, and with the load and/or restraint located on this symmetry axis. For such rings, out-of-plane buckling is uncoupled from in-plane buckling, so out-of-plane buckling can be studied independently. These include Timoshenko and Gere (1961), Goldberg and Bogdanoff (1962), Wah (1967), Trahair and Papangelis (1987) and Teng and Rotter (1987).

There have been three studies which considered out-of-plane buckling of restrained rings. The earliest of these is that by Cheney (1963) which dealt with the buckling of restrained general open section rings. Cheney's (1963) study is restricted to centroidal loading. Teng and Rotter (1988) developed a closed-form solution based on the thin-walled member theory for the buckling of mono-symmetric rings considering the restraint and non-centroidal loading conditions as found at steel silo transition junctions (Fig. 14.1). They also carried out numerical studies of ring buckling behaviour using both their closed-form solution and a linear shell buckling analysis. Teng and Lucas (1994) further extended the work of Teng and Rotter (1988) to the case of general open section. The description given here is based on the studies of Teng and Lucas (1994) and Teng and Rotter (1988).

Teng and Lucas' solution

This solution is for an open section ring of arbitrary form rigidly restrained against in-plane and out-of-plane translations at a single point and elastically restrained at the same point against out-of-plane rotation (Fig. 14.10). The solution models the ring as a thin-walled member, adopting the common assumptions of thin-walled member theory. In addition, the ring is assumed to be thin with a cross-section radial width-to-radius ratio less than about 0.1, so the prebuckling circumferential stresses are taken as uniform. The direction of the load is assumed to remain constant. These assumptions are generally appropriate for transition rings provided that substantial distortion of the ring section does not occur during buckling.

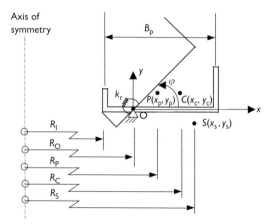

Figure 14.10 Restrained ring geometry and buckling mode.

Under these assumptions, out-of-plane buckling of the ring occurs at the following stress (Teng and Lucas 1994)

$$\frac{\sigma_0}{E} = \left\{ I_x(1 - \beta_s n^2)^2 + I_y n^2 \frac{y_s^2}{R_c^2} + 2I_{xy} n^2 \frac{y_s}{R_c}(2 - \beta_s n^2) \right.$$

$$\left. + \frac{GJ}{E} n^2 (1 - \beta_s)^2 + \frac{I_w n^4}{R_c^2}(1 - \beta_s)^2 + \frac{k_r R_0 R_c}{E} \right\} \Big/ A[r_0^2 n^2 + R_c(x_p - x_c)]$$

$$\tag{12}$$

$$\beta_s = \frac{x_s}{R_c} \tag{13}$$

$$r_0^2 = \frac{I_x + I_y}{A_r} + x_c^2 + y_c^2 \tag{14}$$

where E is the elastic modulus, A_r the ring cross-sectional area, B the ring width, R_c and R_0 the radii of the ring centroid and restraint, respectively, I_x and I_y the second moments of area about the radial and vertical axes of the section, respectively, I_{xy} the product moment of area, G the shear modulus, J the torsional constant, I_w the warping constant, k_r the stiffness of the rotational restraint, x_c and y_c the coordinates of the centroid while x_s and y_s are the coordinates of the shear centre, x_p the x coordinate of the load position and n the number of waves that the ring buckles into. The dimensionless buckling stress, σ_0/E, is a function of the circumferential buckling wave number n for a given ring geometry. A trial and error procedure needs to be applied to find the critical wave number n_{cr} (i.e. critical buckling mode) which gives the minimum buckling stress, and this minimum buckling stress is the critical buckling stress $\sigma_{\theta cr}$.

For a monosymmetric open section ring, $y_s = I_{xy} = 0$, so

$$\frac{\sigma_0}{E} = \frac{I_x(1 - \beta_s n^2)^2 + \frac{GJ}{E} n^2 (1 - \beta_s)^2 + \frac{I_w n^4}{R_c^2}(1 - \beta_s)^2 + \frac{k_r R_0 R_c}{E}}{A[r_0^2 n^2 + R_c(x_p - x_c)]} \tag{15}$$

Equation (15) is the solution derived by Teng and Rotter (1988) for restrained monosymmetric rings.

To apply Eq. (14) or (15), the rotational stiffness k_r provided by the adjacent shell segments needs to be determined. For a general shell junction of radius R where N conical shell segments of different thicknesses t_i and cone apex angles α_i meet (including cylinders, for which $\alpha_i = 0°$), the rotational stiffness can be accurately approximated by (Teng and Barbagallo 1997)

$$k_r = 0.5 \sum_{i=1}^{N} k_{ni}(t_i, \alpha_i) \tag{16}$$

where

$$\frac{k_{ni}}{k_{oi}} = 0.9 \qquad \text{if } \frac{n}{\sqrt{R \cos \alpha_i/t_i}} < 1.18 \tag{17}$$

$$\frac{k_{ni}}{k_{oi}} = \frac{0.76n}{\sqrt{R \cos \alpha_i/t_i}} \qquad \text{if } \frac{n}{\sqrt{R \cos \alpha_i/t_i}} > 1.18 \tag{18}$$

$$k_{oi} = \frac{E t_i^{2.5}}{[3(1 - \nu^2)]^{0.75}\sqrt{R/\cos \alpha_i}} \tag{19}$$

where ν is the Poisson's ratio. For practical design, the following simple expression is sufficient (Teng and Barbagallo 1997):

$$k_{ni} = 0.9k_{oi} \tag{20}$$

Simplified expressions for simply supported T- and angle section rings

To obtain the buckling stress and buckling mode from Eq. (12) or (15), a trial and error procedure needs to be employed to find the critical circumferential mode which gives the critical buckling stress. This process presents some inconvenience if the solution is to be used in practical design. A simplified expression relating the buckling stress to the ring material and geometric properties only is therefore desirable, especially for the practically important T- and angle section rings simply supported and loaded at the inner edge.

For such rings of both T- and angle sections, $x_s = B_p$, $y_s = 0$, $k_r = 0$, $I_w = 0$ and $x_p = 0$, so Eq. (12) reduces to

$$\frac{\sigma_0}{E} = \frac{I_x(1 - \beta n^2)^2 + (GJn^2/E)(1 - \beta)^2}{A_r r_0^2 n^2 - A_r R_c x_c} \tag{21}$$

where

$$\beta = \frac{B_p}{R_c} \tag{22}$$

Using the notation

$$c_1 = \frac{GJ}{EI_x} \tag{23}$$

$$c_2 = \frac{B_p x_c}{r_0^2} \tag{24}$$

Eq. (21) becomes

$$\sigma_0 = \frac{EI_x}{A_r r_0^2}\left[\frac{\beta(1 - \beta n^2)^2 + c_1\beta n^2(1 - \beta)^2}{\beta n^2 - c_2}\right] \tag{25}$$

The critical mode and critical stress can then be obtained by minimizing the buckling stress given in Eq. (25) with respect to the buckling wave number, n. The critical wave number is therefore given by

$$\beta n_{cr}^2 = c_2 \left[1 + \sqrt{\left(1 - \frac{1}{c_2}\right)^2 + \frac{c_1}{\beta c_2}(1 - \beta)^2} \right] = c_3 \tag{26}$$

Introducing Eq. (26) into Eq. (25) leads to

$$\sigma_{\theta cr} = \frac{E I_x}{A_r r_0^2} \left[\frac{\beta(1 - c_3)^2 + c_1 c_3 (1 - \beta)^2}{c_3 - c_2} \right] \tag{27}$$

It is worth noting that Eq. (27) is not only valid for T- and angle section rings but also for eccentrically stiffened rings (Fig. 14.2(d)).

Further simplifications for thin T-section rings

Equation (27) is still rather tedious, so Teng and Rotter (1988) attempted to simplify the solution further by noting that for thin rings for which β is small, it may be assumed that

$$\frac{c_1(1 - \beta)^2}{c_2 \beta} \approx \frac{c_1}{c_2 \beta} \tag{28}$$

and that

$$\left(1 - \frac{1}{c_2}\right)^2 \approx 0 \tag{29}$$

for T-sections that satisfy the following conditions:

$$\frac{B_s T_s}{B_p T_p} > 0.3 \quad \text{and} \quad \frac{B_s}{B_p} < 1.5 \tag{30}$$

With these approximations,

$$c_3 = c_2 + \sqrt{\frac{c_1 c_2}{\beta}} \tag{31}$$

and the buckling stress can be expressed as

$$\sigma_{\theta cr} = \frac{E I_x \sqrt{\beta}}{A_r r_0^2} \left[\frac{\beta(1 - c_3)^2 + c_1 c_3 (1 - \beta)^2}{\sqrt{c_1 c_2}} \right] \tag{32}$$

Finite element buckling analysis

Method of analysis

While the classical buckling theory provides useful information for lightly restrained rings, it becomes less satisfactory as the shell restraint and hence cross-sectional distortion increases. An accurate analysis of such rings can be carried out using the finite element method in which both the ring and the shell walls are modelled using axisymmetric shell elements. The finite element analysis, being based on a thin shell buckling theory, does not suffer from the limitations of the thin-walled member theory. All the design equations presented in this chapter have been developed based on finite element results using the NEPAS program (Teng and Rotter 1989a) or its earlier linear elastic buckling analysis version (Jumikis 1987). The NEPAS program can perform linear and nonlinear bifurcation buckling analysis for elastic–plastic branched axisymmetric shells using a doubly curved isoparametric axisymmetric shell element. A brief summary of the bifurcation analysis procedure can be found in the second section of Chapter 13. Unless otherwise stated, the finite element results presented here are from a small deflection prebuckling analysis, as the effect of prebuckling geometric nonlinearity is small and generally strengthening for this buckling problem (Teng 1997; Teng and Chan 2000, 2001). This means that if buckling occurs in the elastic range, the buckling load presented here is from a linear elastic bifurcation buckling analysis. All plastic buckling loads presented here were obtained using the J_2 deformation theory in the plastic bifurcation analysis, as this offers the most conservative answer, although the differences between the three different plasticity options available in NEPAS (Teng and Rotter 1989a) are not large (Teng 1997; Teng and Chan 2001). The prebuckling behaviour was predicted using the J_2 flow theory as usual.

When the NEPAS program or a similar analysis is applied to predict the buckling stress of isolated rings, it is important to ensure that the prebuckling and buckling restraint conditions are properly specified when the predicted buckling load is compared with that from the classical buckling theory. More specifically, for an inner edge simply supported ring, the prebuckling restraint conditions should allow only radial inward movement of the ring, while during buckling, the inner edge should be restrained against radial and vertical translations. This means that in the case of a simply supported angle section ring, a vertical support and a meridional rotational restraint should be provided during prebuckling analysis so that nearly uniform circumferential compression is induced in the ring section. If the meridional rotational restraint is relaxed, significant rotational deformations result, leading to a linear variation of the circumferential compressive membrane stress over the section height.

It is also worth noting that in all existing finite element studies using NEPAS or its earlier linear elastic version, the circumferential compressive stress at the innermost sampling point of Gauss quadrature was taken to be the inner edge circumferential compressive stress as stresses were output only at Gauss points.

The error so introduced was very small, as the innermost Gauss point was very close to the inner edge. This can be seen in Figs 14.9(a) and (b) where the circumferential compressive stress in the annular plate is plotted from the innermost to the outermost Gauss point.

Finite element assessment of classical solutions

Figure 14.11(a) shows a comparison of the buckling stress between the finite element analysis and the classical solution for a T-section ring simply supported at

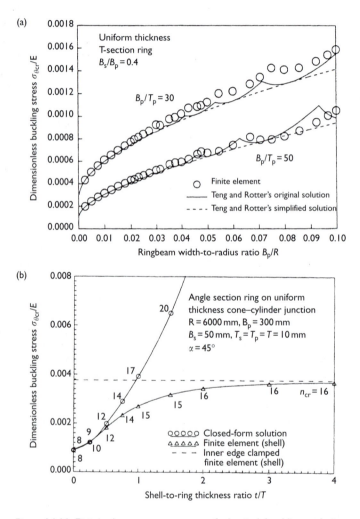

Figure 14.11 Finite element assessment of classical buckling solutions: (a) simply supported T-section ring; (b) angle section ring restrained by shell walls.

the inner edge (Teng and Chan 1999b). The classical buckling stresses from both the original solution (Eq. 15) and the simplified solution (Eq. 32) are shown in Fig. 14.11(a). The curve of the buckling stress predicted by the simplified solution touches the bottom of each cusp of the original solution. Equation (32) is thus a satisfactory simplified version. A very close agreement between the classical solution and the finite element analysis is observed particularly at low β values. For thicker rings (rings with larger β values), the finite element results may be either overestimated or underestimated by the classical solution. The difference is mainly due to cross-section distortion rather than anything else. A similar comparison for a simply supported angle section ring (Teng and Lucas 1994) also found close agreement between the classical solution and the finite element analysis.

Figure 14.11(b) shows a comparison of the buckling stress between the finite element analysis and the classical solution (Eq. 12) for an angle section ring restrained by shell walls (Teng and Barbagallo 1997). For this comparison, a radial ring load was applied at the ring inner edge. When the ratio between the shell wall thickness t and the ring plate thickness T is small (not greater than 0.5), the classical solution matches the finite element analysis well. With further increase in shell wall thickness, the classical solution predicts fast increases in buckling strength, which are, however, not supported by the finite element analysis. As the restraint becomes heavier, the increase in the finite element buckling strength gradually slows down. The finite element analysis predicts that a heavily restrained ring has a strength close to that of a ring which is clamped at its inner edge. The difference in the predicted behaviour from the two approaches stems from the fact that the classical solution is based on the assumptions of thin-walled member theory, and is thus unable to take cross-section distortion into account. The finite element buckling modes (Fig. 14.12) show clearly that cross-section distortion becomes more severe as the rotational restraint becomes stronger.

Remarks

The above comparisons between finite element and classical solutions show that the classical solution provides a satisfactory prediction if the inner edge support provides only a light elastic rotational restraint. If, however, the elastic rotational restraint is significant, then the classical solution overestimates the buckling strength.

Before the development of the classical solution presented in the previous section Jumikis and Rotter (1983) adopted a different approach for the development of a design approximation for the elastic buckling strength of annular plate transition rings. They modelled both the ring and the adjacent shell walls using an axisymmetric shell element, so the assumption of a rigid ring cross-section was not made. To establish a simple design approximation, they first established the buckling strengths of two limiting cases: inner edge clamped annular plates and inner edge simply supported annular plates. The elastic buckling strength of an annular plate

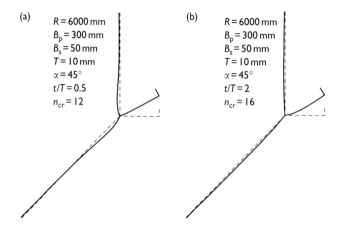

(a)
R = 6000 mm
B_p = 300 mm
B_s = 50 mm
T = 10 mm
α = 45°
t/T = 0.5
n_{cr} = 12

(b)
R = 6000 mm
B_p = 300 mm
B_s = 50 mm
T = 10 mm
α = 45°
t/T = 2
n_{cr} = 16

Figure 14.12 Effect of shell wall restraint on ring buckling mode: (a) flexural–torsional buckling; (b) distortional buckling.

ring with elastic shell wall restraints was then found by a suitable interpolation of the buckling strengths for the two limiting cases. This approach has also been followed by Teng and Chan (2000) in developing a design approximation for the elastic buckling strength of T-section rings, where the elastic buckling strength for the limiting case of a simply supported inner edge is predicted using the simplified classical solution (Eq. 32) with further simplifications. Design approximations for the elastic buckling strengths of rings are presented in the following two sections.

Elastic buckling strength of annular plate rings

Elastic out-of-plane buckling of annular plate rings was first studied by Jumikis and Rotter (1983) and later by Sharma et al. (1987) employing a linear elastic finite element shell buckling analysis. The following simple design equations were developed by Jumikis and Rotter (1983) from their study:

$$\sigma_{\theta cr} = cE \left(\frac{T_p}{B_p} \right)^2 \tag{33}$$

where c can be found from

$$c = \frac{\eta_s c_s + \eta_c c_c}{\eta_s + \eta_c} \tag{34}$$

with

$$c_s = 0.385 + 0.452\sqrt{B_p/R} \tag{35}$$

$$c_c = 1.154 + 0.560(B_p/R) \tag{36}$$

$$\eta_s = 0.43 + \frac{(R/B_p)^2}{4000} \tag{37}$$

$$\eta_c = \frac{1}{2} \left[\frac{\sum_{i=1}^{N} t_i^{2.5}}{T_p^{2.5}} \right] \tag{38}$$

Equation (33) describes the elastic buckling strength of a plate of width-to-thickness ratio B_p/T_p. For a long straight plate with one long edge free, the value of c is 0.385 if the other long edge is simply supported and 1.154 if fixed. The modifications to these coefficients arising from ring curvature are given by $0.452\sqrt{B_p/R}$ and $0.560 B_p/R$, respectively, so that c_s and c_c represent the buckling strength coefficients for simply supported and fixed annular plate rings respectively. Equation (34) represents an accurate, though semi-empirical, assessment of the elastic restraint provided by the cylinder, hopper and skirt against ring buckling. These equations have been included in a design guide (Trahair et al. 1983) and the new European code for the structural design of steel silos (ENV1993-4-1 1999).

The circumferential membrane stress varies slightly across the width of the ring. It should therefore be recognised that Eq. (33) defines the value of the maximum circumferential membrane stress at buckling. This maximum stress is at the inner edge of the ring, and may be found using either the effective section method described earlier (Rotter 1985) or a suitable finite element analysis.

Sharma et al. (1987) later modified the expressions η_s and η_c to

$$\eta_s = 4.22 + \frac{(R/B_p)^2}{408} \tag{39}$$

$$\eta_c = \sqrt{\frac{B_p}{T_p}} \left(\frac{\sum_{i=1}^{N} t_i^{2.5}}{T_p^{2.5}} \right) \tag{40}$$

The use of Eqs (39) and (40) instead of Eqs (37) and (38) leads only to a difference not larger than about $\pm 5\%$ of the predicted buckling stress in most cases (Teng 1997), so the simpler Eqs (37) and (38) are recommended for use here.

In terms of the circumferential compressive force in the junction, the elastic buckling strength is given by

$$F_e = \sigma_{\theta cr}(1 + 0.3 B_p/R) A_e \tag{41}$$

where A_e is the effective section area defined by Eq. (4).

Elastic buckling strength of T-section rings

Geometric limits for distortional buckling

It should be noted that a T-section ring may fail by distortional buckling (Fig. 14.6(a)) or by local buckling (Fig. 14.6(b)). The following buckling strength

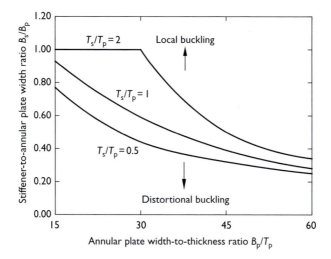

Figure 14.13 Geometric limits for distortional buckling.

formulas are for the mode of distortional buckling. The local buckling mode can be prevented by ensuring that the ring geometry satisfies the geometric limits for distortional buckling as given in the chart of Fig. 14.13 developed by Teng and Chan (1999a).

This chart was developed from a parametric study for elastic clamped T-section rings. A heavily restrained ring is more likely to fail by local buckling than a lightly restrained one, since the inner edge rotational restraint influences the flexural–torsional or distortional buckling load much more strongly than the local buckling load. Consequently, the limits set by Fig. 14.13 are also sufficient for elastic rings with a simply supported or semi-rigid inner edge. Similarly, this chart is sufficient for plastic buckling as yielding always reduces the shell wall rotational restraint first before affecting the stiffness of the ring cross-section (Teng and Rotter 1991c). Although only three values of T_s/T_p are explicitly covered in Fig. 14.13, a linear interpolation may be used for a T_s/T_p value between 0.5 and 2 but not coinciding with one of these three values. The range of T_s/T_p values covered (between 0.5 and 2) is not too restrictive as, in practice, the thicknesses of the annular plate and the stiffener are very often the same or similar. The curve for $T_s/T_p = 2$ can also be used for higher T_s/T_p values as a conservative measure.

Rotter and Jumikis' approximation for rings with a simply supported inner edge

Rotter and Jumikis (1985) developed an empirical fit to their finite element results for the linear elastic buckling of inner edge simply supported T-section rings as follows

$$\sigma_{\theta cr} = \frac{GJ}{A_r r_0^2} + \Gamma \sqrt{\frac{B_p}{R}} \tag{42}$$

where

$$\Gamma = 1.0 \times 10^{-5} \left[23.7 + 0.852 \left(\frac{B_s}{T_s} \right) + 9870 \frac{B_s/T_s}{(B_p/T_p)^2} \right.$$
$$\left. + 30.5a + \frac{15600a}{B_p/T_p} - 215 \frac{(B_s/T_s)^3 a}{(B_p/T_p)^3} \right] \tag{43}$$

and

$$a = \frac{B_s T_s}{B_p T_p} \tag{44}$$

This equation was developed based on finite element results for B_p/R values less than 0.015, which is a rather limited range in comparison with practical ring geometries. In fact, Eq. (42) is inaccurate and unconservative even for low values of B_p/R (e.g. at $B_p/R = 0.01$) for some geometries (Teng and Chan 1999b). Overall, it overestimates the buckling strengths of many ring geometries and is unsuitable for design use for a practical range of ring geometries.

Teng and Chan's approximation for rings with a simply supported inner edge

Teng and Chan (1999b) recently undertook a further study of the buckling strength of inner edge simply supported T-section rings with the aim being the development of a simple design approximation, using Eq. (32) as the starting point. The simplifications of Eqs (28) and (29) were extended by them to

$$c_2 \approx 1 \tag{45}$$
$$(1 - \beta)^2 \approx 1 \tag{46}$$

By implementing these two assumptions, c_3 in Eq. (32) reduces to

$$c_3 = 1 + \sqrt{\frac{c_1}{\beta}} \tag{47}$$

and the buckling stress is then given by

$$\sigma_{\theta cr} = \frac{EI_x}{A_r r_0^2} \left[c_1 + 2\sqrt{c_1}\sqrt{\beta} \right] = \frac{GJ}{Ar_0^2} + \frac{2\sqrt{GJEI_x}}{Ar_0^2} \sqrt{\frac{B_p}{R_c}} \tag{48}$$

Like Rotter and Jumikis' (1985) approximation (Eq. 42), the first term in Eq. (48) represents the buckling stress of a long straight column with an enforced centre of rotation at the ring inner edge, while the second term represents the strength gain due to the ring curvature. It shows that the buckling stress is a linear function of $\sqrt{B_p/R_c}$. The difference between R and R_c is small and may be ignored (Teng and Chan 1999b), so Eq. (48) can be further reduced to

$$\sigma_{\theta cr} = \frac{GJ}{A_r r_0^2} + \frac{2\sqrt{GJEI_x}}{A_r r_0^2}\sqrt{\frac{B_p}{R}} \tag{49}$$

Equation (49) is then of the same form as that of Rotter and Jumikis (1985), but with a much simpler and more rational expression for the coefficient Γ of Eq. (42). Teng and Chan (1999b) showed through comparison with finite element results that Eq. (49) is rather conservative for relatively thick rings. They thus suggested the following modified equation, which provides close predictions of finite element results:

$$\sigma_{\theta cr} = \frac{GJ}{A_r r_0^2} + \frac{2.3\sqrt{GJEI_x}}{A_r r_0^2}\sqrt{\frac{B_p}{R}} \tag{50}$$

In addition, they proposed the following alternative equation which was found to be slightly more accurate than Eq. (50):

$$\sigma_{\theta cr} = \frac{EI_x}{Ar_0^2}\left[c_1 + 0.2\frac{B_p}{R} + 2\sqrt{c_1}\sqrt{\frac{B_p}{R}}\right] \tag{51}$$

Rings with a clamped inner edge

Teng and Chan (1999a) showed that the buckling strength of an inner edge clamped T-section ring is only weakly dependent on the ring width-to-radius ratio B_p/R. This contrasts with that of simply supported T-section rings, for which the elastic buckling strength depends strongly on the B_p/R ratio. This observation means that the buckling stress of a ring with a large radius may be used to approximate the buckling strength of a ring with a smaller radius. Further, the buckling stress of an inner edge clamped ring with a very large radius is similar to that of a longitudinally compressed long rectangular plate of width B_p and thickness T_p with an edge stiffener of width B_s and thickness T_s clamped along the unstiffened edge. Bulson (1970) presented both accurate and approximate solutions for a rectangular plate with an edge stiffener clamped along the unstiffened edge and simply supported along top and bottom edges. His approximate closed-form solution is based on the energy method and the assumption of a single sine half wave in the longitudinal direction for the buckled shape. Teng and Chan (1999a) proposed a modified version of Bulson's solution for predicting the buckling strength of inner edge clamped rings as

$$\frac{\sigma_{\theta cr}}{E} = K_{min} \frac{\pi^2}{12(1-v^2)} \left(\frac{T_p}{B_p}\right)^2 \tag{52}$$

where

$$K_{min} = \frac{\sqrt{0.56[1+4.5(1-v^2)(B_s/B_p)(B_s/T_s)^2(T_s/T_p)^3]} + 0.5}{1+5(B_s/B_p)(T_s/T_p)} \tag{53}$$

Equations (52) and (53) were found to be accurate by comparison with finite element results (Teng and Chan 1999a), but they are still quite complex given that they only form a small part of the elastic buckling strength assessment of a transition ring. Consequently, Teng and Chan (1999a) developed the following alternative approximation to the buckling stress of inner edge clamped rings

$$\frac{\sigma_{\theta cr}}{E} = \psi \left(\frac{T_p}{B_p}\right)^{1.1} \tag{54}$$

with

$$\psi = \chi[0.016 + 0.5\rho - 0.25\rho^2] \tag{55}$$

$$\rho = \frac{B_s}{B_p} \sqrt[3]{\frac{T_s}{T_p}} \tag{56}$$

$$\chi = \frac{1+5\rho}{1+5B_s T_s/(B_p T_p)} \tag{57}$$

It may be noted that for a uniform thickness ring, the above approximation reduces to

$$\frac{\sigma_{\theta cr}}{E} = \left[0.016 + 0.5\left(\frac{B_s}{B_p}\right) - 0.25\left(\frac{B_s}{B_p}\right)^2\right]\left(\frac{T_p}{B_p}\right)^{1.1} \tag{58}$$

Elastic buckling of T-section rings at transition junctions

Teng and Chan (2000) adopted the approach of Jumikis and Rotter (1983) in developing an elastic buckling stress approximation for T-section rings at steel silo transition junctions. That is, the buckling stress of a T-section transition ring is interpolated from the buckling stresses of the ring with the two idealised inner edge conditions: σ_s for a simply supported inner edge and σ_c for a clamped inner edge. This interpolation is given by

$$\sigma_{\theta cr} = \frac{\eta_s \sigma_s + \eta_c \sigma_c}{\eta_s + \eta_c} \tag{59}$$

where η_s and η_c are interpolation parameters that may be functions of all or some of the following eight dimensionless geometric parameters: t_c/T_p, t_h/T_p, t_s/T_p, B_p/R, B_p/T_p, B_s/B_p, T_s/T_p and α, where t_c, t_h and t_s are the thicknesses of the cylinder, the hopper and the skirt, respectively. The expressions for η_s and η_c (Eqs 37 and 38) used by Jumikis and Rotter (1983) for annular plate rings were found by Teng and Chan (2000) in an extensive parametric study to be also appropriate for T-section rings.

Based on a large parametric study, Teng and Chan (2000) recommended that Eqs (50) and (54) for the buckling stresses of inner edge simply supported and inner edge clamped rings, respectively, should be used with Eq. (59) to predict the elastic buckling strength of T-section rings at transition junctions. In terms of the equivalent circumferential compressive force, the elastic buckling strength of a T-section transition ring is given by (Teng and Chan 2000)

$$F_e = \sigma_{\theta cr} A_e \qquad (60)$$

where A_e is the effective section area defined by Eq. (11).

Elastic buckling strength of angle section rings

The design approximations presented in the above two sections for the elastic buckling strengths of annular plate and T-section rings have been developed following a different approach instead of making direct use of the classical solution. However, no similar design approximation has been established for angle section rings. For angle section rings, the only available approach is to calculate the buckling stress using the classical solution (Eq. 12) unless recourse is made to a finite element analysis. As illustrated in Fig. 14.11(b), this classical solution provides a satisfactory prediction only for rings with a light inner edge rotational restraint from the adjacent shell walls, including the limiting case of a simply supported inner edge. If the elastic rotational restraint is significant, the classical solution overestimates the buckling strength.

Plastic collapse strength

For stocky rings, buckling will not occur. Instead, plastic collapse involving the formation of a plastic collapse mechanism controls the strength. The plastic collapse behaviour and strength of these transition junctions have been studied by Rotter (1985, 1987) and Teng and Rotter (1991a,b). This collapse mechanism requires the formation of a plastic hinge circle inside each shell segment as well as a plastic hinge circle at the transition (Fig. 14.5(b)). The first description of transition junction plastic collapse appears to be that of Rotter (1985, 1987). He also produced a simple equation to predict the collapse strength of a uniform thickness junction. Teng and Rotter (1991a) conducted a more complete study of the problem. The plastic collapse strength, in terms of the equivalent circumferential compressive force, can be closely approximated by the following equation developed by Teng and Rotter (1991a) based on the plastic effective section concept already presented in Chapter 13:

$$F_p = f_y A_p = f_y (A_r + A_{ps}) = f_y \left(A_r + \sum_{i=1}^{N} l_{pi} t_i \right) \qquad (61)$$

where A_{ps} is the effective area of the shell segments in resisting the circumferential compression at an equivalent stress of the yield stress f_y at plastic collapse. The plastic effective length l_{pi} for the ith shell segment is given by

$$l_{pi} = 0.975 \gamma_{pi} \sqrt{R t_i / \cos \phi} \qquad (62)$$

where γ_{pi} assumes different values for the two groups of segments above and below the annular plate of the ring respectively and is given by

$$\gamma_{pi} = 1 \qquad \text{for the thinner group} \qquad (63)$$
$$\gamma_{pi} = 0.7 + 0.6\zeta^2 - 0.3\zeta^3 \qquad \text{for the thicker group} \qquad (64)$$

with the equivalent thickness ratio ζ being the same as that defined in Eq. (6). The above method was extended from an earlier simpler proposal by Rotter (1987) for junctions of uniform shell wall thickness. Although it was initially developed for transition junctions with an annular plate ring (Teng and Rotter 1991a), its predictions have also been shown to be accurate for transition junctions with a T-section ring (Teng and Chan 2001). The plastic effective section method for the plastic collapse loads of complex shell junctions presented in the section 'Complex junctions' of Chapter 13 is a generalisation of the above analysis.

Plastic buckling strength

General

Elastic buckling only governs the strength of junctions with a slender ring or with a high yield stress, while the plastic collapse strength of Eq. (61) can overestimate the ultimate strength as plastic buckling is likely to precede plastic collapse. For most transition rings, failure is by plastic buckling.

Plastic buckling of annular plate rings was examined briefly by Rotter (1987), Greiner and Ofner (1991) and Teng and Rotter (1991c), who also considered plastic buckling of T-section rings, but no simple design method was proposed. More recently, Teng (1997) presented a comprehensive study on the plastic buckling of annular plate rings, while a similar study on T-section rings was conducted by Teng and Chan (2001). The description here is based mainly on the latter two studies.

Design approximation for annular plate rings

Plastic buckling results from interaction between elastic buckling and plastic collapse, so the plastic buckling strength is generally formulated in terms of the elastic buckling strength and the plastic collapse strength. To develop a design approximation, it is necessary to define a dimensionless slenderness parameter λ as follows

$$\lambda = \sqrt{F_p/F_e} \qquad (65)$$

For a given structure, the elastic buckling strength F_e and the plastic collapse strength F_p can be found easily using Eqs (41) and (61), so the slenderness parameter can be evaluated without difficulty.

Teng (1997) discovered that the characteristics of interaction between yielding and buckling is controlled by a single parameter, termed the dimensionless ring size parameter k:

$$k = B_p T_p / A_{ps} \qquad (66)$$

This means that for a given value of this parameter, a single function is sufficient to relate the plastic buckling strength to the elastic buckling and plastic collapse strengths, regardless of the actual geometry. Based on this observation, Teng (1997) generated a set of strength curves from finite element analyses for junctions of appropriate geometries corresponding to a range of values of this parameter possibly found in practice (Fig. 14.14). This figure shows that in the region where the interaction between elastic buckling and plastic collapse is the

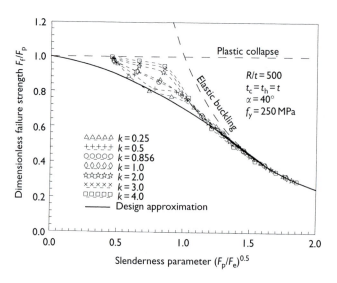

Figure 14.14 Design approximation for annular plate rings.

strongest ($0.5 < \lambda < 1.2$), the lowest strength curve is that given by the smallest k value. Very small rings are not used in practice, so a k value of 0.25 was taken by Teng (1997) as a lower bound to practical k values.

For design use, Teng (1997) suggested a single lower bound approximation in preference to the use of multiple curves corresponding to different ranges of k values, considering the lack of information on the effects of other factors such geometric imperfections and residual stresses. The following equations were suggested by Teng (1997):

$$\frac{F_f}{F_p} = 1 - 0.3\lambda^{1.5}, \quad 0 < \lambda < 1.62 \text{ (plastic buckling region)} \tag{67}$$

$$\frac{F_f}{F_p} = \frac{1}{\lambda^2}, \qquad \lambda > 1.62 \text{ (elastic buckling region)} \tag{68}$$

In Eqs (67) and (68), F_f denotes the ultimate failure strength due to either elastic buckling, plastic buckling or plastic collapse. These equations are seen to provide a lower bound approximation to the plastic buckling region and a close approximation overall (Fig. 14.14). The division into the two regions (elastic and plastic buckling) is based on the strength behaviour, so the elastic buckling region covers purely elastic buckling failures as well as nearly elastic buckling failures occurring after limited yielding which does not reduce the buckling strength appreciably. The above two equations give the same value at the transition point of $\lambda = 1.62$ between the two regions and also have a similar (but not identical) slope at this point.

Design approximation for T-section rings

A similar study was conducted for T-section rings (Teng and Chan 2001). For T-section rings, the characteristics of interaction between yielding and buckling again were found to depend strongly on the dimensionless ring size parameter which is now given by

$$k = \frac{B_p T_p + B_s T_s}{A_{ps}} \tag{69}$$

In addition, the edge stiffener height-to-annular plate width ratio (B_s/B_p) also has a significant influence.

Figure 14.15 shows the buckling strengths corresponding to different values of k and B_s/B_p, where two design approximations are also shown. Approximation P-I consists of three separate functions, while Approximation P-II consists of two separate functions, with one covering the elastic range and the other covering the plastic range. The simpler Approximation P-II was proposed by Teng and Chan (2001)

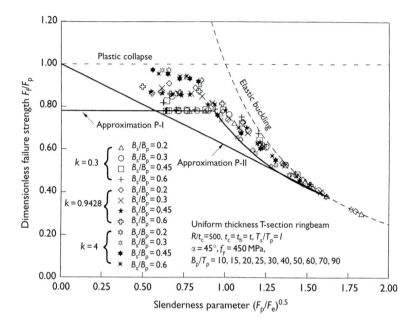

Figure 14.15 Design approximation for T-section rings.

for use in design and is given by the following equations

$$\frac{F_f}{F_p} = 1 - 0.383\,\lambda, \qquad 0 < \lambda < 1.62 \text{ (plastic buckling region)} \qquad (70)$$

$$\frac{F_f}{F_p} = \frac{1}{\lambda^2}, \qquad\qquad \lambda > 1.62 \text{ (elastic buckling region)} \qquad (71)$$

Like the design proposal for annular plate rings (Eqs 67 and 68), the transition between the two regions lies at a value of 1.62. This consistency is certainly a desirable feature. At this transition point, the above equations give the same value.

Experimental behaviour

General

Despite the many theoretical studies, the experimental behaviour of rings at steel silo transition junctions was studied for the first time only recently (Teng *et al.* 2001; Zhao 2001; Zhao and Teng 2003). This lack of experimental results is attributable to the difficulties associated with testing these thin-shell junctions at model scale. The recent experimental results from Zhao and Teng (2003) provided important insights into a number of aspects of the buckling behaviour of these rings, and are summarised in this section.

Overview of experimental programme

The series of buckling experiments on rings at steel silo transition junctions by Zhao and Teng (2003) formed part of a larger experimental program on shell junctions (Zhao 2001). This test series includes five cone–cylinder–skirt–ring junctions subjected to simulated bulk solid loading. The nominal junction dimensions are given in Table 14.1, where L is the length of a given shell segment as defined by the subscript. The width-to-thickness ratio of the ring was the main variable under investigation, and this parameter determines how important buckling of the ring is to the integrity of the junction. All junctions were fabricated using the method of sheet rolling followed by seam welding. Careful fabrication techniques were employed to produce model junctions of high quality (Teng *et al.* 2001). Two types of steel sheets (1 and 2 mm) were used in fabricating these model junctions, with their measured thicknesses being 0.95 and 1.965 mm, respectively. The Young's modulus and the yield stress obtained from tensile tests of coupons cut from the steel sheets are 2.04×10^5 and 253 MPa for the 1 mm sheets and 1.99×10^5 and 165 MPa for the 2 mm sheets, respectively.

An overall view of the experimental set-up is shown in Fig. 14.16. This set-up was built as a multi-purpose test rig for shell buckling experiments with special attention to the testing of steel silo transition junctions (Teng *et al.* 2001). The model junction was filled with sand and a surcharge load was applied through a rigid circular plate which was loaded by a hydraulic jack. This loading method led to realistic pressures on the transition junction without having to build a tall silo.

A laser displacement meter on a measurement frame was rotated around and moved up and down the test specimen to provide a three dimensional map of the shape of the specimen. All the movements were effected using stepper motors controlled by a computer, which also recorded the measurement results. This measurement system provided fast, accurate, non-contact measurements of good repeatability. In addition, strain gauges were installed at selected locations. In particular, one of the models was extensively strain gauged to determine the pressures

Table 14.1 Nominal geometries of model junctions with a ring

Junction	R (mm)	α (degree)	L_c (mm)	L_s (mm)	t_c (mm)	t_h (mm)	t_s (mm)	$B_p \times T_p$ (mm × mm)
CCSR-1	500	40	300	200	1	1	1	20 × 1
CCSR-2	500	40	300	200	1	1	2	30 × 1
CCSR-3	500	40	300	200	1	1	2	20 × 2
CCSR-4	500	40	300	200	1	1	2	40 × 1
CCSR-5	500	40	300	200	1	1	2	30 × 2

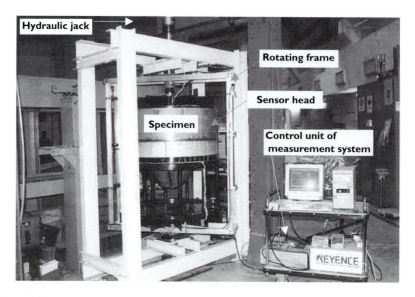

Figure 14.16 Overall view of experimental set-up.

exerted by the sand away from the transition. Further details of the experimental facility are given elsewhere (Teng *et al.* 2001; Zhao 2001).

Typical experimental behaviour

Only the experimental results of junction CCSR-2 are briefly presented here to illustrate the typical buckling behaviour of rings at silo transition junctions. Before the test, a careful three-dimensional survey was conducted on the specimen to obtain the initial imperfect surface. Figure 14.17 shows the initial surface of the example model junction with the annular plate ring excluded. A localised ridge can be seen along each of the two meridional welds. A nearly axisymmetric circumferential weld shrinkage depression can be easily identified at the cylinder-to-skirt welded joint. In addition, short-wave imperfections of smaller amplitude in the critical area near the transition and longer-wave imperfections of greater amplitude away from the transition are seen in the cylinder.

The circumference at 5 mm from the outer edge of the ring was scanned before loading and at each load level. Figure 14.18 shows the results where the dashed line represents the initial shape, with the solid lines being deformed shapes at load levels ranging from 30 to 390 kN with an increment of 20 kN. These results show that most of the short-wave buckles on the ring were amplified from initial geometric imperfections (Fig. 14.18).

At a load of about 170 kN, periodical deformations could be observed on the ring by the naked eye. These deformations continued to grow with further loading,

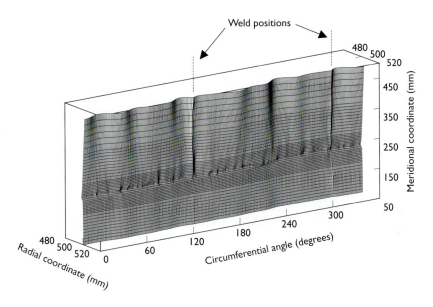

Figure 14.17 Initial surface of junction CCSR-2 excluding the ring.

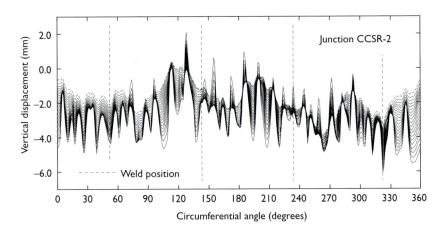

Figure 14.18 Deformed shapes of the ring of junction CCSR-2.

leading to obvious short-wave buckles. Buckles of similar wavelength but smaller amplitude were also found on the bottom of the cylinder. The development of these buckles was not associated with a reduction in the load carrying capacity. That is, the junction displayed a stable postbuckling path. Final failure of the junction occurred at a load of 410 kN, by the formation of a plastic collapse mechanism with nearly uniform plastic deformations over a large part of the circumference.

(a)

(b)

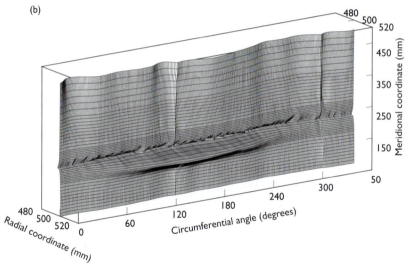

Figure 14.19 Junction CCSR-2 after failure: (a) photograph; (b) three-dimensional plot.

Figure 14.19(a) shows the model junction after final failure with the surcharge load already released, while Fig. 14.19(b) displays the deformed shape plotted from the survey data of the laser meter. The model junction was painted near the transition after test for better contrast in the photograph (Fig. 14.19(a)).

Experimental buckling loads

Due to the stable postbuckling path, the definition of buckling loads is not straight-forward. Zhao and Teng (2003) observed from nonlinear finite element analyses that for a nearly perfect junction, the load–displacement curves of the nodal points within a buckling half wave on the ring are initially coincidental, and then start to differ from each other, and at the same time experience a sudden change in slope at a load close to the bifurcation buckling load of the perfect junction. The buckling load of the corresponding perfect structure can thus be closely defined by examining the trends of these load–displacement curves.

The load–displacement curves of points on the ring of a test junction do not generally show such behaviour due to the presence of significant initial imperfections. However, as a single half wave occupies only a small portion of the ring circumference, if a small nearly perfect portion of the ring can be identified, the development of displacements of this small portion can then be examined to define the buckling load of the perfect ring approximately. Figure 14.20 shows a set of such curves for points within a selected half wave for each of the five model junctions given in Zhao and Teng (2003).

For both junctions CCSR-3 and CCSR-4 (Figs 14.20(c) and (d)), the load–displacement curves of points in the most appropriate half wave almost coincide with each other, before a sudden slope change or reversal occurs in some of the curves. In these curves, sudden divergence in displacements is rather obvious at a certain load level. This load was taken by Zhao and Teng (2003) as the experimental buckling load, which is 340 kN for CCSR-3 and 160 kN for CCSR-4. These loads, shown in Figs 14.20(c) and (d), are believed to be close to the buckling loads of the corresponding perfect junctions (Zhao and Teng 2003).

Half waves with similar curves to those shown in Figs 14.20(c) and (d) could not be found from the displacement measurements of junctions CCSR-2 and CCSR-5. The load–displacement curves shown in Figs 14.20(b) and (e) were judged to be the most suitable for use in buckling load determination for these two junctions by Zhao and Teng (2003), and they show a somewhat different type of behaviour. In these curves, the displacements of different points are initially similar, but diverge gradually with increases of load. A load corresponding to a sudden divergence or an obvious change of slope cannot be identified. Instead, to determine the buckling load, the curve showing the most obvious slope change was identified, and the intercept of the initial slope of this curve and a tangent to a suitable portion of the postbuckling (i.e. post-slope change) part of the curve was taken as the buckling load, as shown in Figs 14.20(b) and (e) for these two junctions, respectively. The buckling loads so obtained are 250 kN for junction CCSR-2 and 325 kN for junction CCSR-5. The two curves used for the determination of buckling loads were among the curves that show the most obvious slope change (Zhao 2001). Closely similar buckling loads could have been obtained by using other curves showing similar slope changes. This approach was first discussed in Teng *et al.* (2001) for Junction CCSR-2.

As Junction CCSR-1 failed by skirt buckling in the initial postbuckling stage of the ring, the determination of its ring buckling load is difficult. Most of the load–displacement curves are nearly linear. A load corresponding to a sudden divergence or an obvious change of slope cannot be identified (Fig. 14.20(a)).

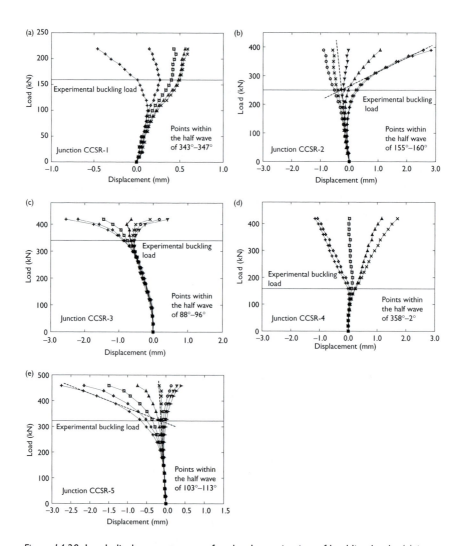

Figure 14.20 Load–displacement curves for the determination of buckling loads: (a) junction CCSR-1; (b) junction CCSR-2; (c) junction CCSR-3; (d) junction CCSR-4; (e) junction CCSR-5.

In addition, the lack of deep postbuckling data means that the intercept of the initial slope of the curve with the most obvious slope change and a tangent to a suitable portion of its postbuckling (i.e. post-slope change) part is unlikely to deliver a reliable prediction of the buckling load. For these reasons, the buckling load was taken as the load when a relatively obvious slope change occurs in two of the curves shown in Fig. 14.20(a), and this load is 160 kN.

Experimental buckling modes

By removing the initial imperfections from the deformed shapes, plots showing only load-induced deformations could be obtained. As the low harmonic modes are believed to have little effect on the development of non-symmetric deformations, the components in harmonic modes 0 and 1 were omitted in these plots. Figure 14.21 shows such a plot for junction CCSR-2. This plot shows a clear process of the development of buckles with loading, and the number of periodical waves counted from this figure is 31. The buckling wave numbers for the other four junctions were obtained using the same approach and the results are listed in Table 14.2. Alternatively, the dominant Fourier harmonic terms preferred for growth during loading can be found by Fourier decomposition of deformation measurements as done by Zhao and Teng (2003). The buckling wave numbers counted from plots such as that shown in Fig. 14.21 are in general agreement with the results of Fourier decomposition.

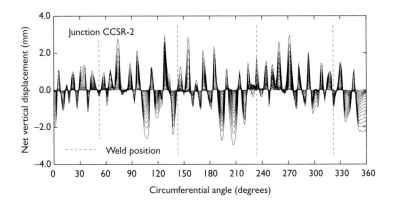

Figure 14.21 Load-induced deformations of the ring of junction CCSR-2.

Table 14.2 Buckling and collapse loads of model junctions

Specimen	Experimental buckling load and mode		Finite element bifurcation load and mode		Experimental collapse load	Finite element collapse load
	P_{cr} (kN)	n_{cr}	P_{cr} (kN)	n_{cr}	P_c (kN)	P_c (kN)
CCSR-1	160	35	213	63	230[a]	267
CCSR-2	250	31	268	34	410	357
CCSR-3	340	20	N/A		434	364
CCSR-4	160	29	198	26	442	389
CCSR-5	325	27	347	43	477	418

Note
a Failure by skirt buckling in the experiment.

Comparison between experimental and finite element results

Experimental buckling loads and buckling wave numbers for all model junctions are shown in Table 14.2, where nonlinear bifurcation loads including the effects of prebuckling deformations and plasticity from the NEPAS program (Teng and Rotter 1989a) are also given. It should be noted that the experimental buckling load for junction CCSR-1 might be unreliable, as this model failed by skirt buckling and thus insufficient data were obtained for the postbuckling path of the ring. In NEPAS analyses, the nominal perfect geometries were used, actual pressures on silo walls determined from one of the tests were adopted, and the effect of welding deposits in these junctions was also included. Further details of the experimental results as well as finite element analyses are available elsewhere (Zhao and Teng 2003). Given the complexity of the problem and the approximation in the determination of the experimental buckling loads, the agreement between the experimental buckling loads and the bifurcation loads as shown in Table 14.2 is believed to be satisfactory. Indeed, it is not unreasonable to expect that the experimental buckling loads, due to the methods of determination employed, are generally lower than the finite element bifurcation loads based on the perfect geometry. The agreement between buckling waves numbers is less satisfactory, but this wave number is believed to be sensitive to the distribution of geometric imperfections.

The postbuckling behaviour of these junctions is stable, so loads higher than the buckling load can be carried by them. The final collapse loads are much higher than the corresponding buckling loads and are also much higher than the theoretical bifurcation loads except for junction CCSR-1, which failed by skirt buckling (Table 14.2). The theoretical final collapse loads from non-linear analyses of the junctions with the effect of welding deposits and measured geometric imperfections taken into account are in close agreement with the experimental collapse loads. The deformed shapes of the ring from finite element analysis and experiment also show a close agreement especially for junctions CCSR-2 and CCSR-4, which had a relatively slender ring (Zhao and Teng 2003). This demonstrates that despite the complex behaviour of these junctions, their behaviour can be properly captured by finite element analyses with all important factors considered.

Implications for design

Zhao and Teng (2003) showed that although the postbuckling deformation is stable, the postbuckling deformations can significantly affect the ultimate load carrying capacity. In Fig. 14.22, nonlinear finite element results of the test junctions based on the measured imperfect shapes and considering both geometric and material nonlinearities are shown, together with the nonlinear bifurcation loads and the axisymmetric collapse loads based on the perfect geometry. In obtaining these results, no welding deposits were assumed as in reality such deposits are small compared to the shell wall thicknesses. The loading was assumed to be a uniform internal pressure with a corresponding frictional traction (coefficient of

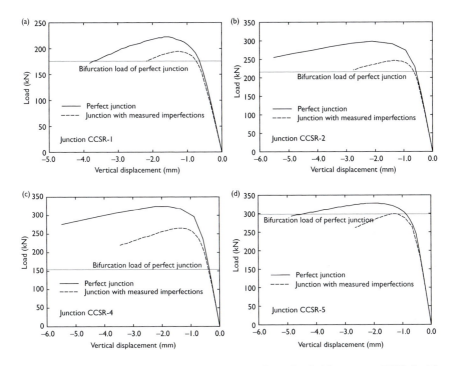

Figure 14.22 Effect of ring buckling on junction collapse load: (a) junction CCSR-1; (b) junction CCSR-2; (c) junction CCSR-4; (d) junction CCSR-5.

friction = 0.5). A comparison of these results shows clearly the effect of ring buckling on the junction collapse load.

For the four model junctions for which bifurcation buckling was found to be possible by NEPAS (Teng and Rotter 1989a), the difference between the collapse load and the bifurcation load is significantly reduced as a result of the detrimental effect of postbuckling deformations. The collapse load of the imperfect geometry remains significantly above the bifurcation load for junctions 1, 2 and 4, so for these junctions, the use of the bifurcation load as the ultimate strength of the junction is conservative. For junction CCSR-5, the collapse load of the imperfect geometry is almost identical to the bifurcation load. While the use of the bifurcation load as the ultimate strength is still safe for this particular junction, it should be borne in mind that for junctions whose axisymmetric plastic collapse load and non-symmetric bifurcation load of the corresponding perfect geometry are similar, the ultimate strength of the real imperfect junction may fall below the bifurcation load due to the effect of postbuckling deformations. It is also worth noting that for junction CCSR-3, although NEPAS found bifurcation buckling to be impossible, the imperfections are seen to still have a significant effect on the collapsed load. This effect is however smaller than that in the other four junctions.

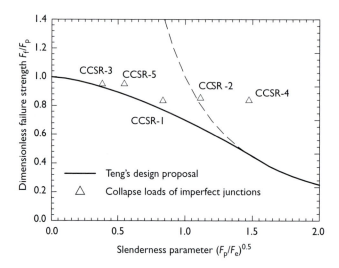

Figure 14.23 Comparison between Teng's design proposal and collapse loads of imperfect junctions.

The collapse loads of the junctions from the nonlinear analyses are plotted in Fig. 14.23 for comparison with the design proposal based on the bifurcation loads of perfect junctions developed by Teng (1997). The design proposal, though based on the bifurcation loads, was also formulated to give conservative predictions when strong interaction between buckling and collapse is likely (Fig. 14.14). It is seen that designs based on this proposal are safe in all cases, but can be rather conservative. This has two implications: (a) the design proposal of Teng (1997) for transition junctions with an annular plate ring presented earlier in this chapter, being derived from numerical analysis based on the perfect geometry, is safe and can be used with confidence; and (b) since the postbuckling behaviour of the ring is stable, it may be possible to develop more liberal design rules in the future so that ring buckling is relegated to a serviceability limit state. For such rules to be developed, it is necessary to carry out a comprehensive study of the effect of ring buckling on the plastic collapse strength of transition junctions. The same implications are also believed to hold for the design proposal of Teng and Chan (2001) for junctions with a T-section ring.

Buckling of rings in column-supported silos

The treatment presented in the preceding sections is for rings at junctions that are uniformly supported, as provided by a skirt or a large number of columns. This treatment cannot be applied to rings which are supported only on a small number of columns.

In theory, a column-supported transition junction should be susceptible to failure by elastic buckling of the ring and plastic collapse of the junction, with the possibility of plastic buckling of the ring between these two cases. One study has examined the conditions under which elastic buckling may be expected (Teng and Rotter 1989b), but the criteria of plastic failure remains unclear at present.

In column-supported silos with only a few columns, the circumferential stresses in the ring vary both around the circumference and over the cross-section (Rotter 1982, 1983; Teng and Rotter 1989b). Rotter (1985) suggested that the buckling strength of an annular plate ring in a column-supported silo might be approximately and conservatively predicted using Eq. (33) if the predicted buckling stress is taken as the highest circumferential compressive stress in the ring. However, this proposal was based on engineering judgement rather than a rational analysis.

In the study of Teng and Rotter (1989b) a closed-form solution was derived based on thin-walled member theory. Its predictions matched finite element results closely. Simplified equations were also presented by them for the strength of a ring simply supported at the inner edge. The buckling stresses at the inner and outer edges of an annular plate ring are given by (Teng and Rotter 1989b)

$$\sigma_{\theta cr}^{I} = \frac{4\sigma_{ucr}}{1 + 3\omega} \tag{72}$$

$$\sigma_{\theta cr}^{O} = \frac{4\omega\sigma_{ucr}}{1 + 3\omega} \tag{73}$$

with

$$\sigma_{ucr} = c_{us} E \left(\frac{T_p}{B_p} \right)^2 \tag{74}$$

$$c_{us} = 0.385 + 0.452 \sqrt{\frac{B_p}{R}} - 0.505 \frac{B_p}{R} \tag{75}$$

in which ω is the ratio of the outer edge to inner edge circumferential stresses and σ_{ucr} is the critical stress for an annular plate under uniform compression. For thin rings ($B_p/R \ll 1$), c_{us} of Eq. (75) reduces to c_s of Eq. (35). Two conclusions from this study may be of particular interest to designers. First, the maximum compressive stress in the ring at buckling is always larger than the buckling stress of the same ring under uniform compression. The section can sustain a much larger maximum stress for some values of ω, and this indicates that the simple approach suggested by Rotter (1985) is a good lower bound but can be very conservative. Second, the load factor against buckling failure varies around the circumference, so a search must be undertaken to find the critical location. Teng and Rotter (1989b) analysed a typical silo supported on different numbers of columns. They showed that the lowest buckling load factor always occurs at the column support position (Fig. 14.24).

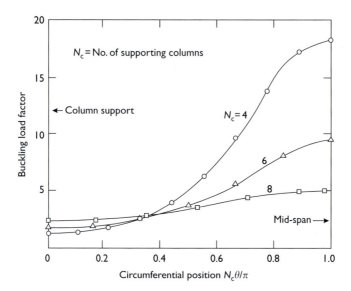

Figure 14.24 Variation of ring buckling load factor between columns.

Rings at other shell junctions

The preceding sections of the chapter specifically deals with rings at steel silo transition junctions. There is, however, much similarity between these rings and rings at other shell junctions, so the design rules presented in this chapter may also be applied to the design of the latter.

For the closely similar problem of ring buckling at transition junctions in elevated liquid storage tanks, the strength rules presented in this chapter are directly applicable, except that no frictional drags exist on the shell walls (e.g. $\mu = 0$ in Eq. 2). The pure internal pressure loading generally leads to a higher stress level in the ring at buckling, so the strength rules presented here are expected to become a little more conservative when applied to tank rings.

The buckling strength of rings at shell junctions in pressure vessels and piping components is expected to be also closely predicted by the strength rules presented in this chapter. This has been proven for the elastic buckling of annular plate rings in a recent study (Teng and Ma 1999). The predictions of the strength rules presented here are expected to be significantly conservative, when the adjacent shell segments are subject to both circumferential and meridional tension.

In general, the strength rules of this chapter can be applied to the design of rings at any shell junctions, provided that the adjacent shell segments are not significantly destabilized by membrane compression in either the circumferential or meridional direction to the degree that its restraint to ring buckling is significantly reduced.

Conclusions

Rings at axisymmetric shell junctions are susceptible to buckling failures when they are subject to circumferential compression. A number of failure modes are possible, including in-plane buckling, out-of-plane buckling and local buckling of the plate/shell elements of a ring. The buckling of rings at shell junctions differs from that of isolated rings, as in the former the adjacent shell walls provide both loading and restraint to the ring. As a result, existing classical theories for the buckling of isolated rings cannot be applied to the stability design of rings at shell junctions.

Over the last two decades, a large amount of research has been carried out on the buckling of rings at the transition junctions of steel silos and tanks. This research has shown that the in-plane buckling mode is generally restrained by the adjacent shells which provide substantial resistance to any periodical radial deformations, while the possibility of local buckling should be suppressed by limiting the slenderness of the plate/shell elements of the ring. Extensive research has thus been conducted on the out-of-plane buckling mode of rings, so this chapter has been focused on this failure mode.

The extensive existing research has included closed-form solutions, finite element studies, and experiments. Closed-form solutions have been developed using the thin-walled member theory in which the section is assumed to retain its shape during buckling, an assumption which has been shown by finite element analysis to be invalid in general for junction rings because of the heavy rotational restraint from the adjacent shell walls. Instead, the buckling mode is generally distortional, involving significant shape changes of the section. The closed-form solutions are nevertheless accurate for the limiting case of rings with no or light rotational restraint at the inner edge, and can be exploited in design with this limitation in mind.

Most of the information required for the development of design proposals was generated by finite element analysis as a shell buckling problem. The design proposals were developed based on the finite element bifurcation buckling loads of the perfect geometry. A complete design proposal is available for both annular plate rings and T-section rings, but not for angle section rings.

Recent experiments on annular plate rings in steel silo transition junctions have been described together with complementary nonlinear finite element modelling. This recent work showed that the postbuckling path of rings is stable, with the ultimate failure generally being the plastic collapse of the junction. There is adverse interaction between buckling and plastic collapse, so the collapse load of a real junction can be significantly lower than the axisymmetric collapse load of a perfect junction. This work also showed that the theoretically based design proposals are safe for practical application, but can be rather conservative for some junctions, due to the postbuckling strength reserve. Further research on the effect of ring buckling on junction final collapse load is required to develop a more liberal rule for the exploitation of postbuckling strength reserve.

Almost all existing research has been on rings at steel silo transition junctions uniformly supported around circumference. Towards the end of the chapter, the applicability of existing information on these rings to rings at other similar junctions was briefly addressed. Furthermore, much research is required on rings in column-supported silos, on which there is very little information.

Finally, it should be borne in mind that in the design of steel shell junctions, adverse interaction between the buckling of the ring or collapse of the junction and the buckling of the shell body away from the junction should be avoided by appropriating proportioning of shell wall thicknesses. In the treatment of ring buckling provided in this chapter, such interaction has not been included.

Acknowledgements

The authors are grateful to The Hong Kong Research Grants Council and The Hong Kong Polytechnic University for their financial support in the last few years for some of the work covered in this chapter.

Notation

A_e	effective ring area for elastic analysis
A_p	effective ring area for plastic analysis
A_{ps}	effective area of shell segments for plastic analysis
A_r	cross-sectional area of ring
a	stiffener-to-annular plate area ratio
B	width of ring
B_p	width of annular plate
B_s	width of vertical stiffener for an annular plate
c	coefficient defined by Eq. (34)
c_c	coefficient defined by Eq. (36)
c_s	coefficient defined by Eq. (35)
c_{us}	coefficient defined by Eq. (75)
c_1	parameter defined by Eq. (23)
c_2	parameter defined by Eq. (24)
c_3	parameter defined by Eq. (31)
E	elastic modulus
F	circumferential compressive force in a junction
F_e	circumferential compressive force in a junction at elastic buckling
F_f	ultimate failure strength
F_0	circumferential compressive force in a junction due to meridional tension in hopper only
F_p	circumferential compressive force in a junction at plastic collapse
f_y	yield stress
G	shear modulus
I_w	warping constant of ring section

I_x	second moment of area about the radial axis of ring section
I_y	second moment of area about the vertical axis of ring section
I_{xy}	product moment of area of ring section
J	torsional constant of ring section
K	number of shell segments in a given group in effective section analysis
$K_{\min}$	buckling stress coefficient
k	dimensionless ring size parameter
k_r	stiffness of rotational restraint to ring from shell walls
k_{ni}	rotational stiffness of ith shell segment in mode n
k_{oi}	rotational stiffness of ith shell segment in axisymmetric deformation
l_c	effective length of cylinder
l_h	effective length of hopper
l_s	effective length of skirt
l_{ei}	effective length of ith shell segment for elastic analysis
l_{pi}	effective length of ith shell segment for plastic analysis
N	total number of shell segments in a junction
$N_{\phi h}$	meridional tension at top of hopper
n	number of buckling waves
n_{cr}	critical number of buckling waves
p	internal pressure
p_{c1}	pressure defined by Fig. 14.8
p_{c2}	pressure defined by Fig. 14.8
p_{h1}	pressure defined by Fig. 14.8
p_{h2}	pressure defined by Fig. 14.8
R	radius of cylinder middle surface or junction
R_c	radius of ring centroid
R_0	radius of ring restraint
r_0^2	$[(I_x + I_y)/A_r] + x_c^2 + y_c^2$
T	plate thickness of ring
T_p	thickness of annular plate
T_s	thickness of vertical stiffener for an annular plate
t	shell wall thickness
t_c	thickness of cylinder
t_h	thickness of hopper
t_s	thickness of skirt
t_{eq}	equivalent thickness of shell wall
t_i	thickness of ith shell segment
x_c	x coordinate of ring centroid
x_p	x coordinate of load position of ring
x_s	x coordinate of ring shear centre
y_c	y coordinate of ring centroid
y_p	y coordinate of load position of ring
y_s	y coordinate of ring shear centre

α	cone apex half angle
α_i	apex half angle of ith shell segment
β	B_p/R_c
β_s	x_s/R_c
Γ	coefficient defined by Eq. (43)
γ_{ei}	modification coefficient for shell segment effective length for elastic analysis
γ_{pi}	modification coefficient for shell segment effective length for plastic analysis
ζ	equivalent thickness ratio
η_s	parameter defined by Eq. (37) & Eq. (39)
η_c	parameter defined by Eq. (38) & Eq. (40)
λ	dimensionless ring slenderness parameter
ν	Poisson's ratio
ρ	parameter defined by Eq. (56)
σ_c	buckling stress of a T-section ring with a clamped inner edge
σ_s	buckling stress of a T-section ring with a simply supported inner edge
σ_θ	(maximum) circumferential compressive stress at ring inner edge
σ_0	circumferential stress at bucking of a ring under uniform circumferential compression in a given buckling mode
$\sigma_{\theta cr}$	circumferential compressive stress at ring buckling
$\sigma_{\theta cr}^{I}$	circumferential compressive stress at buckling at the inner edge of an annular plate ring in a column-supported silo
$\sigma_{\theta cr}^{O}$	circumferential compressive stress at buckling at the outer edge of an annular plate ring in a column-supported silo
σ_{ucr}	circumferential compressive stress at buckling of an annular ring under uniform compression
ϕ	α for the cone; 0 for the cylinder
ψ	parameter defined by Eq. (55)
χ	parameter defined by Eq. (57)
ω	outer edge to inner edge stress ratio in an annular plate ring in a column-supported silo

References

Bulson, P.S. (1970). *The Stability of Flat Plates*. Chatto and Windus, London.

Chen, J.F. and Rotter, J.M. (1998). Effective cross-sections of asymmetric rings for cylindrical shells. *Journal of Structural Engineering*, ASCE **124**(8), 1074–1080.

Cheney, J.A. (1963). Bending and buckling of thin-walled open section rings. *Journal of the Engineering Mechanics Division*, ASCE **89**(5), 17–34.

ENV 1993-4-1 (1999). *Eurocode 3: Design of Steel Structures, Part 4–1: Silos*. European Committee for Standardisation, Brussels.

Gaylord, E.H. and Gaylord, C.N. (1984). *Design of Steel Bins for Storage of Bulk Solids*. Prentice Hall, Englewood Cliffs, NJ.

Goldberg, J.E. and Bogdanoff, J.L. (1962). Out-of-plane buckling of I-section rings. *International Association for Bridges and Structural Engineering* **22**, 73–92.

Greiner, R. and Ofner, R. (1991). Elastic plastic buckling at cone cylinder junctions of silos. In *Buckling of Shell Structures on Land, in the Sea and in the Air* (ed. J.F. Jullien). Elsevier Applied Science, London, pp. 304–312.

Jumikis, P.T. (1987). Stability problems in silo structures. PhD Thesis, University of Sydney, Australia.

Jumikis, P.T. and Rotter, J.M. (1983). Buckling of simple ringbeams for bins and tanks. *Proceedings of the International Conference on Bulk Materials Storage, Handling and Transportation*, IEAust, Newcastle, Australia, August, pp. 323–328.

Rotter, J.M. (1982). Analysis of ringbeams in column-supported bins. *Proceedings of the 8th Australasian Conference on the Mechanics of Structures and Materials*, University of Newcastle, Australia, August, pp. 33.1–33.6.

Rotter, J.M. (1983). Ringbeams for elevated bins and silos. *Proceedings of the Conference on Metal Structures*, IEAust, Brisbane, Australia, May, pp. 111–116.

Rotter, J.M. (1985). Analysis and design of ringbeams. In *Design of Steel Bins for the Storage of Bulk Solids*, School of Civil and Mining Engineering, University of Sydney, Australia, March, pp. 164–183.

Rotter, J.M. (1987). The buckling and plastic collapse of ring stiffeners at cone/cylinder junctions. *Proceedings of the International Colloquium on the Stability of Plate and Shell Structures*, Gent, Belgium, April, pp. 449–456.

Rotter, J.M. and Jumikis, P.T. (1985). Elastic buckling of stiffened ringbeams for large elevated bins. *Proceedings of the Metal Structures Conference*, IEAust, Melbourne, Australia, May, pp. 104–111.

Sharma, U.C., Rotter, J.M. and Jumikis, P.T. (1987). Shell restraint to ringbeam buckling in elevated steel silos. *Proceedings of the 1st National Structural Engineering Conference*, IEAust, Australia, August, pp. 604–609.

Smith, C.V., Jr. and Simitses, G.J. (1969). Effect of shear and load behaviour on ring stability. *Journal of the Engineering Mechanics Division*, ASCE **95**(3), 559–569.

Teng, J.G. (1997). Plastic buckling approximation for transition ringbeams in steel silos. *Journal of Structural Engineering*, ASCE **123**(12), 1622–1630.

Teng, J.G. and Barbagallo, M. (1997). Shell restraint to ring buckling at cone–cylinder intersections. *Engineering Structures* **19**(6), 425–431.

Teng, J.G. and Chan, F. (1999a). Out-of-plane distortional buckling of T-section ringbeams clamped at inner edge. *Engineering Structures* **21**(7), 615–628.

Teng, J.G. and Chan, F. (1999b). New buckling approximation for T-section ringbeams simply supported at inner edge. *Engineering Structures* **21**(10), 889–897.

Teng, J.G. and Chan, F. (2000). Elastic buckling strength of T-section transition ringbeams in steel silos and tanks. *Journal of Constructional Steel Research* **56**(1), 69–99.

Teng, J.G. and Chan, F. (2001). Plastic buckling strength for T-section ringbeams in steel silos. *Engineering Structures* **23**(3), 280–297.

Teng, J.G. and Lucas, R.M. (1994). Out-of-plane buckling of restrained thin rings of general open section. *Journal of Engineering Mechanics*, ACSE **120**(5), 929–947.

Teng, J.G. and Ma, H.W. (1999). Elastic buckling of ring-stiffened cone–cylinder intersections under internal pressure. *International Journal of Mechanical Sciences* **41**(11), 1357–1383.

Teng, J.G. and Rotter, J.M. (1987). Unrestrained out-of-plane buckling of monosymmetric rings. *Journal of Constructional Steel Research* **7**, 451–471.

Teng, J.G. and Rotter, J.M. (1988). Buckling of restrained monosymmetric rings. *Journal of Engineering Mechanics, ASCE* **114**(10), 1651–1671.

Teng, J.G. and Rotter, J.M. (1989a). Non-symmetric bifurcation of geometrically non-linear elastic–plastic axisymmetric shells subject to combined loads including torsion. *Computers and Structures* **32**(2), 453–477.

Teng, J.G. and Rotter, J.M. (1989b). Buckling of rings in column-supported bins and tanks. *Thin-Walled Structures* **7**(3&4), 251–280.

Teng, J.G. and Rotter, J.M. (1991a). Collapse behaviour and strength of steel silo transition junctions – Part I. Collapse mechanics. *Journal of Structural Engineering, ASCE* **117**(12), 3587–3604.

Teng, J.G. and Rotter, J.M. (1991b). Collapse behaviour and strength of steel silo transition junctions – Part II. Parametric study. *Journal of Structural Engineering, ASCE* **117**(12), 3605–3622.

Teng, J.G. and Rotter, J.M. (1991c). Plastic buckling of rings at steel silo transition junctions. *Journal of Constructional Steel Research* **19**, 1–18.

Teng, J.G., Zhao, Y. and Lam, L. (2001). Techniques for buckling experiments on steel silo transition junctions. *Thin-Walled Structures* **39**(8), 685–707.

Timoshenko, S.P. and Gere, J.M. (1961). *Theory of Elastic Stability*, 2nd ed. McGraw-Hill, New York.

Trahair, N.S., Abel, A., Ansourian, P., Irvine, H.M. and Rotter, J.M. (1983). *Structural Design of Steel Bins for Bulk Solids*. Australian Institute of Steel Construction, Sydney.

Trahair, N.S. and Papangelis, J.P. (1987). Flexural–torsional buckling of monosymmetric arches. *Journal of Structural Engineering, ASCE* **113**(10), 2271–2288.

Wah, T. (1967). Buckling of thin circular rings under uniform pressure. *International Journal of Solids and Structures* **3**, 967–974.

Wozniak, R.S. (1979). Steel tanks. In *Structural Engineering Handbook*, 2nd ed. (eds E.H. Gaylord and C.N. Gaylord). McGraw-Hill New York, NY.

Yoshida, S., Tomiya, M., Daimaruya, M., Nishida, K. and Kobayashi, H. (1998). Free vibration finite element analysis of fluid-filled double-decker cylindrical tanks. *Proceedings of the 1998 Joint ASME/JSME Pressure Vessel and Piping Conference*, PVP **370**, 157–163.

Zhao, Y. and Teng, J.G. (2003). Buckling experiments on steel silo transition junctions, to be published.

Zhao, Y. (2001). Stability and strength of steel silo transition junctions, PhD Thesis, The Hong Kong Polytechnic University, Hong Kong, China.

Chapter 15

A probabilistic approach to design shell structures

J. Arbocz and A.R. Stam

Introduction

Because of their favourable strength over weight ratios thin-walled stiffened or unstiffened, metallic or composite shells are widely used structural elements in many applications in the aerospace, the off-shore, the maritime, the process-technology and other civil engineering fields. Unfortunately, thin-walled shells are also often very prone to buckling instabilities.

In the 1960s, the engineering community in the United States was facing the urgent problem of having to devise reliable design procedures for both the ballistic missile program of the US Air Force and the launch vehicles of the NASA space program. In these weight-critical applications, the structural optimization schemes often lead to buckling-critical designs. Facing a multiplicity of very complicated decisions the different governmental agencies started large research programs dealing with shell stability.

Looking back at the results of each of these programs separately, one has to classify them as successful. The Lower Bound Design Philosophy (Anonymous 1968), which was based on the results of the experimental programs, has provided generations of shell designers with a useful tool for successful shell designs.

The many publications dealing with different aspects of the imperfection sensitivity theory (Koiter 1945, 1967; Budiansky and Hutchinson 1964) have improved our understanding of the different effects causing the sometimes wide scatter of buckling loads observed experimentally. With the introduction of ever faster workstations and supercomputers the current generation of the shell of revolution codes such as BOSOR-5 (Bushnell 1976) and the general shell codes such as STAGS-2 (Anonymous 1994) provide us with excellent tools needed for an accurate modelling of the collapse behaviour of thin-walled shells.

However, it is very disturbing that, when it comes to determining the load carrying capability of a thin-walled shell structure then, also in 2000 the accounting for the uncertainties involved in every design is usually done by the above mentioned lower bound design philosophy, a very conservative deterministic procedure that has already been in use 70 years ago. That is, the uncertainties are accounted for by using an empirical 'knockdown factor', which is so chosen that when it is

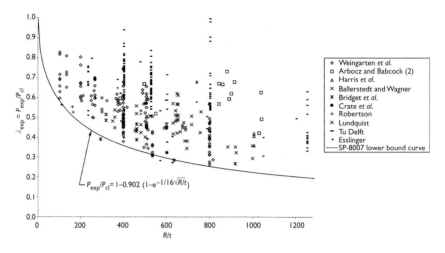

Figure 15.1 Test data for axially compressed isotropic shells – (R/t < 1300, all L/R) – from Doup (1997).

multiplied with the calculated perfect shell buckling load a 'lower bound' to all existing experimental data is obtained (see Fig. 15.1).

Deterministic versus probabilistic design procedures

In principle, the use of empirical knockdown factors to account for the damaging effect of as yet unknown causes was and is a good engineering solution to a pressing problem. However, the question that immediately comes to one's mind is, where has the scientific community failed? How come that today, in the new millennium, after so many years of concentrated research effort one cannot do any better, and this despite the enormously increased computational facilities provided by today's high-powered computers?

Lately, probabilistic design procedures have been proposed as a viable alternative (Ryan and Townsend 1993). It is felt, that quantifying and understanding the 'problem uncertainties' and their influence on the design variables provides an approach which will ultimately lead to a better engineered, better designed and safer structure.

However, the vast majority of practising engineers agree that true reliability must be demonstrated and not simply estimated from analysis. It is the authors' opinion, that before the engineering community will begin in large numbers to accept the current generation of probabilistic analysis tools, two conditions must be satisfied. First, there must be test-constructed data bases which can help in mapping the input parameter uncertainties into probability density functions. In addition, there must be failure and failure rate data bases, which can be used for test verification of the probabilistic failure load predictions.

Turning to the most frustrating case, the collapse problem of axially compressed isotropic cylinders can best be formulated in terms of a response (or limit state) function

$$g(\mathbf{X}) = \Lambda(\mathbf{X}) - \lambda \tag{1}$$

where λ is a suitable normalised load parameter ($= P/P_c$, say), Λ is the random collapse load of the shell, and the vector $\mathbf{X}$ represents the random variables of the problem. The components of the random vector, X_i, may be Fourier coefficients of the initial imperfections and other parameters quantifying the uncertainties in the specified boundary conditions, the constitutive equation used to describe the nonlinear material behaviour, thickness distribution, residual stresses, etc. Notice that the evaluation of the response function thus involves the solution of a complicated nonlinear structural analysis problem, represented by a detailed and possibly large finite element model. However, with today's computational resources this, in itself, should not pose any unsurmountable difficulties.

Clearly, the response function $g(\mathbf{X}) = 0$ separates the variable space into a 'safe region' where $g(\mathbf{X}) > 0$, and a 'failure region' where $g(\mathbf{X}) \leq 0$. The reliability $R(\lambda)$ – or the probability of failure $P_f(\lambda)$ can then be calculated as

$$R(\lambda) = 1 - P_f(\lambda) \tag{2}$$

where

$$P_f(\lambda) = \text{Prob}\{g(\mathbf{X}) \leq 0\} = \int \cdots_{g(\mathbf{X}) \leq 0} \int f_{\mathbf{X}}(\mathbf{x}) \, d\mathbf{x} \tag{3}$$

and $f_{\mathbf{X}}(\mathbf{x})$ is the joint probability density function of the random variables involved.

The credibility of this approach depends on two factors, the accuracy of the mechanical model used to calculate the limit state function and the accuracy of the probabilistic techniques employed to evaluate the multi-dimensional integral.

Comparing the two approaches in short, the deterministic design procedure recommends the use of the following buckling formula

$$P_a \leq \frac{\gamma}{\text{FS}} P_c \tag{4}$$

where P_a is the allowable applied load, P_c the lowest buckling load of the perfect structure, γ the 'knockdown factor' and FS the factor of safety. The empirical knockdown factor γ is shown for axially compressed isotropic shells in Fig. 15.1 as the lower bound curve to all available experimental data.

In the proposed probabilistic design procedure, improvements with respect to the deterministic lower bound based design procedure are sought by using a more rational approach for the definition of the 'knockdown' factor γ. The proposed new shell design procedure can be represented by the following formula

$$P_a \leq \frac{\lambda_a}{FS} P_c \tag{5}$$

where λ_a is the reliability-based improved (less conservative) scientific 'knockdown' factor, to be determined from the appropriate reliability curve, P_c the buckling load for a perfect shell and the factor of safety FS is used to account for possible load uncertainties. In the following the steps involved in the derivation of such a reliability-based improved (less conservative) scientific 'knockdown' factor λ_a will be discussed.

Characteristic imperfection distributions

The introduction of a statistical imperfection sensitivity analysis depends heavily on the availability of so-called characteristic initial imperfection distributions associated with the different manufacturing processes.

In the beginning, initial imperfection surveys were carried out on laboratory-scale shells by the Caltech group (Arbocz and Babcock 1969; Singer et al. 1971) and others (Singer et al. 1978; Walker and Sridharan 1979). This was followed by complete imperfection surveys on large- and full-scale structures in different parts of the world (Horton 1977; Arbocz and Williams 1977; Sebek 1981). Nowadays practically all designers are aware of the fact that in shell buckling problems initial imperfections play an important role. Carrying out initial imperfection surveys has become fashionable. However, when carrying out initial imperfection measurements one should always remember that Koiter's work (1945, 1963) has shown that the degree of imperfection sensitivity depends not only on the magnitude but also on the shape of the initial imperfections. Thus, it is not sufficient to spot check the shell surface for the maximum imperfection amplitude by carrying out selected circumferential and/or axial scans. One must always provide for sufficient cross-reference data, so that later the individual scans can be pieced together to a complete surface map of the measured structure via numerical data reduction programs. The need for providing cross-reference data sometimes poses added complications but as has been shown by Arbocz (1982) on several examples in most cases the difficulties can be overcome rather routinely.

In all cases, where one attempts to measure the exact shape of a shell, before one can determine the initial imperfections one has to define the perfect cylinder. This is done by finding numerically the best-fit cylinder to the measured surface map assembled using the initial imperfection scans. Using the method of least-squares data reduction programs (Arbocz and Babcock 1969) can compute the eccentricities Y_1 and Z_1, the rigid body rotations ε_1 and ε_2 and the mean radius R (see Fig. 15.2) rather routinely. Finally, the measured radial displacements are recomputed with respect to the newly found 'perfect' cylinder. These values are

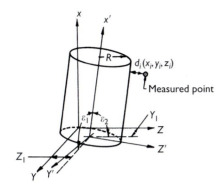

X, Y, Z   Reference axis of traversing pick-up
X', Y', Z' Reference axis of best fit cylinder
d_i Normal distance from measured point
         to best fit cylinder

Figure 15.2 Definition of the 'perfect' cylinder (Arbocz and Babcock 1969).

then used to prepare the three-dimensional plots of the initial imperfections shown later in this chapter.

In all cases, the measured initial imperfections can be represented by one of the following two double Fourier series

$$\bar{W}(x, y) = t \sum_{k,\ell=0}^{N} \cos k\pi \frac{x}{L} \left(A_{k\ell} \cos \ell \frac{y}{R} + B_{k\ell} \sin \ell \frac{y}{R} \right) \tag{6}$$

referred to as the half-wave cosine representation, and

$$\bar{W}(x, y) = t \sum_{k,\ell=0}^{N} \sin k\pi \frac{x}{L} \left(C_{k\ell} \cos \ell \frac{y}{R} + D_{k\ell} \sin \ell \frac{y}{R} \right) \tag{7}$$

called the half-wave sine representation. Here R, L and t are shell radius, length and wall thickness, respectively, x and y are axial and circumferential coordinates; k and ℓ are integers denoting the number of axial half waves and the number of full waves in the circumferential direction, respectively. In all cases the measured imperfections are referred to the so-called 'best fit' cylinders.

The adoption of a standard representation for the measured initial imperfections is necessary in order to be able to compare the different imperfection distributions that are associated with the different fabrication processes. Notice that in all cases the Fourier coefficients are normalised by the corresponding wall thicknesses of the shell bodies.

At the moment, there are two Initial Imperfection Data Banks in existence, one at the Delft University of Technology (Arbocz and Abramovich 1979) and one at the Technion in Haifa (Singer *et al*. 1978).

For a successful application of reliability-based design methods, the existence of data bases containing such experimentally measured initial imperfections, is very helpful, whereby the critical question is:

> Can one associate characteristic initial imperfection distributions with a specified manufacturing process?

That the answer to this question is an unconditional yes has been demonstrated very clearly by Arbocz (1982) and Arbocz and Hol (1991), where characteristic imperfection distributions for different fabrication processes are shown and analysed.

Shells fabricated with welded seams

Thus, in 1977 Arbocz and Williams (1977) published the results of detailed imperfection surveys carried out on a large (radius: 1534.0 mm, length: 2387.6 mm, wall thickness: 2.54 mm) integrally ring and stringer-stiffened shell. Figure 15.3 shows the test specimen with the equipment used to carry out the imperfection

Figure 15.3 Initial imperfection survey instrumentation at NASA Langley (Arbocz and Williams 1977).

scans visible on the side of the shell. The technique employed measured the deviation of the cylinder outer surface relative to an imaginary cylindrical reference surface. Physically, this was accomplished with a 3.0 m long aluminium guide rail supported on the outside diameter of the two steel end rings, which were machined so accurately that they matched to within 0.15 mm. A trolley car carrying a direct current differential transformer instrument was slowly driven along the guide rail by an electric motor. The position of the car was electronically measured using a potentiometer. Accuracy of the displacement measurements was to within ±0.05 mm. The guide rail was moved stepwise around the cylinder circumference in 5° increments thus yielding a total of 72 discrete scans for the complete cylinder. Discrete digital data was recorded every 6.0 mm along the cylinder length, thus yielding 343 data points for a typical 2057.4 mm scan. This digitised data were recorded on a magnetic tape and used later for data reduction.

In the above, the radial deviations from the imaginary cylindrical reference surface, defined by the two very accurately machined steel end rings and the rigid aluminium rail connecting them, were measured.

Next, as described earlier, the data reduction program of Arbocz (1968) was used to find the best-fit cylinder to the measured data of the initial imperfection scans.

Finally, the measured displacements were recomputed with respect to the newly found 'perfect' cylinder. These values were then used to prepare the three-dimensional plot of the initial imperfections shown in Figure 15.4. The three welded seams are clearly visible in the longitudinal direction.

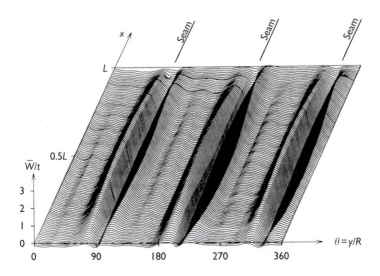

Figure 15.4 Measured initial shape of Langley shell LA-1 (Arbocz and Williams 1977) (radius: 1524.0 mm, length: 2057.4 mm, wall thickness: 2.54 mm).

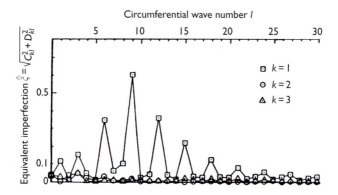

Figure 15.5 Circumferential variation of the half-wave sine Fourier representation – shell LA-1 (Arbocz and Williams 1977).

Results of the imperfection surveys can be displayed in several different ways. Figure 15.5 shows the variation of the calculated half-wave sine Fourier coefficients as a function of the circumferential wave number ℓ for selected axial half-wave numbers k. By computing the respective mean imperfection amplitudes $\hat{\xi}$ the phase shift in the circumferential direction is eliminated. Notice that a characteristic imperfection distribution produced by the three axial welded seams is clearly evident, since the plot is dominated by Fourier coefficients, which are multiples of 3. The amplitudes of the Fourier harmonics with a single half wave in the axial direction have a distinct maximum at nine circumferential waves. All Fourier coefficients with more that a single half wave in the axial direction are comparatively much smaller. Also, because the heavy end rings have been machined to very close tolerances the shell exhibits practically no ovalization (i.e. the amplitude of the Fourier coefficients with $\ell = 2$ is very small).

In the 1970s, the Caltech group carried out imperfection surveys on flight hardware at different aerospace companies.

Figure 15.6 shows a three-dimensional plot of the measured initial imperfections referenced to the 'best fit' cylinder from a large (radius: 1212.1 mm, length: 6454.1 mm, wall thickness: 1.549 mm) integrally stiffened shell constructed out of three pieces. The lines are drawn in the circumferential direction but the data were collected from axial scans. The three welded seams are clearly visible in the imperfection plot. The disturbance at the center results from a ring frame at this location which tends to minimise the weld imperfection. The waviness is caused by the pockets of the stiffened shell. A better view of this is shown in Figure 15.7, where the regular pattern of the pockets is clearly seen. The large jumps at the end of the shell are a result of the end domes of the shell being force-fitted inside the cylindrical section. In all three-dimensional plots, positive imperfections are outward. An exception to this rule are Figs 15.6 and 15.7 where

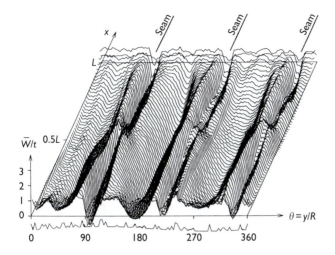

Figure 15.6 Measured initial shape of aerospace shell X-1 (Arbocz 1982) (radius: 1212.1 mm, length: 6454.1 mm, wall thickness: 1.549 mm).

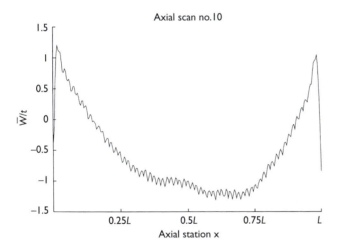

Figure 15.7 Typical meridional shape of aerospace shell X-1.

positive imperfections are pointing inward. This change in orientation became necessary because of the form of the measured initial imperfections near the two edges. A closer look at the axial cross-plot shown in Fig. 15.7 reveals that the initial imperfections consist of three main components, namely a large step function like component due to the uniform radial displacement produced by the force-fitted end domes, a large half-wave sine component in the axial direction plus a small short-wave imperfection component, which accounts for the waviness. Since for

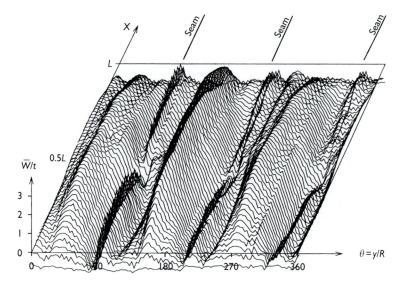

Figure 15.8 Measured initial shape of aerospace shell X-1 with edge zones removed.

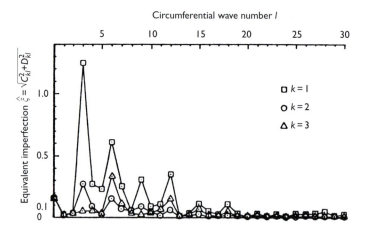

Figure 15.9 Circumferential variation of the half-wave sine Fourier representation (aerospace shell X-1).

stability calculations the uniform radial expansion is of no consequence the edge zones have been eliminated from the data. Thus, the Fourier decomposition of the measured data was done with the initial imperfections shown in Fig. 15.8. Here positive imperfections are pointing outward as usual.

Figure 15.9 shows the variation of the half-wave sine Fourier representation for this large integrally stiffened aerospace shell which is also assembled out of three

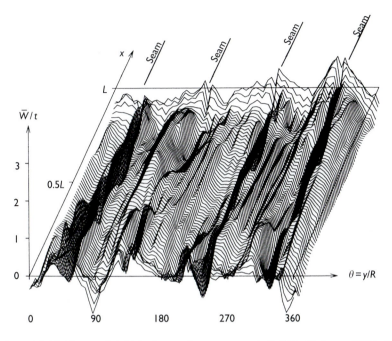

Figure 15.10 Measured initial shape of aerospace shell X-2 (Arbocz 1982) (radius: 1527.4 mm, length: 6047.7 mm, wall thickness: 2.629 mm).

curved parts. In this case, the amplitudes of the Fourier harmonics with a single half wave in the axial direction have a distinct maximum at $\ell = 3$, which corresponds to the number of longitudinal welds. Only harmonics that are integer multiples of 3 have significant amplitudes. The amplitudes of the Fourier coefficients decay with increasing wave numbers both in the axial and in the circumferential direction. Further, since the circular end domes are force-fitted into the cylindrical part, the shell possesses a negligibly small $\ell = 2$ component (ovalisation).

The measured initial imperfections referenced to the 'best fit' cylinder from still another large (radius: 1527.4 mm, length: 6047.7 mm, mean wall thickness: 2.629 mm) stringer stiffened shell made out of four pieces are shown in Fig. 15.10. The four longitudinal welded seams are once again clearly visible. Also, since this shell has only longitudinal stiffeners the short-wave waviness caused by the intersecting transverse stiffeners observed on the previous shell is absent. On the other hand, this shell has more waviness in the circumferential direction.

Figure 15.11 shows the variation of the half-wave sine Fourier representation for this large integrally stiffened aerospace shell, which is assembled out of four curved parts. In this case, the end domes were but-welded to the cylindrical part. This resulted in a relatively large ovalization. As a matter of fact the Fourier harmonics with a single half wave in the axial direction have a distinct maximum

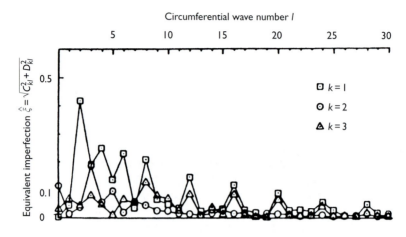

Figure 15.11 Circumferential variation of the half-wave sine Fourier representation (aerospace shell X-2).

at $\ell = 2$ (out-of-roundness). Besides $\ell = 2$ and $\ell = 6$, only those harmonics that are integer multiples of 4 have significant amplitudes. Also, in this case the amplitudes of the Fourier coefficients decay with increasing wave numbers both in the axial and in the circumferential direction.

Considering the results presented so far one can state that if the longitudinal seams, needed to assemble the curved parts, are welded then the resulting imperfection distributions are rather complicated. The dominant imperfection harmonics have a single half wave in the axial direction but the number of full waves in the circumferential direction are obviously influenced by the number of full length panels out of which the shell is assembled. Further, the welding procedure used has also a strong influence on the resulting number of full waves. Thus, both the Langley shell LA-1 and the first aerospace shell X-1 (radius $R = 1212.1$ mm) consist of three full length curved panels. Both have negligibly small out-of-roundness (the $\ell = 2$ Fourier coefficients are very small), however, for the Langley shell LA-1 the largest Fourier coefficient has nine full waves in the circumferential direction (three times the number of welded seams), whereas for the first aerospace shell X-1 the largest Fourier coefficient has three full waves (equal to the number of welded seams). Apparently the welding procedure used by the two aerospace companies were similar because also the second aerospace shell X-2 (radius $R = 1527.4$ mm) has a large Fourier coefficient with the same number of full waves in the circumferential direction as the number of curved panels out of which it is assembled, namely four. However, the initial imperfection distribution of this shell has also comparatively large harmonics with $\ell = 2$ (out-of-roundness) and $\ell = 6$ full waves in the circumferential direction, besides the harmonics that are integer multiples of 4, the number of welded seams. Thus, for shells assembled out of a fixed number of curved parts by welded seams, an important question remains to be answered, namely: when is the number of full waves of the dominant Fourier coefficient in the circumferential direction equal to the number of welded seams, and when will it be equal to the number of welded seams times an integer?

Shells fabricated with riveted seams

In the early 1970s, Horton and his co-workers carried out a series of shell buckling tests at the Georgia Institute of Technology (Horton 1977). Both small- and large-scale specimens were tested. The large shells (radius: 945.8 mm, length: 2743.2 mm, wall thickness: 0.643 mm) were made of 7075-T6 aluminium alloy, each consisting of six identical panels. On the inside they were reinforced by 312 closely spaced Z-shaped stringers. One edge of each panel was joggled, and two stringers were riveted along each joint line. The remaining stringers were attached to the sheet with adhesive. The shells were held circular by two heavy end rings, which were rolled out of bracket-shaped extruded sections, and by seven equally spaced Z-shaped rings on the outside. On two of the large shells complete imperfection surveys consisting of 32 equally spaced circumferential scans were carried out. For this purpose stiff end-plates with central bearings were attached to the specimen, which then was mounted in a heavy framework, with its axis horizontal in such a fashion that it could be rotated about its axis. A 3 m long precision straight edge was positioned parallel to the shell and served as a reference beam. A linear voltage differential transducer (LVDT) was then attached to the straight edge at different axial positions. The shell was rotated about its axis and the variation of the profile recorded at equal intervals along the length of the shell. A three-dimensional view of the measured initial imperfections referenced to the previously described 'best-fit' cylinder is shown in Fig. 15.12.

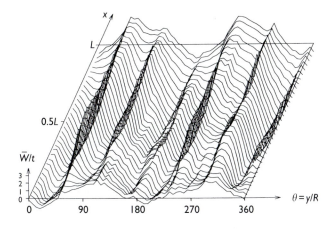

Figure 15.12 Measured initial shape of Horton's shell HO-1 (Horton 1977).

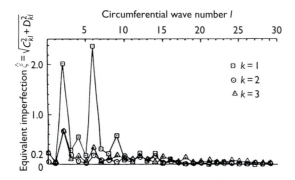

Figure 15.13 Circumferential variation of the half-wave sine Fourier representation (Horton's shell HO-1).

Figure 15.13 shows the variation of the half-wave sine Fourier representation for this shell. The amplitudes of the Fourier harmonics with a single half-wave in the axial direction have two distinct maxima, one at $\ell = 2$ (out-of-roundness) and another at $\ell = 6$ (which corresponds to the number of panels the shell is assembled from). The Fourier coefficients with more than a single half wave in the axial direction are comparatively much smaller.

The Solid Mechanics Group of the Faculty of Aerospace Engineering of Delft University of Technology used the one piece portable rail system shown in Fig. 15.14 to carry out imperfection surveys on the Ariane interstage I/II and II/III shells (Sebek 1981). Here the shell is positioned upright on a two-piece turntable, with the reference beam placed parallel to it on an adjustable tripod. There are three LVDT pick-ups installed on the reference beam. The fixed ones on the top and on the bottom are bearing against the machined end rings. The third one is installed on a carriage which is moved along the beam by an electric drive to record the shape of the corresponding shell generator. Next, the shell is rotated to a new position followed by another axial scan. The process is continued until the whole surface has been surveyed and recorded. The exact shape of the reference beam has been measured optically and it is removed from the measured data during the data reduction step. The axial position of the carriage is recorded by an electro-optical device which scans a strip with equally spaced cut-outs. The resulting square-shaped signal is used to digitise the data in intervals of 10 mm. Special care is taken to record possible random rigid body displacements of the shell assembly with respect to the fixed position of the reference beam during rotating the turntable to a new circumferential position. This is accomplished by monitoring the planar displacements of a calibrated circular ring placed in the centre of the turntable. These displacements are then also removed from the measured data during reduction (see Sebek 1981 for a detailed description).

Figure 15.14 Initial imperfection survey system from the TU-Delft (Sebek 1981).

The corresponding measured initial imperfections are displayed in Fig. 15.15. Figure 15.16 shows the variation of the coefficients of the half-wave sine Fourier representation for this shell. The amplitudes of the Fourier harmonics with a single half wave in the axial direction have a distinct maximum at $\ell = 8$ (which corresponds to the number of panels the shell is assembled from). There is also a sizable $\ell = 2$ (out-of-roundness) component. All other Fourier coefficients are comparatively much smaller.

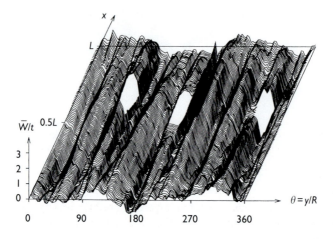

Figure 15.15 Measured initial shape of an ARIANE interstage II/III shell (Sebek 1981) (radius: 1300.0 mm, length: 2730.0 mm, wall thickness: 1.2 mm).

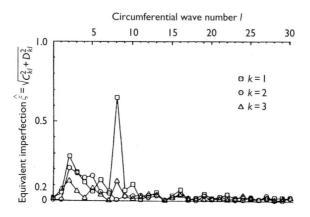

Figure 15.16 Circumferential variation of the half-wave sine Fourier representation (aerospace shell AR23-3 from Sebek 1981).

It appears from the results presented in this section that for full-scale aerospace shells built up out of a fixed number of curved panels, the initial imperfection distributions will be dominated by two components only, if the joints are riveted. Using the half-wave sine representation both components will have a single half-wave in the axial direction and 2 and ℓ full waves in the circumferential direction, where ℓ is the number of full length panels out of which the shell is assembled. By using accurately machined rigid end rings, the $\ell = 2$ out-of-roundness component can be significantly reduced in size. Thus, the variation of the measured harmonics shown in Figs 15.13 and 15.16 can be considered as the characteristic initial imperfection distribution for this particular type of fabrication process.

Stochastic stability analysis

It has been demonstrated in the preceding section that, indeed, one can associate characteristic initial imperfection distributions with the different fabrication processes. The question then arises:

> Given a characteristic initial imperfection distribution, how does one proceed to incorporate this knowledge into a systematic design procedure?

Since initial imperfections are obviously random in nature, the probabilistic design procedure, described earlier by Eqs (1)–(3), is one way to introduce the results of the experimentally measured initial imperfections into the analysis.

The proposed approach is based on the notion of a reliability function $R(\lambda)$, where by definition

$$R(\lambda) = \text{Prob} \ (\Lambda > \lambda) \tag{8}$$

and λ is the normalised load parameter ($= P/P_c$) and Λ the normalised random buckling load.

As can be seen from Fig. 15.17, the knowledge of the reliability function permits the evaluation of the allowable load, defined as the load level λ_a for which the

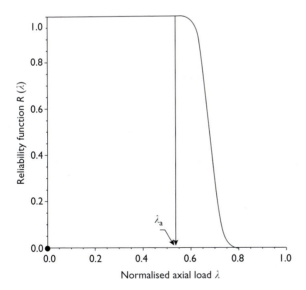

Figure 15.17 Reliability curve calculated via the FOSM method using Koiter's modified three-mode imperfection model, for the group of seven A-shells (Arbocz 1968).

specified reliability is achieved, for a whole ensemble of similar shells produced by a given manufacturing process. Notice that the allowable load level λ_a is identical to the improved (higher) scientific knockdown factor introduced in Eq. (5).

Turning now to the collapse problem of axially compressed cylinders with random initial imperfections one is faced with the evaluation of the multi-dimensional integral given by Eq. (3).

Whereas with today's advanced nonlinear finite element codes such as STAGS (Anonymous 1994) and ABAQUS (Anonymous 1998) the limit state function $g(\mathbf{X})$ (if so desired) can be determined with great accuracy, the evaluation of the multidimensional integral, where the domain of integration depends on the properties of the limit state function, is still a subject of detailed investigations. Since an exact numerical evaluation of this multidimensional integral is considered impractical, in the following it will be shown that by using the first-order, second-moment (FOSM) method (Karadeniz et al. 1982; Elishakoff et al. 1987) it is possible to develop a simple but rational method for checking the reliability of axially compressed shells using some statistical measures of the imperfections involved, and to provide an estimate of the structural reliability, whereby also the specified boundary conditions are rigorously enforced.

The use of the FOSM method involves linearisation of the response function $g(\mathbf{X})$ at the mean point and knowledge of the distribution of the random vector $\mathbf{X}$. Calculations are relatively simple if $\mathbf{X}$ is normally distributed. If $\mathbf{X}$ is not normally distributed, an appropriate normal distribution has to be substituted instead of the actual one. In the present case, one wants to obtain the reliability $R(\lambda)$ of the structure at any given load λ; that is, one wants to know

$$R(\lambda) = \mathrm{Prob}(\Lambda_s > \lambda) = \text{Probability that } \Lambda_s > \lambda \tag{9}$$

In this case, the limit state (or failure) surface has the form given by Eq. (1)

$$Z(\lambda) = \Lambda_s - \lambda = \psi(X_1, X_2, \ldots, X_n) - \lambda = 0 \tag{10}$$

where Λ_s is the random buckling load and λ is the applied non-dimensional deterministic load.

To combine the use of numerical codes with the mean value FOSM method, one needs to know the lower order probabilistic characteristics of Z. In the first approximation, the mean value of Z is

$$
\begin{aligned}
E(Z) &= E(\Lambda_s) - \lambda \\
&= E[\psi(X_1, X_2, \ldots, X_n)] - \lambda \\
&\approx \psi[E(X_1), E(X_2), \ldots, E(X_n)] - \lambda
\end{aligned} \tag{11}
$$

where the value of

$$\psi[E(X_1), E(X_2), \ldots, E(X_n)] \tag{12}$$

is calculated numerically with the code that was chosen for the numerical work. It corresponds to the deterministic collapse load of the imperfect shell with mean imperfection amplitudes.

The variance of Z is given by

$$\text{Var}(Z) = \text{Var}(\Lambda_s) \approx \sum_{j=1}^{n} \sum_{k=1}^{n} \left(\frac{\partial \psi}{\partial \xi_j} \right) \left(\frac{\partial \psi}{\partial \xi_k} \right) \text{cov}(X_j, X_k) \tag{13}$$

where $\text{cov}(X_j, X_k)$ is the variance–covariance matrix. The calculation of the derivatives $\partial \psi / \partial \xi_j$ (or $\partial \psi / \partial \xi_k$) is performed numerically by using the following numerical differentiation formula evaluated at values of $\xi_j = E(X_j)$ (or $\xi_k = E(X_k)$)

$$\frac{\partial \psi}{\partial \xi_j} = \frac{\psi(\xi_1, \ldots, \xi_{j-1}, \xi_j + \Delta \xi_j, \xi_{j+1}, \ldots, \xi_n) - \psi(\xi_1, \xi_2, \ldots, \xi_n)}{\Delta \xi_j} \tag{14}$$

Having obtained the quantities $E(Z) = E(\Lambda_s) - \lambda$ and $\text{Var}(Z)$, one can proceed to estimate the probability of failure $P_f(\lambda)$ as

$$P_f(\lambda) = \text{Prob}(Z < 0) = F_Z(0) = \int_{-\infty}^{0} f_Z(t) \, dt \tag{15}$$

where $F_Z(t)$ is the probability distribution function and $f_Z(t)$ is the probability density function of Z.

Assuming that the limit state function is normally distributed, then

$$f_Z(t) = \frac{1}{\sigma_Z \sqrt{2\pi}} \exp\left[-\frac{1}{2} \left(\frac{t - a}{\sigma_Z} \right)^2 \right] \tag{16}$$

where $a = E(Z)$ and $\sigma_Z = \sqrt{\text{Var}(Z)}$. Further,

$$F_Z(0) = \int_{-\infty}^{0} f_Z(t) \, dt = \frac{1}{2} + \text{erf}\left(\frac{-a}{\sigma_Z} \right) = \frac{1}{2} - \text{erf}\left(\frac{a}{\sigma_Z} \right) = \phi(-\beta) \tag{17}$$

where $\beta = a/\sigma_Z$ is the reliability index and $\phi(\beta)$ is the standard normal probability distribution function. The error function $\text{erf}(\beta)$ is defined as

$$\text{erf}(\beta) = \frac{1}{\sqrt{2\pi}} \int_{0}^{\beta} e^{-t^2/2} \, dt \tag{18}$$

Finally, the reliability $R(\lambda)$ will be estimated as

$$R(\lambda) = 1 - P_f(\lambda) = 1 - \text{Prob}(Z < 0) = 1 - F_Z(0) = \tfrac{1}{2} + \text{erf}(\beta) = \phi(\beta) \tag{19}$$

As can be seen from Eqs (11) and (13), one must know the mean values and the variance-covariance matrix of the input random variables X_i in order to be able to evaluate the mean value and the variance of the limit state function Z. Since in this case the input random variables X_i represent the Fourier coefficients of the initial imperfections, the above statistical measures can only be evaluated if a sufficiently detailed initial imperfection data bank (Arbocz and Abramovich 1979; Abramovich *et al.* 1981) is available.

Numerical results

In the previous section, the solution of the stochastic stability problem of Eq. (3) has been reduced to a series of $n + 1$ deterministic buckling load analysis. Several of the currently available structural analysis computer codes (Anonymous 1994, 1998) have the capability of calculating the effect of initial imperfections on the buckling load. However, as has been pointed out by Arbocz and Babcock (1976), the success of such deterministic buckling load analysis depends very heavily on the appropriate choice of the model used, which in turn requires considerable knowledge by the user of the physical behaviour of the imperfect shell. This knowledge can best be acquired by first using the series of imperfection sensitivity analyses of increasing complexity that have been published in the literature (Budiansky and Hutchinson 1964; Koiter 1974; Arbocz and Babcock 1976, 1980).

For the statistical calculations, the data associated with the seamless copper electroplated isotropic shells tested at Caltech in 1967, the so-called A-shells, are used initially (Arbocz 1968). The shell properties and test results are given in Table 15.1. The properties of shell A-8 are used for the numerical computations.

Table 15.1 Geometric and material properties, and experimental and theoretical buckling loads of the A-shells (Arbocz 1968)

	R (mm)	t (mm)	L (mm)	$E \times 10^{-5}$ (N/mm^2)	$P_{c\ell}$ (N)	P_{exp} (N)	λ_{exp}	γ	λ_s
A-7	101.6	0.1140	203.2	1.0411	5158.6	3036.4[a]	0.589	0.238	0.668
A-8	101.6	0.1179	203.2	1.0480	5535.8	3673.8	0.664	0.242	0.677
A-9	101.6	0.1153	203.2	1.0135	5125.2	3724.9	0.727	0.239	0.717
A-10	101.6	0.1204	203.2	1.0273	5662.6	3196.9[b]	0.565	0.245	0.729
A-12	101.6	0.1204	209.55	1.0480	5776.9	3853.0	0.667	0.245	0.665
A-13	101.6	0.1128	196.85	1.0411	5035.4	3108.9	0.617	0.236	0.631
A-14	101.6	0.1110	196.85	1.0894	5103.9	3442.9	0.675	0.234	0.707

Notes
a Stable local buckling at $P_{local} = 2819.7^N (\lambda_{local} = 0.547)$.
b Stable local buckling at $P_{local} = 3051.0^N (\lambda_{local} = 0.539)$.
   For all shells $\nu = 0.3$.

The measured initial imperfections are represented by the following double Fourier series

$$\bar{W}(x, y) = t \sum_{i=1}^{n_1} A_{io} \cos i\pi \frac{x}{L}$$

$$+ t \sum_{k=1, \, \ell=2}^{n_1} \sum_{}^{n_2} \sin k\pi \frac{x}{L} \left(C_{k\ell} \cos \ell \frac{y}{R} + D_{k\ell} \sin \ell \frac{y}{R} \right) \qquad (20)$$

It is proposed to use the root mean square (RMS) value of the measured initial imperfections as an approximate measure of their size. By definition the RMS value is

$$\Delta_{RMS}^2 = \frac{1}{2\pi RL} \int_0^{2\pi RL} \int_0^{} [\bar{W}(x, y)]^2 \, dx \, dy$$

$$= \frac{t^2}{4} \left\{ 2 \sum_{i=1}^{n_1} A_{io}^2 + \sum_{k=1, \, \ell=2}^{n_1} \sum_{}^{n_2} (C_{k\ell}^2 + D_{k\ell}^2) \right\} \qquad (21)$$

Thus

$$\left(\frac{\Delta_{RMS}}{t} \right)^2 = \frac{1}{2} \sum_{i=1}^{n_1} A_{io}^2 + \frac{1}{4} \sum_{k=1, \, \ell=2}^{n_1} \sum_{}^{n_2} (C_{k\ell}^2 + D_{k\ell}^2) = \Delta_{axi}^2 + \Delta_{asy}^2 \qquad (22)$$

Relying on the results of earlier investigations of the buckling behaviour of the imperfect shell A-8 (Arbocz and Babcock 1976; Arbocz et al. 1998), it was decided to use the following modified three-mode imperfection model of Koiter (1974) for the collapse load calculations.

$$\frac{\bar{W}}{t} = \bar{\xi}_1 \cos i_{c\ell}\pi \frac{x}{L} + \left(\bar{\xi}_2 \sin k_1 \pi \frac{x}{L} + \bar{\xi}_3 \sin k_2 \pi \frac{x}{L} \right) \cos \ell \frac{y}{R} \qquad (23)$$

where

$$\bar{\xi}_1 = \Delta_{asy} \frac{k_1}{\sqrt{2} i_{c\ell}}; \quad \bar{\xi}_2 = \Delta_{asy}; \quad \bar{\xi}_3 = \frac{k_1}{k_2} \Delta_{asy}; \quad i_{c\ell} = \frac{L}{\pi} \sqrt{\frac{2c}{Rt}};$$

$$c = \sqrt{3(1 - v^2)}$$

and k_1 and k_2 are roots of the equation which defines the so-called Koiter (1945) circle

$$k^2 \frac{Rt}{2c} \left(\frac{\pi}{L} \right)^2 + \ell^2 \frac{Rt}{2c} \left(\frac{1}{R} \right)^2 - k \sqrt{\frac{Rt}{2c}} \left(\frac{\pi}{L} \right) = 0 \qquad (24)$$

Table 15.2 Values of equivalent Fourier coefficients and
the sample mean vector

	$X_1(=\bar{\xi}_1)$	$X_2(=\bar{\xi}_2)$	$X_3(=\bar{\xi}_3)$
A-7	0.014090	0.67747	−0.020529
A-8	0.013231	0.63617	−0.019278
A-9	0.009782	0.47037	−0.014254
A-10	0.008974	0.43149	−0.013075
A-12	0.014210	0.68327	−0.020705
A-13	0.018545	0.89171	−0.027022
A-14	0.010929	0.52548	−0.015924
$E(\ )$	0.012823	0.61657	−0.018684

The Koiter circle is the locus of a family of modes belonging to the lowest
eigenvalue $\lambda_c = 1.0$, where by definition

$$\lambda_c = \frac{N_o}{N_{c\ell}}; \qquad N_{c\ell} = \frac{Et^2}{cR} \tag{25}$$

and N_o is the applied axial compressive stress resultant. The circumferential wave
number ℓ is chosen such that $k_1 = 1$. This yields the first Koiter triad (Arbocz
and Babcock 1976). The ratios of the amplitudes of the imperfections are chosen
such that the direction of the postbuckling path coincides with the path of steepest
descent, which yields the most adverse imperfection shape (Koiter 1974).

In order to apply the FOSM method, the mean buckling load has to be calculated first. Using the imperfection model of Eq. (23) with the mean values of the
corresponding equivalent imperfection amplitudes listed in Table 15.2, the result
of the calculation is

$$E(\Lambda_s) = 0.681599$$

The derivatives $\partial \psi / \partial \xi_j$ are calculated as follows. For the increment of the random
variable in Eq. (14), 1% of the original mean value of the corresponding equivalent
Fourier coefficient is used, so that $\Delta \xi_j = 0.01 E(X_j)$. The calculated derivatives
are listed in Table 15.3. In this study, the increments of the path parameter are
chosen in such a way that the limit loads are found accurate to within 0.01%.
Next, using the sample variance–covariance matrix displayed in Table 15.4, one
can evaluate the mathematical expectation and the variance of Z. The results of
these calculations are

$$E(Z) = 0.681599 - \lambda \quad \text{and} \quad \text{Var}(Z) = 0.0012425$$

Finally, the reliability is calculated directly from Eq. (19) and is plotted in
Fig. 15.17. Notice that for a reliability of 0.99997, one obtains an improved

Table 15.3 Derivatives of ψ with respect to the equivalent
Fourier coefficients of the A-shells

	X_j	$\partial \psi / \partial X_j$
$\bar{\xi}_1 = \bar{W}_{34.0}$	1	−3.58730
$\bar{\xi}_2 = \bar{W}_{1.9}$	2	−0.00748
$\bar{\xi}_3 = \bar{W}_{33.9}$	3	2.47270

Table 15.4 Sample variance–covariance matrix of the A-shells (Arbocz 1968)

	1	2	3
1	0.10681×10^{-4}		Symmetric
2	0.51358×10^{-3}	0.24695×10^{-1}	
3	-0.15563×10^{-4}	-0.74835×10^{-3}	-0.22678×10^{-4}

scientific 'knockdown' factor $\lambda_a = 0.54$. As can be seen from the results listed in Table 15.1, the reliability-based improved scientific 'knockdown' factor λ_a always yields a safe estimate of the load carrying capacity of the shells tested. That is, the normalised experimental collapse loads λ_{exp} are always greater than $\lambda_a = 0.54$. Still, the value of the probabilistic $\lambda_a = 0.54$ is, on the average, more than twice as high as the overly conservative allowable load γ obtained from the lower bound design curve recommended in the current version of NASA SP-8007 (Anonymous 1968).

Conservativeness of the FOSM method for buckling problems

The FOSM method is a so-called approximate probabilistic method. The method is based on calculating the probability of failure via a local approximation of the response probability density. The transformation of the random input variables to the random output variable is performed with a local linearisation of the response surface. The linearised response surface is used as a transformation function of the probability density function of the input variables. Usually, the assumption of normality of the input variables is used which generally seems quite legitimate. For linear transformation functions the normal probability density function transforms linearly, that is, it translates and scales but does not change type.

For buckling problems, in general, the limit-state function $g(\mathbf{X})$ is nonlinear and the curvature of the limit-state curve/surface strongly depends on the magnitude of the imperfection amplitude and the mode of the imperfection shape. However, it turns out that predictions made by the FOSM method with a simple imperfection model are often conservative because of the convexity of the limit-state surface.

Estimations of the probability of failure with the FOSM method are only valid under the following conditions.

- The probability distributions of the input variables are assumed to be normal. Since the transformation is linear the output distributions are also normal.
- The transformation of the input probability density function to the output probability density function is based on a linearisation around the mean.
- The mean response is approximated under the assumption that the probability density function has a low coefficient of variation.

The advantage of using the FOSM method is that it requires only a small number of response surface function evaluations. A disadvantage is that, depending on the curvature of the response surface, the approximation of the limit-state function in comparison with the exact limit state function could be quite crude, as will be shown. The results obtained for quite a number of buckling calculations show that in most cases the FOSM method gives conservative predictions of the probability of failure.

As an example, in the following a two-mode imperfection model

$$\frac{\bar{W}}{t} = \bar{\xi}_1 \cos i\pi \frac{x}{L} + \bar{\xi}_2 \sin k\pi \frac{x}{L} \cos \ell \frac{y}{R} \tag{26}$$

in combination with the Donnell type nonlinear equations will be used. This two-mode imperfection model captures elementary buckling phenomena, such as coupling of modes. Although quite simple the mechanical model will, in the following sections, be represented by this two-mode imperfection model.

Response surface geometry

Before obtaining solutions from the rather complex design space, it is a good practice to navigate through the design space in order to be able to detect anomalies. A simple and efficient tool for this exercise is to span the response surface in the design space.

Using the response surface, $\lambda = H(\bar{\xi})$, with the Fourier expressions from Eq. (6) or (7) it can be demonstrated that the response surface is convex for the dimensions represented by the axisymmetric imperfection components. Whereas the asymmetric imperfection components turn the response surface in certain dimensions in a concave shape. It is the combination of these two effects that can cause anomalous behaviour.

The response surface can be derived from the regular response equations $\mathbf{F} = \mathbf{0}$ by simply adding the criterion that the obtained solution, i.e. every solution, should be either a limit point or a bifurcation point, $\det(\mathbf{J}) = 0$, in which $\mathbf{J}$ is the Jacobian of the regular response equilibrium equations.

In order to be able to properly interpret this phenomenon one could best visualise the shape of the response surface. The response surface function is a set of nonlinear

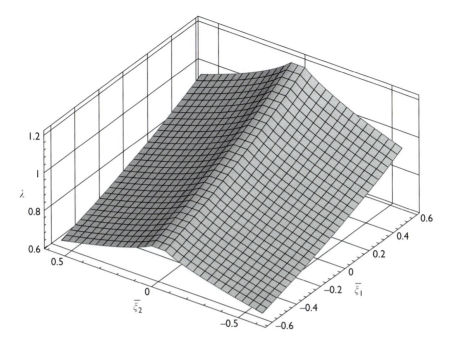

Figure 15.18 Response surface of the two-mode imperfection model.

algebraic equations in which both the regular, deterministic, response variables as well as the probabilistic response parameters are varying. In Fig. 15.18, the geometry of the response surface is shown. Notice that for the design variable $\bar{\xi}_2 = 0$, there is no asymmetric imperfection, and the buckling equations become singular since only symmetric bifurcation points can be found. Next to that, the buckling behaviour of the shell in the chosen model is insensitive to the sign of the asymmetric imperfection mode, that is, symmetry. The behaviour of such a system with symmetry can best be described by a series system. Therefore, we have to add the calculated probabilities that belong to the critical modes.

$$\beta_g = -\Phi^{-1}(\Phi(-\beta_1) + \Phi(-\beta_2)) \tag{27}$$

It should also be noticed that for increasing values of $\bar{\xi}_1$, the buckling load increases. This implies that for positive values of the axisymmetric imperfection mode, the centre section of the shell bulges outward and provides a stiffening effect on the load carrying capacity of the shell. When levels of equal buckling loads are projected on a plane, it can be seen in Fig. 15.19 that the limit-state function for a particular level of λ will not be linear. This is confirmed by Fig. 15.20, in which the curvature of the limit-state function is plotted with respect to $\bar{\xi}_2$. It should be

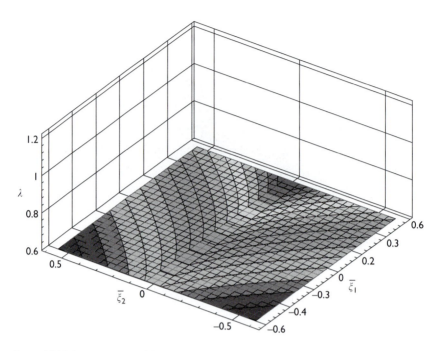

Figure 15.19 Iso-contours of the response surface of the two-mode imperfection model at specified load levels.

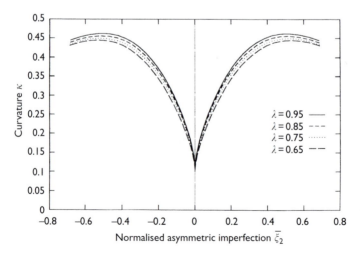

Figure 15.20 Curvature of the limit-state function with respect $\bar{\xi}_2$.

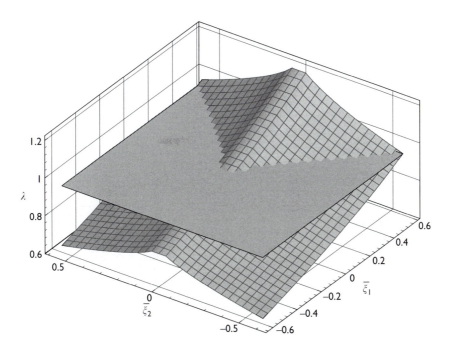

Figure 15.21 Definition of failure when a deterministic load λ is applied.

mentioned that the shape of the response surface, especially its curvature, is highly dependent on the type of shell one is investigating.

As can be seen in Fig. 15.21, when a specified deterministic load level λ (say, $\lambda = 0.9$) is applied, for a safe structure the required stiffness should be above this particular load level. Therefore, all combinations of imperfection amplitudes, resulting in a stiffness level lower than the applied load level, will result in failure. These two states are separated by the limit-state function $g(\mathbf{X}) = \Lambda(\mathbf{X}) - \lambda = 0$, indicated by the intersection between the yellow surface, representing the applied load λ, and the red-coloured space curve, the response surface $\Lambda(\mathbf{X})$. The probability of failure $P_f(\lambda)$ is calculated by evaluating the probability integral of Eq. (3) over the visible part (yellow coloured) of the horizontal surface in Fig. 15.21.

For other load levels, the intersection contours also have a shape as presented in Fig. 15.19. It is clear that this response surface is convex. It is this property of the exact response surface that provides a handle to the suggestion that probability of failure predictions are conservative.

In order to demonstrate this rationale the approximated response surface will be plotted with the exact response surface. The combined set of surfaces is presented in Fig. 15.22. The mean value of the input variables are $\bar{\xi}_1 = -0.31$ and $\bar{\xi}_2 = -0.31$. It is to be expected that the linearised response surface coincides with the exact

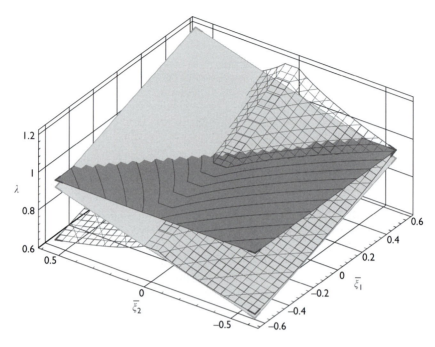

Figure 15.22 Approximation of failure by the FOSM method in the design space.

response surface at that particular point. As can be seen from Fig. 15.20, the curvature of the limit-state function is at any location on the response surface positive. It is this feature of positive curvature that positions the approximated limit-state function closer to the origin than the exact limit-state function. In Fig. 15.22, this situation is graphically explained. Although the resulting probability of failure turns out to be conservative in this analysis, there are actually two mechanisms working that influence the probability of failure calculation.

First, the symmetry of the response surface is neglected in the probability density function transformation. This is quite easy to imagine. For $\bar{\xi}_2 = 0$, there is a discontinuity in the response surface that for a certain applied load level creates two sub-domains because the mechanical model behaves as a series system with respect to the asymmetric imperfection mode. So the presence of a two-part probability of failure contribution is not taken into account. The mechanism that provides the conservatism in the solution is the fact that the linearisation largely overestimates the probability of failure. This overestimation will always be larger than the sum of the two 'real' probabilities of failure, because the volume under the probability density function decreases exponentially with the coordinates.

There are, however, also factors that can reduce the conservativeness of this computational model. One is the magnitude of the coefficient of variation of the input random variables, especially with respect to the position of the mean value.

Second, the response surface seems to become less curved when the shell is made of composites, that is, the imperfection sensitivity is reduced. These factors reduce the margin of overpredicting the probability of failure with this approximate model.

More accurate methods

Results of the FOSM method can be verified with more accurate methods. Its attractiveness for design purposes is mainly based on simplicity and efficiency of the calculation, not on its accuracy. It is clear that more accurate results require more specific model information and require therefore more computational resources.

The first order reliability method (FORM) is also a linearisation of the response surface. However, the linearisation is done in standard normal design space. The main idea behind this method is that it transforms the possibly correlated input probability density functions, to the uncorrelated standard normal design space, with the result that every design dimension is assigned equal weight. This implies that the limit-state function in the original design space $g(\mathbf{X})$ is transformed to the standard normal space $g(\mathbf{U})$ via scaling, rotation and translation. Although the limit-state function could be symmetric in the original design space, the limit-state function in the standard normal design space will not necessarily have to be symmetric. This is highly dependent on the type of probability density functions involved. One of the advantages is that this method does not put constraints on the type of input probability density functions.

For low-dimensional probabilistic design models as the two-mode imperfection model used FORM can be explained rather simply. In the standard normal space, the linearisation point is chosen as the point with the highest probability density anywhere on the limit state function. Following the limit state function as a path through the design space and continuously checking the value of the probability density function will show easily, that for the two-mode imperfection model the point with the highest probability density is

$$\beta = \min \|\mathbf{u}\| = \min \sqrt{u_1^2 + u_2^2} \tag{28}$$

This is the geometric reliability index. In the standard normal space the distance from the origin to the limit-state function is a measure for the reliability, since all variables have identical weight. The distance from the origin to the limit-state function can be used as the integration limit for the one dimensional integration for the probability of failure because of the hypersphere representation of the transformed design space.

For the determination of the geometric or generalised reliability index in the most general case an optimisation algorithm is needed to obtain the β-point. In case of the two-mode imperfection model it turned out that a re-formulation of the path following method offered the possibility of following the nonlinear solution trace in the standard normal space.

The evaluation of the generalized reliability index is, in the most general case, a cumbersome operation. This is because the failure domain in the standard normal space has to be integrated with respect to the limit state. This can be formulated as

$$\beta_g = -\Phi^{-1}(P_f) \tag{29}$$

where

$$P_f = \int_{-\infty}^{\infty} \left[\int_{-\infty}^{\bar{\xi}_2 = H(\bar{\xi}_1, \lambda)} f_{\bar{\Xi}}(\bar{\xi}_1, \bar{\xi}_2) \, d\bar{\xi}_2 + \int_{\bar{\xi}_2 = H(\bar{\xi}_1, \lambda)}^{\infty} f_{\bar{\Xi}}(\bar{\xi}_1, \bar{\xi}_2) \, d\bar{\xi}_2 \right] d\bar{\xi}_1 \tag{30}$$

and H describes the functional relation between $\bar{\xi}_1$ and $\bar{\xi}_2$ for a constant level of λ. The value of the generalised reliability index is not just a distance measure, it depends also on the curvature of the limit state function. For details see Stam (1996). In most cases, however, the limit state has only a slight curvature. Therefore a sufficiently accurate estimate of the reliability can be obtained with the geometric reliability index, that is, an estimation with a linearised limit-state function.

In this analysis, the axial compressive load is assumed to be deterministic. Thus, failure of the cylindrical shell will occur when for a certain value of the axisymmetric imperfection amplitude $\bar{\xi}_1$ the asymmetric imperfection satisfies both the conditions $\bar{\xi}_1 \leq \bar{\xi}_1^*$ and $\bar{\xi}_2 \leq \bar{\xi}_2^*$, where $\bar{\xi}_2^*$ is the asymmetric imperfection amplitude that corresponds to the axisymmetric imperfection amplitude for a level of $\lambda = $ constant.

The most accurate method for evaluating the multi-dimensional probability integral Eq. (3) is the Monte Carlo Method (Elishakoff 1983). To simulate the large number of initial imperfection profiles needed for the Monte Carlo method, first the mean values and the variance–covariance matrices of the measured initial imperfections must be determined. For the two-mode imperfection of Eq. (26) this involves the evaluation of the following ensemble averages for a sample of experimentally measured initial imperfections.

$$\bar{A}^{(e)} = \frac{1}{M} \sum_{m=1}^{M} A^{(m)}; \qquad \bar{C}^{(e)} = \frac{1}{M} \sum_{m=1}^{M} C^{(m)} \tag{31}$$

$$K_{AA}^{(e)} = \frac{1}{M-1} \sum_{m=1}^{M} \left[A^{(m)} - \bar{A}^{(e)} \right] \left[A^{(m)} - \bar{A}^{(e)} \right]$$

$$K_{CC}^{(e)} = \frac{1}{M-1} \sum_{m=1}^{M} \left[C^{(m)} - \bar{C}^{(e)} \right] \left[C^{(m)} - \bar{C}^{(e)} \right] \tag{32}$$

$$K_{AC}^{(e)} = K_{CA}^{(e)} = \frac{1}{M-1} \sum_{m=1}^{M} \left[A^{(m)} - \bar{A}^{(e)} \right] \left[C^{(m)} - \bar{C}^{(e)} \right]$$

where M is the number of sample shells and m the serial number of the shells.

The variance–covariance matrix

$$[V^{(e)}] = \begin{bmatrix} K_{AA}^{(e)} & K_{AC}^{(e)} \\ K_{CA}^{(e)} & K_{CC}^{(e)} \end{bmatrix} \tag{33}$$

is positive-semidefinite and can be uniquely decomposed in the form

$$\lfloor V^{(e)} \rfloor = [G][G]^T \tag{34}$$

where $[\ ]^T$ means transpose and $[G]$ is a lower triangular matrix found by the Cholesky decomposition algorithm. The vectors of the Fourier coefficients of the two-mode imperfection model are simulated as follows

$$\begin{bmatrix} A \\ C \end{bmatrix} = [G][B] + \begin{bmatrix} \bar{A}^{(e)} \\ \bar{C}^{(e)} \end{bmatrix} \tag{35}$$

where B is a random vector of length $2M_s$, the elements of which are normally distributed, statistically independent with zero means and unit variance (for further details see Elishakoff 1982; Elishakoff and Arbocz 1985). Notice that M_s is the number of simulated shells.

Next, one can proceed to the second step of the Monte Carlo method: the evaluation of the buckling load for each of the 'simulated' shells. Here this is done with the code MIUTAM (Arbocz and Babcock 1976). Once the critical buckling load for the large number of simulated shells (here $M_s = 1089$) are available, one can then calculate the histogram of the buckling loads and determine the corresponding reliability function.

Finally, in Fig. 15.23 the results obtained by the above mentioned methods are displayed. Notice that for the desired high reliability (say, $R(\lambda) = 0.999$) the FOSM method of Eq. (19) predicts an improved scientific knockdown factor of $\lambda_a \cong 0.540$. The more accurate FORM method of Eq. (30) predicts an improved scientific knockdown factor of $\lambda_a = 0.555$, whereas the most accurate Monte Carlo method (for 1089 realisations) predicts $\lambda_a = 0.576$. It should be stressed, however, that the accuracy of the Monte Carlo method strongly depends on the number of realisations used for the simulation procedure.

Probabilistic criterion for preliminary shell design

The proposed new probabilistic approach to identify the scientific 'knockdown' factor λ_a for preliminary shell design (see Eq. 5) consists of the following steps

1   Establishment of Initial Imperfection Data Bank(s) – This step is necessary to identify the characteristic initial imperfection distributions of nominally identical shells, produced by the same fabrication process and possessing similar wall-constructions. Quality of shells is initially quantified by the RMS

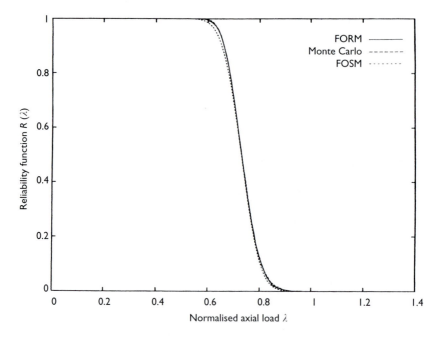

Figure 15.23 Reliability curves computed by different methods for the same probability density functions.

value of the measured initial imperfections, whereby one distinguishes the axisymmetric Δ_{axi} and the asymmetric Δ_{asy} RMS amplitudes (see Eq. 22).

2   Theoretical buckling analysis of the proposed shell structure – This step is necessary to identify the critical (lowest) eigenvalues and the corresponding buckling modes. A hierarchical approach (see Arbocz *et al.* 1999) is recommended, where the effect of prebuckling edge restraint and the specified boundary conditions is included in the analysis. The results of this investigation will guide the investigator to identify the simple imperfection model, which describes the collapse behaviour of the imperfect structure accurately (see Arbocz and Babcock 1976, 1980).

3   FOSM analysis to identify the 'scientific knockdown factor λ_a' – This step is necessary to identify the scientific 'knockdown' factor λ_a, which replaces the empirical 'knockdown' factor γ in Eq. (4), when one calculates the allowable applied load P_a.

It has been demonstrated by Arbocz *et al.* (1999, 2000) that a data base containing both information about the buckling tests and experimentally measured initial imperfections can be used successfully to calculate improved, less conservative allowable normalised buckling loads with the same reliability as the current deterministic, lower-bound based, more conservative traditionally recommended allowable buckling loads.

It is anticipated that for applications where the total weight of the structure is one of the critical parameters (i.e. aerospace structures), there will be a chance for definite improvement in the design process with the help of the proposed new probabilistic design procedure. It is felt, that the small added cost involved in systematically carrying out the required initial imperfection surveys will be fully justified by the overall cost-savings and by producing improved, less conservative and more reliable shell structures.

In order to make the proposed probabilistic design procedure for buckling-sensitive structures fully operational, there is a need for additional, systematic combined experimental and analytical or numerical results. Only then can the benefits be reaped from the many years of concentrated shell buckling research of the late 1960s and early 1970s. It is the authors' opinion that the technology now exists for such an undertaking.

Acknowledgement

The research reported in this paper was supported in part by NASA Grant NAG 1 1826. This aid is gratefully acknowledged.

References

Anonymous (1968). Buckling of thin-walled structures, NASA space vehicle design criteria (structures), SP-8007.

Anonymous (1994). *STAGS Manual*, Version 2.0, Lockheed Palo Alto Research Laboratories, Palo Alto, CA.

Anonymous (1998). *ABAQUS Manual*, Version 5.8, Hibbit, Karlsson & Sorensen Inc., 1080 Main Street, Pawtucket, RI, 02860-4847, USA.

Abramovich, H., Singer, J. and Yaffe, R. (1981). Imperfection characteristics of stiffened shells – Group 1, TAE Report 406, Technion, Haifa, Israel.

Arbocz, J. (1968). The effect of general imperfections on the buckling of cylindrical shells. PhD Thesis, California Institute of Technology, Pasadena, CA.

Arbocz, J. (1982). The Imperfection Data Bank, a mean to obtain realistic buckling loads. *Proceedings Buckling of Shells – A-State-of-the-Art Colloquium* (ed. E. Ramm). Springer Verlag, Berlin, pp. 535–567.

Arbocz, J. and Abramovich, H. (1979). The Initial Imperfection Data Bank at the Delft University of Technology – Part I, Report LR-290, Delft University of Technology, The Netherlands.

Arbocz, J. and Babcock, C.D., Jr (1969). The effect of general imperfections on the buckling of cylindrical shells. *Journal of Applied Mechanics* **36**(1), 28–38.

Arbocz, J. and Babcock, C.D. (1976). Prediction of buckling loads based on experimentally measured initial imperfections. *Proceedings IUTAM Symposium: Buckling of Structures* (ed. B. Budiansky). Springer Verlag, Berlin, pp. 291–311.

Arbocz, J. and Williams, J.G. (1977). Imperfection surveys on a 10 ft diameter shell structure. *AIAA Journal* **15**(7), 949–956.

Arbocz, J. and Babcock, C.D. (1980). The buckling analysis of imperfection sensitive structures. NASA CR-3310.

Arbocz, J. and Hol, J.M.A.M. (1991). Collapse of axially compressed cylindrical shells with random imperfections. *AIAA Journal* **29**(12), 2247–2256.

Arbocz, J., Starnes, J.H. and Nemeth, M.P. (1998). Towards a probabilistic criterion for preliminary shell design. *Proceedings 39th AIAA/ASME/ASCE/AHS/ASC Structures, Structural Dynamics and Materials Conference*, 20–23 April, Long Beach, CA, pp. 2941–2955.

Arbocz, J., Starnes, J.H. and Nemeth, M.P. (1999). A hierarchical approach to buckling load calculations. *Proceedings 40th AIAA/ASME/ASCE/AHS/ASC Structures, Structural Dynamics and Materials Conference*, 12–15 April, St. Louis, MO, pp. 284–299.

Arbocz, J., Starnes, J.H. and Nemeth, M.P. (2000). A comparison of probabilistic and lower bound methods for predicting the response of buckling sensitive structures. *Proceedings 41st AIAA/ASME/ASCE/AHS/ASC Structures, Structural Dynamics and Materials Conference*, 3–5 April, Atlanta, GA, ISBN # 1-56347-435-2.

Budiansky, B. and Hutchinson, J.W. (1964). Dynamic buckling of imperfection sensitive structures. *Proceedings of the 11th IUTAM Congress* (ed. H. Görtler). Springer-Verlag, Berlin, pp. 636–651.

Bushnell, D. (1976). BOSOR 5 – program for buckling of elastic–plastic complex shells of revolution including large deflections and creep. *Computers and Structures* **6**, 221–239.

Doup, M.R. (1997). Probabilistic analysis of the buckling of thin-walled shells using an imperfection database and a two-mode analysis. Ir. Thesis (also Memorandum M-808), Delft University of Technology, Faculty of Aerospace Engineering, The Netherlands.

Elishakoff, I. (1982). Simulation of an Initial Imperfection Data Bank, Part I: Isotropic shells with general imperfections. TAE Report 500, Department of Aeronautical Engineering, Technion-Israel Institute of Technology, Haifa, Israel.

Elishakoff, I. (1983). *Probabilistic Methods in the Theory of Structures*. Wiley-Interscience, New York, pp. 433–468.

Elishakoff, I. and Arbocz, J. (1985). Reliability of axial compressed cylindrical shells with general nonsymmetric imperfections. *Journal of Applied Mechanics*, **52**(1), 122–128.

Elishakoff, I., van Manen, S., Vermeulen, P.G. and Arbocz, J. (1987). First-order second-moment analysis of the buckling of shells with random imperfections. *AIAA Journal* **25**(8), 1113–1117.

Horton, W.H. (1977). On the elastic stability of shells. NASA CR-145088.

Karadeniz, H., van Manen, S. and Vrouwenvelder, A. (1982). Probabilistic reliability analysis for the fatigue limit state of gravity and jacket type structures. *Proceedings of the Third International Conference on Behavior of Off-Shore Structures*. McGraw-Hill, London, pp. 147–165.

Koiter, W.T. (1945, 1967). On the stability of elastic equilibrium. PhD Thesis (in Dutch), TH-Delft, The Netherlands, H.J. Paris, Amsterdam, 1945; English translation NASA TTF-10, pp. 1–833.

Koiter, W.T. (1963). Elastic stability and postbuckling behavior. *Proceedings Symposium Nonlinear Problems*. University of Wisconsin Press, Madison, pp. 257–275.

Koiter, W.T. (1974). Personal communication. California Institute of Technology.

Ryan, R.S. and Townsend, J.S. (1993). Application of probabilistic analysis/design method in space programs. The Approaches, the Status and the Needs. In *Proceedings 34th AIAA/ASME/ASCE/AHS/ASC Structures, Structural Dynamics and Materials Conference*, 15–22 April, La Jolla, CA, AIAA Paper No. 93-1381.

Sebek, R.W.L. (1981). Imperfection surveys and data reduction of ARIANE interstages I/II and II/III. Ir. Thesis, TH-Delft, Department of Aerospace Engineering.

Singer, J., Arbocz, J. and Babcock, C.D., Jr. (1971). Buckling of imperfect stiffened cylindrical shells under axial compression. *AIAA Journal* **9**(1), pp. 68–75.

Singer, J., Abramovich, H. and Yaffe, R. (1978). Initial imperfection measurements of integrally stringer-stiffened shells. TAE report No. 330, Technion, Haifa, Israel.

Stam, A. (1996). Stability of imperfect cylindrical shells with random properties. *Proceedings 37th AIAA/ASME/AHS/ASC Structures, Structural Dynamics and Materials Conference*, 18–19 April, Salt Lake City, UT, pp. 1307–1315.

Walker, A.C. and Sridharan, S. (1977). Buckling of compressed, longitudinal stiffened cylindrical shells. BOSS '79. *Proceedings 2nd International Conference on the Behaviour of Off-shore Structures*, Imperial College, London, Vol. 2, pp. 341–356.

Index

analysis: geometrically and materially nonlinear 17–18, 91, 98, 112, 121, 269–81; geometrically and materially nonlinear with imperfections 17–18, 71–2, 91, 107, 112–17, 121, 124, 271–4, 281; geometrically nonlinear 17, 91, 112, 268–75, 281; linear bifurcation 17–18; materially nonlinear 17, 91, 112–14

axisymmetric: buckling 48, 49–53, 73, 75, 81, 302, 326; prebuckling 451

bending: general 135–6, 142, 375; local 48, 52, 65, 77, 273, 395

bifurcation: axisymmetric 48–53, 73, 81, 302, 326; critical mode 18, 52, 56, 63–4, 73, 81, 379, 402, 422, 479; event 46, 65, 140; linear 1, 13, 17–18, 47, 210; mode 43–8, 52–8, 62–4, 73–4, 81, 89, 103, 112, 165–71, 191, 201–5, 214, 219, 225, 231–7, 240–58, 263, 268–73, 283, 294, 302, 306, 312–14, 326, 330, 344–7, 355–61, 372–5, 380, 392, 395, 411–14, 419–21, 426–8, 443, 449–52, 486; mode change 57, 85; non-symmetric 46–7, 58, 63, 179, 231, 246, 421, 462; point of 320, 478–9; wavelength 62–3, 68, 231, 382, 389

boundary conditions: clamped 3, 26, 49, 120, 158–60, 188, 204, 214–18, 230–2, 245, 251–63, 272, 297, 300–12, 413, 424–32, 452–3; end effects 50; free 3, 49, 64, 94, 158–64, 188, 202, 208, 213–15, 245, 264–6, 427; general 3, 49–50, 56–8, 98, 158–63, 204, 306, 312, 320, 326–32; simply supported 1, 3,

78–9, 155–8, 204–5, 262–6, 280, 307, 311–12, 362, 421–32, 447–53

collapse 109–14, 371, 376–7, 387, 393, 398–403, 433, 445

cyclic loading 230, 243–4, 255–8

ductile behaviour 254, 371

dynamic phenomena 5, 10–13, 46, 231–4, 241–3, 248–54, 258–9

ECCS recommendations 6, 96, 162, 172, 213, 263, 265, 276, 277–8, 280

eigenmode 18–19, 53–6, 60–8, 106–9, 201–2, 237, 245–56, 262, 272–3, 320

eigenvalue 47, 70, 80, 204, 209, 214–15, 247, 262, 266, 281, 320, 476, 486

elastic–plastic 7, 70, 71, 77, 109, 119, 123–4, 210–11, 215, 232, 237, 263, 271, 317–19, 349, 363, 372, 376, 394, 398, 415

elephant's foot buckling 10, 13, 77–80, 112, 114, 201, 218, 231–5, 263

Eurocode 6, 13–19, 24–5, 57, 67–72, 83, 91, 97, 126, 151, 200–5, 211, 231, 258, 393, 452

Eurocode 3 Part 1.6 for shell structures 10–12, 19, 24, 57, 67–70, 72–80, 83, 91, 138, 151, 205, 218–25, 258–63, 278–83

Eurocode 3 Part 4.1 for silo structures 10–14, 24, 137–8, 151, 258

Eurocode 3 Part 4.2 for tank structures 13, 25

geometric imperfection forms: dent 66; dimple 70–1, 144, 147–51, 220, 272; eigenmode 56; mode 19–23, 57–60, 106, 114, 471–86; seam 220–1, 437;